Electrostatic Discharge

Electrostatic Discharge

Kenneth L. Kaiser

Kettering University
Flint, Michigan

CRC Press
Taylor & Francis Group
Boca Raton London New York

CRC Press is an imprint of the
Taylor & Francis Group, an **informa** business

A TAYLOR & FRANCIS BOOK

This material was previously published in the *Electromagnetic Compatibility Handbook* © CRC Press LLC 2004.

CRC Press
Taylor & Francis Group
6000 Broken Sound Parkway NW, Suite 300
Boca Raton, FL 33487-2742

First issued in paperback 2019

© 2006 by Taylor & Francis Group, LLC
CRC Press is an imprint of Taylor & Francis Group, an Informa business

No claim to original U.S. Government works

ISBN-13: 978-0-367-39209-3 (pbk)

Library of Congress Card Number 2005049277

Library of Congress Cataloging-in-Publication Data

Kaiser, Kenneth L.
  Electrostatic discharge / Kenneth L. Kaiser.
    p. cm.
  Includes bibliographical references and index.

  1. Electrostatics. I. Title.

QC571.K35 2005
537.2--dc22

2005049277

Visit the Taylor & Francis Web site at
http://www.taylorandfrancis.com

and the CRC Press Web site at
http://www.crcpress.com

Preface

Although I have been interested in electromagnetics my entire adult life, my concentration on applied electrostatics and electrostatic discharge began in graduate school. During this time, I was introduced by Walter L. Weeks, Ph.D. (Purdue University) and Ed C. Escallon (Terronics Development Corporation), to a fascinating research project on electrostatic spraying. A consequence of this introduction is that I dedicated nearly 4 years to researching a project on this topic and to learning everything I could about applied electrostatics. While experimenting using kV-level voltages during my research, I had an enjoyable time and developed the proper respect for electrostatic discharges. After reading this book, it will become apparent why electrostatics is not only a fun field but also an important field.

The electrostatic discharge related topics contained in this book were selected from my previously published reference and textbook published by CRC Press entitled *Electromagnetic Compatibility Handbook*. I hope this "spinoff" book, being only a small fraction of the *Handbook*, can be more conveniently used and more easily owned by individuals interested mainly in electrostatic discharge. One of the main purposes of this particular book is to demystify electrostatic discharge and to help explain the source and limitations of the approximations, guidelines, models, and rules-of-thumb seen in this field. For further reference and personal edification, many of the examples contained in this book were written by me to document the answers to questions I had about the subject matter. Although the chapters in this book are fairly self contained, it is assumed that the reader has a rudimentary background in electromagnetics, including vector analysis.

I have tried to be diligent in crediting all of the sources that were used in solving a problem, generating tabled results, or understanding an unfamiliar or a confusing concept, and I apologize in advance for any oversights. (If by chance you locate some material in this book not appropriately referenced, which is possible considering the book's length and the number of editing iterations, please e-mail me. I will include the addition on my web site and in any future editions of this book.)

So as not to burden the reader with frequent citation interruptions, in most cases the references are grouped together at the end of each problem statement and located between brackets, [ ]. The references are listed by last name roughly from most used to least used. (The date of publication is also given when necessary to avoid confusion with other authors with the same surname or different publications by the same author.) In some cases, as with many of the tables, specific references are provided for particular equations or results. When possible, I have tried to use "original" sources.

The program Mathcad was used to generate most of the plots and solve many of the equations. The major reason this program was selected is that it is easily understood even with little or no prior experience with Mathcad. Although it is somewhat unorthodox, entire Mathcad programs have been provided in many cases. This allows all of the variable assignments, assumptions, equations, and possible mistakes to be clearly seen. I used the program VISIO, sometimes with an embedded Mathcad output, to generate *all* of the figures.

An enormous amount of time was devoted to crafting the tables in this book. A brief comment concerning the accuracy of the many expressions contained in this book, including in the tables, is appropriate. What is not seen in this book are the many derivations and checks. Although most of the equations in this book were personally derived, obviously, not every equation in this book has been personally obtained from basic principles. Extensive use has been made of the years of hard work of myriad scholars. It can be stated, however, that nearly every expression has been either derived or checked sometimes through much effort. A typical analytical and numerical check might involve taking a limit on one or more variables in an expression and then comparing it to a reliable result. The number of approximate expressions contained in this book prohibited an indication of their relative error. Sets of approximations are provided, in part, to help show the relationship of the individual expressions to each other.

As in the preface of *Electromagnetic Compatibility Handbook*, I again thank my friend, mentor, and colleague Professor James C. McLaughlin (Kettering University) for his many suggestions and his steadfast support. Also, I thank Mr. Ed C. Escallon, Dr. Johannes C. Almekinders, and Dr. Caner U. Yurteri of Terronics Development Corporation for their review of these enormous chapters.

Your comments, suggestions, and corrections are most welcomed and encouraged. They are invaluable to the current edition and any possible future editions of this book. At the site http://www.klkaiser.com, student problems, errata, and a few extras are provided for the *Electromagnetic Compatibility Handbook*. I sincerely hope this book on electrostatic discharge is helpful and inspirational to you. Unless specified otherwise, Système International (SI) units are used throughout this book. Enjoy.

Kenneth L. Kaiser
Electrical and Computer Engineering Department
Kettering University

The Author

Kenneth L. Kaiser's interest in electrical engineering began in high school with his involvement in amateur radio. While obtaining a solid theoretical background in a number of fields in electrical engineering, he obtained additional inspiration and practical experience working in several nonacademic positions. Since obtaining a Ph.D. in electrical engineering from Purdue University, he has focused his attention on the researching of topics of personal and industrial interest, on effective teaching methods, and on the writing of this book. He is currently at Kettering University (formerly GMI Engineering & Management Institute), continuing with his research and stimulating his students to excel.

Contents

1

Air Breakdown

The breakdown of the air between two charged conductors is a function of many factors including the voltage difference across the conductors, maximum possible current in the breakdown path, shape of the conductors, and atmospheric conditions. This breakdown is a source of electrical noise. For circuits containing switches, various networks can be added to reduce the effects of or eliminate the breakdown.

1.1 Breakdown Voltage

What is the minimum voltage necessary to break down the air between two parallel electrical contacts at room temperature and standard pressure? [Bell, '71; Holm; Somerville; Khalifa; Spangenberg; Cruft]

In introductory courses in electrical engineering, it is emphasized that no charge flows through air since the conductivity of air is zero.[1] Although any gas, liquid, or solid has a nonzero conductivity, in many instances, the conductivity of air and many other insulators is assumed zero since the current through the medium is extremely small. Since a major topic of this book is modeling the real behavior of devices, it should not be surprising that significant current can pass through air given the right conditions.

The presence of current implies the movement of free charge. In addition to the inherent free charge in a medium, this free charge can be due to impurities (e.g., dirt), frictional charging, ionization, and electron emission. Impurities can significantly raise a medium's conductivity, which is especially true for insulating materials such as Teflon or pure water. Frictional charging, also referred to as tribocharging, is the transfer of charge between two objects when they contact each other or are rubbed together.

Ionization is the process of tearing away one or more outer electrons from an atom thereby leaving the atom positively charged. There are several methods of ionizing: irradiation by rays of the appropriate frequency, strong electric fields, and impact from fast moving particles. The current due to background radioactivity and cosmic radiation is usually quite low (e.g., much less than a μA). When an atom is exposed to a strong electric field, however, free charges produced from the tearing away of the electron from their mother (nucleus) can be important. These freed electrons can acquire sufficient velocity and energy in the electric field to release other electrons if they collide with other atoms. The number of these collisions can avalanche producing very high current levels.

The production of free charges due to electron emission, unlike ionization, does not require atoms or molecules in the medium between two electrodes. There are a number of ways that electrons can be emitted or liberated from a metal surface or electrode. Again, electron emission can occur in a vacuum and does not require "resident" atoms in the medium. Electrons can be emitted from a metal surface if the metal is exposed to waves of sufficient energy, which is referred to as surface irradiation. For example, light of sufficient energy can cause electrons to be emitted from the metal, which is also

---

[1]The terms air and free space are frequently used interchangeably, even in this book. Free space, though, is generally considered to correspond to a perfect vacuum. The conductivity of air is not quite zero, and the conductivity of a perfect vacuum is zero. The relative permittivity and permeability of air (and free space) are both considered one in this book.

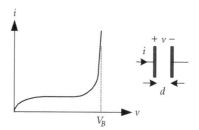

FIGURE 1.1 Townsend curve.

referred to as photoelectric emission. Electrons can be emitted from a metal surface if its temperature is sufficiently high, which is referred to as thermionic emission. The kinetic energy of the electrons is then great enough to escape the tug of the nucleus. Electrons can be emitted from a metal surface if hit with a high-velocity electron or ion, which is referred to as secondary emission. The energy required to release the electrons from the metal surface is obtained from the striking electron. Finally, electrons can be emitted from a metal surface when the electric field at the surface is very high, which is referred to as high-field or field emission. For a cold electrode (i.e., room temperature), a field of about 10^9 V/m is required for field emission. To obtain such large fields, using a reasonable voltage, often the cathode or electrode electron emitter is very sharp. Furthermore, the cathode material generally has a low work function, a high melting point, and high mechanical strength. (Tungsten, although it has a high work function, has a high melting temperature and can be pulled into a sharp point.) When the cathode is surrounded by air, rather than a vacuum, electron emission is not common, at least not for a very long time. Usually, in air, ionization of the surrounding air generates free charge that moves toward and impacts the cathode, and the cathode melts or vaporizes. Sharp pointed electrodes will become dulled from this ionized charge bombarding the cathode. Consequently, the electric field for a given voltage will drop, and the field emission will cease unless the voltage is increased. Of course, as in vacuum tubes, the field emission can be assisted by increasing the temperature of the cathode, as well as reducing the work function of the surface.

Before delving into the details of breakdown, it is instructive to examine the current-voltage relationship for closely spaced parallel plate electrodes with air between them. The electric field between the plates (away from the edges) is approximately uniform. The *i-v* curve for these electrodes, known as the Townsend current/voltage characteristics of a gas, is shown in Figure 1.1. The small initial current present at low voltage levels may be due to external radiation. As the voltage increases, the electric field increases and greater ionization current by impact or collision occurs. The uniform electric field between two parallel electrodes separated by a distance d is proportional to the voltage:

$$\left| \vec{E} \right| = \frac{v}{d} \tag{1.1}$$

Eventually, avalanching occurs and the gap breaks down: the change in the current is so large that the space (or a channel) between the electrodes appears like a short circuit or high-conductivity medium. This breakdown voltage, V_B, is a function of the gas, temperature, pressure, and contact separation distance, d. The general trend of the breakdown voltage vs. the product of the pressure, p, and the separation distance, d, is shown in Figure 1.2. (This sketch is based on the Paschen relationship.) The breakdown voltage is a function of the pressure, as well as the separation distance. The pressure determines the separation distance between the molecules and, hence, the average path length between collisions. At low pressures, the path is long and the probability of collision is low. At high pressures, the path is short, but it is more difficult to acquire sufficient kinetic energy over this short path to ionize another atom. After the product $(pd)_{min}$ is exceeded, if d is fixed and the pressure is increased, the voltage required

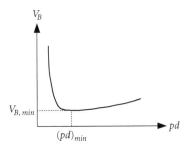

FIGURE 1.2 Relationship between the breakdown voltage of a gas and the pressure-distance product.

for breakdown increases: at higher pressures the gas density is greater, and the mean or average distance an electron can travel before colliding with a molecule is smaller.

For low values of pd, below $(pd)_{min}$, it is difficult to confirm experimentally the breakdown voltage. For values of pd greater than $(pd)_{min}$ an empirical (first-order) relationship for this breakdown voltage for uniform fields in air at atmospheric pressure is given by

$$V_B = 320 + 7 \times 10^6 d \qquad (1.2)$$

where d is the separation distance in meters. This equation is often used for "small" gaps less than approximately 100 μm but not too close to $(pd)_{min}$. For example, the minimum breakdown voltage, $V_{B,min}$, in air at one atmosphere is about 330 V at $d = 7.5$ μm, which is not obtained from (1.2). For distances greater than approximately 7.5 μm, the breakdown voltage is larger than 330 V. When the separation distance corresponds to a "large" gap, which is greater than approximately 100 μm, the guideline used is that the electric field should be greater than 3 MV/m. This corresponds to a voltage of

$$V_B = 3 \times 10^6 d \qquad (1.3)$$

At 1 mm, the breakdown voltage is 3 kV, and at 0.5 cm the voltage is 15 kV. Both of these voltages are greater than 330 V.

As seen, there are essentially two criteria for breakdown due to ionization. One is based on the voltage and the other on the electric field. Generally, both criteria must be satisfied for breakdown to occur. For large gaps at standard conditions, the electric field must exceed about 3 MV/m or 30 kV/cm ($\approx$ 76 kV/inch), while for small gaps, the voltage must exceed *about* 320 V.

Obviously, very close contacts or air gaps can break down for voltages less than 320 V. This breakdown is due to (or at least is initiated by) field emission. If the electric field at the surface of the electrode is very large, around 50 MV/m for contaminated surfaces (200 MV/m for clean surfaces), and the voltage is greater than the arcing voltage, V_A, the contact can break down. The arcing voltage is around 12 V. For breakdown due to electron field emission, therefore, the voltage across the gap, v, must satisfy both of the following inequalities:

$$50 \times 10^6 < \frac{v}{d} \quad \text{and} \quad v > V_A \qquad (1.4)$$

For voltages less than about 12 V, a gap is unlikely to break down. (This minimum arcing voltage varies with the contact material. Contacts constructed of certain gold alloys, for example, can have an arcing voltage less than 12 V. If the two electrodes are of different materials, the minimum arcing voltage is determined mainly by the cathode electrode.) Field emission is not visible unless the electrons strike some solid or gas so that light is emitted.

To summarize, to break down an air gap, large voltages (greater than about 320 V) are usually required. However, if the gap spacing is very small, breakdown can occur at lower voltages if the field is sufficiently intense. To sustain breakdown, as will be discussed shortly, a minimum current must be available to "feed" the breakdown.

1.2   Glows, Arcs, Coronas, and Sparks

What is the difference between a glow discharge, an arc, a corona, and a spark? [Bell, '71; Saums; Somerville; Khalifa; Holm; Copson; Browne]

The breakdown of air is a complex, definitely nonlinear, process. There are several different types of breakdown: glow, long arc, short arc, corona, and spark. If the voltage across a gap exceeds the critical breakdown voltage previously discussed, the current can increase rapidly through the gap. If the current is available, a glow or long arc discharge can be produced and sustained. Whether it is sustained is a function of the electrical circuit connected to the gap and the nature of the discharge path (e.g., length and cleanliness). Sometimes the term flashover is used to describe the arcing between two objects (before the objects make contact).

For a glow discharge, faint glowing areas or bands are present due to the ionization and excitation processes. In addition, for a glow discharge, as the avalanching process stabilizes, the voltage drop across the gap *decreases* from the Paschen limit given in (1.2) to (for palladium cathodes)

$$V_G = 280 + 1,000\, d \tag{1.5}$$

where d is the width of the gap in meters and V_G is the glow voltage. In this region, the current becomes independent of the voltage. Most of this voltage drop occurs near the cathode (i.e., the more negative terminal). The current for a glow discharge, I_G, varies from mA to A. As long as the current through the gap exceeds a minimum glow current and the voltage across the gap exceeds V_G, the glow discharge will be sustained. This minimum glow current is in the mA range. If the switch or gap is connected to a low-voltage high-resistance circuit, the minimum glow current and voltage may not be available, and the glow discharge may be extinguished. A neon tube is an example of a device that utilizes the glow discharge.

If current from high mA to A is available, the glow discharge can turn into a long arc (or an electric arc), also referred to as a luminous discharge. The voltage across the gap for an arc condition, V_A, is around 12 V and is a function of the electrode material and its level of contamination. The voltage does not change much with current. Note that the current required for an arc, I_A, is much greater than the current required for a glow discharge, but the voltage required is much less. The current during an arc can become so great that the electrodes melt or partially evaporate. An arc welder is a device that utilizes the arc discharge. Since the voltage for an arc is much less than for a glow discharge, it is possible to produce an arc while not producing a glow discharge. For example, the voltage across a switch may be only 20 V, but if the distance between the contacts is very small (e.g., during the initial opening of a switch), the electric field is very intense. If the current available to the switch is sufficient, an arc could be produced across the switch. An arc generated by field emission over a very short gap is referred to as a short or metal-vapor arc. The electrons can travel almost directly from one electrode to the other for a short arc. If a short arc is formed and then d is increased, a long arc can be produced and is said to be "drawn." In the arc region, the voltage drop across the arc or electrodes is small since the number of charge carriers or the conductivity in the arc is so large (at least compared to that of free space without any arcing or breakdown). The voltage-current relationship for these breakdown regions is sketched in Figure 1.3. When plotting or sketching several breakdown regions on the same set of axes, the current axis (and sometimes the voltage axis) is logarithmic. There are usually many orders of magnitude between the current well before a glow discharge and after an arc.

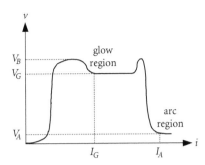

FIGURE 1.3 Approximate relationship between the voltage and current during breakdown.

Corona is a local, beautiful discharge near sharply curved surfaces such as wires. Since the electric field near objects where the curvature is high can be intense and nonuniform, the voltage required to begin corona is less than V_G, which assumed a uniform field in the previous discussions. The voltage required to produce an arc or a sparkover between electrodes is greater (and sometimes considerably greater) than the corona voltage. Corona is often the glow seen in dark conditions (e.g., at night) near those locations where the electric field diverges or converges.[2] In some situations, it is also referred to as St. Elmo's fire. For corona, breakdown occurs only in those regions where the electric field is strong and not everywhere across the gap. For this reason, it is referred to as a partial discharge. Corona is a function of the sign or polarity of the voltage applied to the electrode. Corona is also a source of power loss via heat and radiation (visual and radio frequencies). On high-voltage lines, corona is a source of power loss, acoustical noise, and electrical noise. The "snapping" due to corona can sometimes be detected on an AM automobile radio. In some applications, corona is desirable such as in electrostatic powder coating and smoke precipitators.

A spark discharge is a short duration, high electric field event. It is a transitory phenomenon. An example of this breakdown process is the sparking that sometimes occurs when metal objects such as conductive plates or utensils are used inside a microwave oven. The microwaves in the oven are reflected off the metal and the resultant electric field near the object can exceed the breakdown strength of the surrounding air. When the object is smooth without sharp edges, the likelihood of sparking is less. The effect of the sharpness of an object on the intensity of the field will be briefly discussed in a later section. (Although usually not recommended by the manufacturer because of the magnetron's cathode sensitivity to reflected energy and the intense sparking and heating which can occur, sometimes aluminum foil is wrapped around parts of a food item for selective heating control.) Again, a spark is a transient or temporary phenomenon. Frequently, though, the term spark is used by the public and engineers as any visible electrical discharge between two objects.

1.3 Nonuniform Fields and Time-Varying Arc

A metallic crane was approximately 1.5 ft from an overhead conductor[3] (12.5 kV rms line-to-line). Workers claim that an arc occurred between the crane and conductor without any direct contact. Is this possible? If the workers did actually notice an arc across this 1.5 ft span, what most likely happened? [Khalifa; Stevenson; Eaton; Greenwald; Somerville; Bazelyan]

---

[2]The color of the glow can vary with the polarity of the applied voltage and available energy to the discharging electrode. With a negative voltage, the glow is bluish, while with a positive voltage, the color is purple-like. Some observations indicate that positive corona can also be blue-like.

[3]It is common in the power field to use the terms line and conductor interchangeably. In other fields, a line is considered a collection of two or more conductors.

Before delving into this problem, background information on the distribution of power will be presented.[4] The magnitude of the voltage on power transmission lines between generating stations (e.g., hydroelectric plant) and distribution systems is quite high (e.g., 69 to 765 kV rms line-to-line). These high voltages allow for lower current levels and lower I^2R losses. However, there are costs associated with these high voltage levels that include increased likelihood of corona, larger spacing between conductors to prevent arcing, and greater breakdown strength of the insulators. The conductors supported by massive metal and wood structures, often located near highways and low-population regions, are these high-voltage lines. These high voltages are fed to bulk-power substations where the voltage level is reduced to 34.5 to 138 kV rms line-to-line. (Some industrial consumers that require high power levels may need these high voltages.) These voltages are then fed to distribution substations where they are reduced to 4 to 34.5 kV rms line-to-line. The voltages leaving these substations feed residential, commercial, and most industrial users. The portion of the power system from these substations to the pole transformers is referred to as the primary distribution system. The poles, many of which are wood, carry the primary distribution lines. The primary line voltage is stepped down at the pole transformer to 120, 240, or 480 V rms line-to-ground when needed by a customer. The system from the pole transformer to the actual user is referred to as the secondary distribution system. (In some locations, the primary lines are buried and pad-mounted, surface-level transformers are used.)

A curious student, after reading the previous discussion, may decide to study the power lines near the highway to reduce boredom on a long trip. It will be immediately apparent that there is not one standard way of arranging the power lines on the structures or support lattices. Since primary distribution lines are more common near high-population regions where accidents are more likely to occur, only these will be discussed. The power from the substation is usually three phase. The three hot conductors, for example, may be 7.2 kV rms (line-to-ground) at 0°, 7.2 kV rms at 120°, and 7.2 kV rms at −120°. If the source were Y-connected, a neutral or return conductor would also be present, while if the source were Δ-connected, the neutral conductor would be absent. The pole transformer and lines to the consumer (the secondary distribution) are located below these three phase lines. One possible line layout is shown in Figure 1.4.

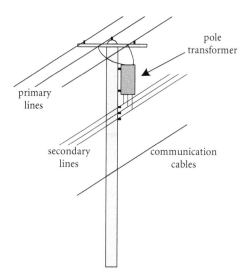

FIGURE 1.4   One possible layout of power and communication lines on a pole. Sometimes a smaller diameter lightning "shield" conductor is seen above the primary lines.

---

[4]Although this and the next few paragraphs are relevant background material, this information is not required to answer the given question. It was included since few engineering graduates have an understanding of these important power-related topics.

The three hot lines are located on the horizontal cross arm, and the pole transformer is mounted on the side of the pole. Fuses or breakers, which are not shown, are located on the primary side of the transformer. Communication cables (e.g., cable TV and telephone lines) can be located below the power lines on the same pole.

On some poles, only two primary lines are present. This represents one phase of the three-phase system from the substation. For a two-wire single-phase system, one line is hot and the other line is the neutral or return. On other poles, even though three primary lines are present, the system is actually two phase: two hot lines are 90° out of phase while the remaining line is the neutral. The two hot lines, for example, may be 7.2 kV rms (line-to-ground) at 90° and 7.2 kV rms at 0°.

The secondary lines for residential users are usually single phase even though there are three conductors present. The center conductor is the neutral or ground wire while the two remaining lines are hot and 180° out of phase. The two hot lines, for example, may be 120 V rms at 0° (line-to-ground) and 120 V rms at 180° (= −120 V rms at 0°). Notice that the difference between the two hot voltages is 240 V rms. The two hot voltages are in phase (but of opposite sign).[5] The 240 V rms is used for high-electrical load devices and appliances. To obtain these two hot voltages, the center-tap of the secondary of the pole transformer is the ground or neutral while the two ends of the secondary are the hot voltages.

With this background information, the effect of a time-varying nonuniform field on the production of an arc can be explored: the voltage on an ac line is time varying, not dc, and the electric field is definitely nonuniform near the conductors. Since the breakdown process occurs in a very small period of time (0.01 to 1 μsec) compared to one-half of a cycle (1/120 sec ≈ 8.3 msec) of the line frequency, the breakdown mechanism for the 60 Hz power frequency is similar to dc. Of course, the direction of the ions change as the polarity of the field changes. There is one important difference between ac and dc. With an ac signal, the current to the arc varies and drops to zero twice in every period. Thus, the ionization production drops to a small value when the ac current is at zero but the deionization processes continues (e.g., recombination of the negative particles with the positive particles). Obviously, the current in a stable arc is constant for dc. Whether the arc is sustained for an ac source, after the current passes through zero, is dependent on the rate of ionization and rate of the rise in the voltage across the gap. The voltage across the gap is a function of the circuit that is connected to the gap.

How the electrode responds after the current passes through zero (for ac voltages) is a function of the electrode material. For electrode materials with high melting points such as carbon or tungsten, thermal emission will occur at the electrodes at high field strengths, and the arc will be sustained even for low voltages (approximate amplitude of 15 to 20 V), assuming adequate current is available. The electrodes can become sufficiently hot, without melting, to emit thermally electrons (a burning-spot-free arc). However, for electrode materials with low melting points such as copper, a much greater voltage (approximate amplitude of 300 to 400 V) is required across the gap to sustain the arc after the current passes through zero. This breakdown is a burning-spot arc. Note that copper contacts are usually used in electrical receptacles and the peak amplitude of the outlet voltage is $120\sqrt{2} \approx 170 < 300$ V.

In a strong nonuniform electric field, ionization may occur in those regions where the electric field strength is large but not in those regions where it is small. In other words, corona can occur in the high-field regions. A guideline seen for nonuniform field discharges is that if the maximum field at an electrode is less than five times the average field, the discharge is similar to a uniform field. Near a high-voltage conductor, the field is not uniform. The actual electric field is a function of the distance from the conductor and the transmission line configuration. Referring to the original question posed, even if the electric field were relatively uniform at a distance 1.5 ft from the conductor, the crane itself would distort the field. The degree of distortion is dependent on the physical shape of the crane. For example, if the crane tip was modeled as a conducting floating cylinder parallel to the high-voltage conductor, the maximum field at the surface of the crane could be twice the uniform field level.

---

[5]It is traditional to refer to this as a single-phase system even though there are two hot voltages of opposite sign present (i.e., 120 V rms and −120 V rms).

A crude estimate for the electric field near the crane can be obtained by assuming that the electric field is mainly due to the nearest high-voltage conductor and that the crane is grounded. The voltage in this situation is much greater than the minimum 320 V required for a glow discharge, and for separation distances as large as 1.5 ft, it is probably best to use the electric field guideline of 3 MV/m. The voltage of 12.5 kV rms line-to-line is a common primary distribution voltage. The electric field across two parallel electrodes spaced 1.5 ft apart with a corresponding peak line-to-ground voltage is

$$\left| \vec{E} \right| = \frac{\dfrac{12.5 \times 10^3 \sqrt{2}}{\sqrt{3}}}{1.5 \times 0.305} \approx 22 \, \text{kV/m} \ll 3 \, \text{MV/m}$$

The $\sqrt{2}$ multiplier converts the rms voltage to a peak voltage, and the $1/\sqrt{3}$ multiplier converts the line-to-line voltage to a line-to-ground voltage. It appears unlikely that air breakdown occurred across the 1.5 ft span. Of course, environmental factors, such as whether it was foggy, can affect the breakdown potential of the air.

It is more likely that the crane itself touched (or was in very close proximity to) the high-voltage line and then pulled away from the line. In this case, a short arc would have been initially produced between the crane and line since the field intensity would have been large. (Many of the conductors have no external insulation and are constructed of aluminum.) Then, a long arc could have been drawn from this short arc as the crane was pulled away. If sufficient current were available to feed the long arc, the workers would have seen a long arc between the crane and the conductor. Whether an arc of this length could have been produced or sustained is dependent on factors such as the water content of the air and the composition, shape, and condition of the crane and line conductor.

1.4   Ideal Switching of Simple Loads

When the load in series with a switch is purely resistive, inductive, or capacitive, what is the voltage across the switch after it is opened? Assume that no breakdown occurs. Why is noise produced? [Eaton]

The voltage across a gap is a major factor in determining whether a gap will break down. The voltage across the gap when the load consists of a single resistor, single inductor, or single capacitor will be determined in this discussion. Situations that allow for breakdown and more general loads are discussed in later examples.

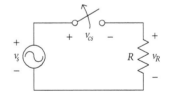

FIGURE 1.5   Switching a purely resistive circuit.

Assume that a switch interrupts a time-varying current through a purely resistive load as shown in Figure 1.5. Using Kirchoff's voltage law, the voltage across the contacts (or switch) is

$$v_{cs} = v_s - v_R = A\cos(\omega t) - v_R$$

where the source voltage is assumed sinusoidal. *Before* the switch is opened, the voltage across the resistor is equal to the voltage of the source, since the voltage across the ideal switch is zero. Also, the current is in phase with the supply voltage: when the current through the resistor is zero, the voltage across the resistor is zero. The voltage across the switch *after* it opens is

$$v_{cs} = A\cos(\omega t) - 0(R) = A\cos(\omega t)$$

The voltage source appears entirely across the switch, and the current through the resistor is zero. If the switch is opened when the ac current is zero, the voltage of the source is zero, and the voltage across the switch both immediately before and after the opening is also zero. There is no sudden change in voltage across the switch if the switch is opened during a zero-current crossing. This reduces the likelihood of breakdown across the

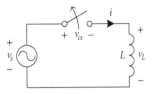

FIGURE 1.6 Switching a purely inductive circuit.

contacts of the switch. The change in the voltage after a zero-current crossing is determined solely by the nature of the voltage source. If the switch is opened when the current is not zero, the voltage jump across the switch at the time of the opening is $A\cos(\omega t)$. The maximum voltage drop across the contacts is thus A. In this and the following scenarios, no breakdown is assumed so the current through the switch and load is zero after the opening.

When the load is purely inductive, the approach is similar, but the results are more interesting. This load could be representing a relay coil or dc motor. The voltage across the contacts shown in Figure 1.6 is

$$v_{cs} = v_s - v_L = A\cos(\omega t) - v_L$$

Before the switch is opened, the voltage across the inductor is equal to the source voltage and

$$v_s - v_L = A\cos(\omega t) - L\frac{di}{dt} = 0$$

Unlike the resistive load, the current through the inductor is 90° out of phase with the voltage across it:

$$\frac{di}{dt} = \frac{A}{L}\cos(\omega t) \quad \Rightarrow \quad i(t) = \frac{A}{L}\int_0^t \cos(\omega t)\, dt + i(0) = \frac{A}{L\omega}\sin(\omega t) + i(0)$$

Therefore, if the voltage across the inductor is zero *before* opening the switch, the current through it is not zero (assuming $i(0) = 0$). If the switch is opened during a zero-current crossing point, the voltage across the contacts immediately after opening the switch is

$$v_{cs} = A\cos(\omega t) - L(0) = A\cos(\omega t)$$

The current is zero before and after the switch is opened, and the rate of change in current through the inductor is zero. However, unlike the resistive load, the voltage across the switch is not zero but $\pm A$ at the zero-current crossings. The voltage and current are 90° out of phase. (One zero-current crossing occurs at $\omega t = 0$, if $i(0) = 0$, but the voltage is $A\cos(0) = A$ at this time.) This voltage across the switch increases the likelihood of breakdown. Now, if the switch is opened during a time when the current is not zero, the inductor will "kickback." The current through an inductor cannot change instantaneously. The voltage across the switch is

$$v_{cs} = A\cos(\omega t) - L\frac{di}{dt} \approx A\cos(\omega t) - L\frac{\Delta i}{\Delta t} = A\cos(\omega t) - L\frac{0 - i}{\Delta t}$$

$$= A\cos(\omega t) + L\frac{i}{\Delta t}$$

(1.6)

where i is the current before the switch was opened and Δt is the time required for the current to drop to zero. Since this time is very short, the "kickback" voltage across the inductor is quite large. (Using this over simplified model, when Δt is zero, the kickback voltage is infinite. A better model for this circuit would allow for breakdown and include parasitic capacitance. This would prevent an infinite voltage transient across the inductor.) This

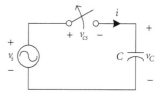

FIGURE 1.7   Switching a purely capacitive circuit.

large voltage definitely increases the likelihood of breakdown across the switch. Opening a switch at a zero-current crossing is obviously advantageous. Switch packages are available that include the necessary electronics to wait for a zero-current crossing before opening.

The final load that will be examined is a purely capacitive load. While inductors resist current change, capacitors resist voltage change. The voltage across the switch shown in Figure 1.7 is

$$v_{cs} = v_s - v_C = A\cos(\omega t) - v_C$$

Before the switch is opened, the voltage across the capacitor is equal to the source voltage as with the resistive and inductive loads. However, since the voltage across the capacitor cannot change instantaneously, after the switch is opened, the voltage across the contacts is zero:

$$v_{cs} = A\cos(\omega t) - A\cos(\omega t) = 0$$

However, the voltage across the contacts does not remain at zero. Note that before the switch is opened, the current and voltage are 90° out of phase since

$$i = C\frac{dv_C}{dt} = C\frac{d}{dt}[A\cos(\omega t)] = -CA\omega\sin(\omega t)$$

If the switch is opened during a zero-current crossing, the voltage across the capacitor is not zero because of this phase difference. At $\omega t = 0$, $i = -CA\omega\sin(0) = 0$ while $v_C = A\cos(0) = A$. If the switch is opened during a zero-current crossing, the magnitude of the voltage across the capacitor both immediately before and after the switching is A. Although the voltage across the switch is initially zero after the opening, the voltage across the switch will swing from 0 to $-2A$ as time progresses to a zero-current crossing:

$$v_{cs}(t) = A\cos(\omega t) - A \quad \text{if the switch is opened at } t = 0 \tag{1.7}$$

This large voltage will increase the likelihood of a *delayed* breakdown, which may be more difficult to locate. If the initial voltage across the capacitor is $-A$ immediately prior to the switching, the voltage across the switch swings from 0 to $+2A$ after the switch opens. If the switch is opened during a nonzero-current crossing, the maximum magnitude of the voltage across the switch will be less than $2A$. Depending on the voltage across the capacitor at the time of the switching, assumed at $t = t_o$, the switch voltage is

$$v_{cs}(t) = A\cos(\omega t) - v_C(t_o) \tag{1.8}$$

The voltage across the capacitor remains constant after the switch opens since no closed path is available to charge or discharge the ideal capacitive load.

When the current in the circuit suddenly changes in value, high-frequency noise is produced. The second break frequency of a trapezoidal waveform is $1/(\pi\tau_f)$ Hz. As the fall time, τ_f, of the signal decreases, the high-frequency spectral content of the signal increases. As will be seen, noise is also produced when the switch begins to break down.

As shown, there are advantages to interrupting an ac circuit during a zero-current crossing. If a switch is opened at any time, not just during a zero-current crossing, a purely inductive load is likely to generate the largest voltage across the switch's contacts because of its tendency to resist current change. A purely capacitive load can generate a voltage of twice the amplitude of the supply voltage across the switch, but not at the time of the opening of a switch. A purely resistive load will generate a maximum voltage equal to the amplitude of the supply voltage across the switch. The time when this maximum voltage occurs for a resistive load is dependent on the time of opening.

1.5 Ideal Switching of Complex Loads

When an *RL*, *RC*, or *RLC* load is in series with a switch, what is the voltage across the switch after it is opened? Assume that no breakdown occurs. [Eaton]

In this discussion, several standard combinations of resistors, inductors, and capacitors will be analyzed as loads. As will be seen, a series *RL* or series *RC* load can reduce the voltage across an opening switch, while *RLC* loads can produce oscillations and additional noise. The switch could be representing a circuit breaker, light switch, or thermostat.

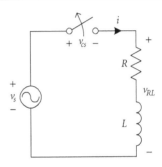

FIGURE 1.8 Switching of a series *RL* circuit.

A switch in series with a series *RL* load is shown in Figure 1.8. The voltage across the switch is

$$v_{cs} = v_s - v_{RL} = A\cos(\omega t) - v_{RL}$$

Before the switch is opened, the voltage across the load is equal to the voltage of the source. The current from the source is not in phase with the voltage of the source, unless $L = 0$. If $R = 0$, then the current and voltage are 90° out of phase. The current through the inductor, working in the frequency domain, is

$$I_s = \frac{A\angle 0°}{R + j\omega L} = \frac{A\angle 0°}{\sqrt{R^2 + (\omega L)^2}\angle \tan^{-1}\left(\dfrac{\omega L}{R}\right)} = \frac{A}{\sqrt{R^2 + (\omega L)^2}}\angle - \tan^{-1}\left(\frac{\omega L}{R}\right)$$

or in the time domain

$$i = \frac{A}{\sqrt{R^2 + (\omega L)^2}}\cos\left[\omega t - \tan^{-1}\left(\frac{\omega L}{R}\right)\right]$$

The magnitude of the phase difference between the current through and voltage across this load is $\tan^{-1}(\omega L/R)$. If the switch is opened during a zero-current crossing, for example, when

$$\omega t - \tan^{-1}\left(\frac{\omega L}{R}\right) = \frac{\pi}{2} \quad \text{or} \quad \omega t = \frac{\pi}{2} + \tan^{-1}\left(\frac{\omega L}{R}\right) \tag{1.9}$$

then the voltage across the switch immediately after the opening is

$$v_{cs} = A\cos(\omega t) - iR - L\frac{di}{dt} = A\cos\left[\frac{\pi}{2} + \tan^{-1}\left(\frac{\omega L}{R}\right)\right] - (0)R - L(0)$$

$$= A\cos\left[\frac{\pi}{2} + \tan^{-1}\left(\frac{\omega L}{R}\right)\right]$$

$$(1.10)$$

In this case, the current through the inductor (after and immediately before the switch is opened) is zero and $di/dt = 0$. The voltage across the switch immediately after the opening has a magnitude less than A. Thus, the resistance reduces the maximum voltage that can appear across the switch as compared to a purely inductive load. When $R \gg \omega L$, $v_{cs} \approx 0$ immediately after the zero-current crossing opening. If $R \ll \omega L$, $v_{cs} \approx -A$ immediately after the opening. If the switch is not opened at a zero-current crossing, the voltage across the switch can be approximated as

$$v_{cs} = A\cos(\omega t) - iR - L\frac{di}{dt} \approx A\cos(\omega t) - (0)R - L\frac{0-i}{\Delta t} = A\cos(\omega t) + L\frac{i}{\Delta t}$$

$$(1.11)$$

(Obviously, this expression does not consider many real factors such as arcing across the switch and the contact's nonlinear time-dependent resistance.) The current through the inductor cannot change instantaneously. The variable Δt is the time required for the current to drop from i to zero. The kickback voltage from the inductor is proportional to L and i and inversely proportional to Δt. This voltage can be substantial. The voltage across the switch during the Δt interval will change since the current through the resistor decreases. The voltage across the resistor decreases from iR to 0 over this interval. Often,

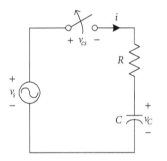

FIGURE 1.9 Switching of a series *RC* circuit.

the voltage across the inductor will be much larger than the voltage across the resistor. It is important to state that during switching operations energy is lost through radiation. This form of energy loss during breakdown is not modeled in this book.

A switch in series with a series *RC* load is shown in Figure 1.9. This load could be representing a piezoelectric buzzer. As with the *RL* load, the current from the source is not in phase with the voltage of the source unless the capacitor is replaced with a short circuit. If $R = 0$, then the current and voltage are 90° out of phase. Therefore, the voltage across the load is not zero if the current through the load is zero. For an *RC* load, the magnitude of the phase difference between the current through and voltage across this load is $\tan^{-1}[1/(\omega RC)]$. Working in the frequency domain, the current through the switch *before* it is opened is

$$I_s = \frac{A\angle 0°}{R + \frac{-j}{\omega C}} = \frac{A\angle 0°}{\sqrt{R^2 + \left(\frac{1}{\omega C}\right)^2}\angle\tan^{-1}\left(\frac{-1}{\omega RC}\right)} = \frac{A}{\sqrt{R^2 + \left(\frac{1}{\omega C}\right)^2}}\angle\tan^{-1}\left(\frac{1}{\omega RC}\right)$$

Therefore, the voltage across the capacitor is

$$V_{Cs} = I_s\frac{-j}{\omega C} = \frac{A}{\sqrt{(\omega RC)^2 + 1}}\angle\left[\tan^{-1}\left(\frac{1}{\omega RC}\right) - \frac{\pi}{2}\right]$$

The corresponding current and voltage in the time domain are

$$i = \frac{A}{\sqrt{R^2 + \left(\dfrac{1}{\omega C}\right)^2}} \cos\left[\omega t + \tan^{-1}\left(\frac{1}{\omega RC}\right)\right]$$

$$v_C = \frac{A}{\sqrt{(\omega RC)^2 + 1}} \cos\left[\omega t + \tan^{-1}\left(\frac{1}{\omega RC}\right) - \frac{\pi}{2}\right] \tag{1.12}$$

The voltage across the switch *after* it is opened ($i = 0$) is then

$$v_{cs} = A\cos(\omega t) - iR - v_C$$

$$= A\cos(\omega t) - \frac{A}{\sqrt{(\omega RC)^2 + 1}} \cos\left[\omega t + \tan^{-1}\left(\frac{1}{\omega RC}\right) - \frac{\pi}{2}\right] \tag{1.13}$$

If the switch is opened during a zero-current crossing, for example, when

$$\omega t + \tan^{-1}\left(\frac{1}{\omega RC}\right) = \frac{\pi}{2} \quad \text{or} \quad \omega t = \frac{\pi}{2} - \tan^{-1}\left(\frac{1}{\omega RC}\right) \tag{1.14}$$

then the voltage across the switch immediately after the opening is

$$v_{cs} = A\cos\left[\frac{\pi}{2} - \tan^{-1}\left(\frac{1}{\omega RC}\right)\right]$$

$$- \frac{A}{\sqrt{(\omega RC)^2 + 1}} \cos\left[\frac{\pi}{2} - \tan^{-1}\left(\frac{1}{\omega RC}\right) + \tan^{-1}\left(\frac{1}{\omega RC}\right) - \frac{\pi}{2}\right] \tag{1.15}$$

$$= A\cos\left[\frac{\pi}{2} - \tan^{-1}\left(\frac{1}{\omega RC}\right)\right] - \frac{A}{\sqrt{(\omega RC)^2 + 1}}$$

The voltage across the switch immediately *after* the opening has a magnitude less than A. When $R \gg 1/(\omega C)$, $v_{cs} \approx -A/(\omega RC)$ immediately after the opening. If $R \ll 1/(\omega C)$, $v_{cs} \approx 0$ immediately after the opening. Compared to a purely capacitive load, the resistance increases the voltage across the switch immediately after the opening. However, the maximum magnitude of the voltage over time after the opening, will be less with a nonzero value for R. For times after the switch opening, when $R = 0$ the maximum magnitude of the voltage across the switch is $2A$. Referring to (1.12), the voltage across the capacitor, which remains constant after the switch opens, decreases with increasing R. The dc offset of the voltage across the switch is this capacitor voltage. As the dc offset decreases, the maximum magnitude decreases (the peak-to-peak voltage swing is not affected by the dc offset).

The final load to be analyzed, shown in Figure 1.10, is a series RL circuit in shunt with a capacitance. This C could be representing the parasitic capacitance across the RL load. Before the switch is opened, the voltage across the entire load is equal to the source voltage. Since the capacitor is in parallel with the

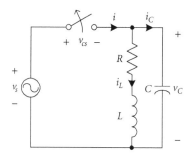

FIGURE 1.10   Switching of an *RLC* circuit.

source voltage and its voltage cannot change instantaneously, the voltage across the switch both immediately before and after the switch is opened is

$$v_{cs} = v_s - v_C = A\cos(\omega t) - A\cos(\omega t) = 0$$

After the switch is opened, the voltage across the switch will vary according to the frequency of the source and ringing frequency of the *RLC* load. To determine the expression for the transient voltage across the load, the initial current through the inductor and the initial voltage across the capacitor must be known. The initial voltage across the capacitor is

$$v_C(t_o) = A\cos(\omega t_o)$$

where t_o corresponds to the time the switch is opened. The initial current through the inductor is

$$i_L(t_o) = \frac{A}{\sqrt{R^2 + (\omega L)^2}} \cos\left[\omega t_o - \tan^{-1}\left(\frac{\omega L}{R}\right)\right]$$

The standard solution for the voltage across the capacitor in a series *RLC* circuit is (for the underdamped case assuming that the circuit is source free)

$$v_C(t) = E e^{-\alpha t} \cos(\omega_d t) + F e^{-\alpha t} \sin(\omega_d t)$$

where

$$\alpha = \frac{R}{2L}, \quad \omega_d = \sqrt{\omega_o^2 - \alpha^2} = \sqrt{\frac{1}{LC} - \left(\frac{R}{2L}\right)^2}$$

The constants *E* and *F* can be determined by setting

$$v_C(t = t_o) = A\cos(\omega t_o) \quad \text{and} \quad i_L(t_o) = -C\frac{dv_C(t)}{dt}\bigg|_{t=t_o}$$

After the switch is opened, the current through the inductor and capacitor are identical since they are in series. Thus, the initial current through the capacitor is equal to the initial current through the inductor, which cannot change instantaneously. The constants E and F can be determined using these initial conditions:

$$v_C(t_o) = Ee^{-\alpha t_o}\cos(\omega_d t_o) + Fe^{-\alpha t_o}\sin(\omega_d t_o) \qquad (1.16)$$

$$i_L(t_o) = -C\frac{d}{dt}[Ee^{-\alpha t}\cos(\omega_d t) + Fe^{-\alpha t}\sin(\omega_d t)]\Big|_{t=t_o}$$

$$= -C\begin{bmatrix} -\alpha Ee^{-\alpha t_o}\cos(\omega_d t_o) - \omega_d Ee^{-\alpha t_o}\sin(\omega_d t_o) \\ -\alpha Fe^{-\alpha t_o}\sin(\omega_d t_o) + \omega_d Fe^{-\alpha t_o}\cos(\omega_d t_o) \end{bmatrix} \qquad (1.17)$$

Because of its importance, the voltage across the switch for the zero-current crossing case will be determined. To simplify the equations, it is assumed that the zero-current crossing occurs at $t = t_o = 0$. This requires that the original current and voltage conditions be modified to (since $v_C(0) = A\cos(0) = A$ and $i_L(0) \neq 0$)

$$i_L(t) = B\cos(\omega t + 90°)$$

$$v_C(t) = i_L(t)R + L\frac{di_L(t)}{dt} = B\sqrt{R^2 + (\omega L)^2}\cos\left[\omega t + \tan^{-1}\left(\frac{R}{-\omega L}\right)\right] \qquad (1.18)$$

For this set of equations, $i_L(0) = 0$ and

$$v_C(0) = B\sqrt{R^2 + (\omega L)^2}\cos\left[\tan^{-1}\left(\frac{R}{-\omega L}\right)\right] = B\sqrt{R^2 + (\omega L)^2}\,\frac{-\omega L}{\sqrt{R^2 + (\omega L)^2}} = -B\omega L$$

(Note the quadrant of the argument of the arctangent function.) These initial conditions can now be used in (1.16) and (1.17):

$$-B\omega L = E$$

$$0 = -C(-\alpha E + \omega_d F)$$

Solving for E and F and substituting into the voltage expression yields

$$v_C(t) = -B\omega Le^{-\alpha t}\left[\cos(\omega_d t) + \frac{\alpha}{\omega_d}\sin(\omega_d t)\right]$$

The voltage across the switch is therefore

$$v_{cs}(t) = B\sqrt{R^2 + (\omega L)^2}\cos\left[\omega t + \tan^{-1}\left(\frac{R}{-\omega L}\right)\right] - v_C(t)$$

$$= B\sqrt{R^2 + (\omega L)^2}\cos\left[\omega t + \tan^{-1}\left(\frac{\omega L}{R}\right) + \frac{\pi}{2}\right] \qquad (1.19)$$

$$+ B\omega Le^{-\alpha t}\left[\cos(\omega_d t) + \frac{\alpha}{\omega_d}\sin(\omega_d t)\right]$$

This expression assumes that the switch is opened during a zero-current crossing. A plot of this equation for one set of *RLC* values is given in Mathcad 1.1. Often the ringing frequency of the *RLC* load (e.g., 5 kHz) is

$$R := 1 \quad L := 1 \cdot 10^{-3} \quad C := 1 \cdot 10^{-6} \quad \omega := 377 \quad B := 10$$

$$\alpha := \frac{R}{2 \cdot L} \quad \omega_d := \sqrt{\frac{1}{L \cdot C} - \left(\frac{R}{2 \cdot L}\right)^2} \qquad \frac{\omega_d}{2 \cdot \pi} = 5.032 \times 10^3$$

$$t := 0, 0.000001 .. 0.01$$

$$v_s(t) := B \cdot \sqrt{R^2 + (\omega \cdot L)^2} \cdot \cos\left(\omega \cdot t + \operatorname{atan}\left(\frac{\omega \cdot L}{R}\right) + \frac{\pi}{2}\right)$$

$$v_c(t) := -B \cdot \omega \cdot L \cdot e^{-\alpha \cdot t} \cdot \left(\cos(\omega_d \cdot t) + \frac{\alpha}{\omega_d} \cdot \sin(\omega_d \cdot t)\right)$$

$$v_{cs}(t) := v_s(t) - v_c(t)$$

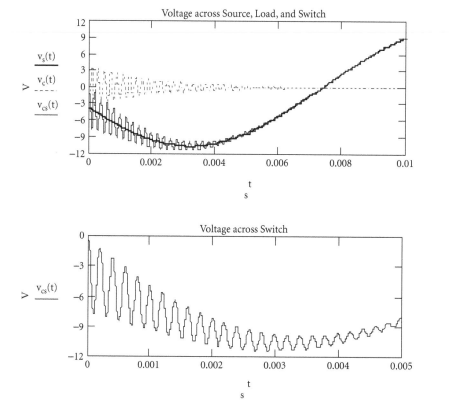

MATHCAD 1.1 Voltage of the source, voltage across the *RLC* load, and voltage across the opened switch.

much greater than the frequency of the source (e.g., 60 Hz). The oscillatory nature of the voltage across the switch is clearly evident in this example. This oscillation is a source of electrical noise.

1.6   Switching and Breakdown

When can breakdown occur across a closing switch? [Paul, '92(b); Bell, '71]

It is clear from the previous discussions that the various types of breakdown are voltage and current dependent. This dependency is clearly seen by analyzing the simple resistive-load dc circuit shown in Figure 1.11. Recall that arcing can occur at low voltages and gas discharge can occur at high voltages.

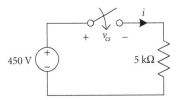

FIGURE 1.11 Switching of a purely resistive load.

When the distance between the contacts is greater than about 100 μm (a large gap), the breakdown guideline is that the electric field should be greater than 3 MV/m. As given in (1.3), this corresponds to a voltage of

$$V_B = 3 \times 10^6 d$$

across the switch. Of course for a real switch, the contacts are not flat parallel electrodes and the electric field can be more intense at the ends or at sharp points. That is, the shape of the contacts will affect the breakdown voltage. When the distance between the contacts is $d = 150$ μm, the required voltage across the switch is 450 V. Thus, it is possible at this distance that breakdown may be initiated. However, to sustain the breakdown, sufficient current must be available. If all 450 V is dropped across the switch, no voltage is available for the resistor, and the current is (450 − 450 V)/5 kΩ = 0. The breakdown will not be sustained. If the distance between the contacts drops to 110 μm, $V_B = 330$ V, and the electric field is 3 MV/m. The current available is now (450 − 330 V)/5 kΩ = 24 mA. Typical current values to sustain a glow discharge, I_G, are 10 to 600 mA. In this example, a glow discharge may or may not be sustained at this point. As the switch continues to close, the distance between the contacts will be less than 100 μm and the likelihood of breakdown increases.

As the contact distance becomes smaller, the glow discharge may develop into an arc. For an arc to develop and be sustained, the electric field must be greater than 50 MV/m (for contaminated surfaces):

$$\frac{v}{d} > 50 \times 10^6$$

Also, the voltage across the switch must be greater than the minimum arcing voltage,

$$v > V_A$$

and the current must be greater than the minimum arcing current, I_A. The minimum arcing voltage is around 12 V, which is much less than the 450 V supply voltage. In order for the electric field to be greater than 50 MV/m, the distance between the contacts must be (450 V)/(50 MV/m) = 9 μm. Finally, if the arc is developed, the available current is (450 − 12 V)/5 kΩ ≈ 88 mA. Minimum arc-sustaining currents typically range from 50 mA to 1 A. (The minimum arcing current and voltage are a function of the contact material. It is an interesting observation that the minimum arcing voltage appears to increase with the innate hardness of a metal.) It is not obvious whether an arc will be initiated or sustained in this case. Notice that the voltage required across a switch to produce an arc is much less than for a glow discharge, however, the required current is much greater. If the current is not sufficient to sustain an arc, the breakdown will revert to a glow discharge. Actually, the situation is even more complex than stated. For example, the current required to initiate an arc is sometimes greater than the current required to sustain an arc. Finally, at $d = 0$ the current rises to 450 V/5 kΩ = 90 mA.

1.7 Showering Arc

What is a showering arc, and why does it occur? [Paul, '92(b); Bell, '71; Holm; Horowitz]

If the arc current and voltage requirements are met, an arc will be sustained. It is common, however, for an arc to be ignited, extinguished, reignited, and reextinguished in an oscillatory fashion. This oscillating breakdown is referred to as arc showering and is possibly a consequence of the parasitic capacitance and inductance associated with the contacts and connecting circuitry. It was shown previously that a load consisting of a series *RL* circuit in parallel with a capacitance, *C*, could produce oscillations. These oscillations can cause the current and voltage to rise above and fall below the minimum arc sustaining levels. This produces the showering arc. There are two types of showering arc: arc-controlled and circuit-controlled. An arc-controlled shower is dependent mainly on the properties of the metal contacts. A circuit-controlled arc is dependent mainly on the circuit connected to the switch. Arc showering is a source of electrical noise since the current is oscillatory.

The physical bouncing of the switch can also cause this showering effect. Switches when closed do not just make contact and stop: they make and break contact several times. The frequency of this mechanical bouncing is usually much lower than the frequency due to the arc-controlled and circuit-controlled showering. Sometimes, several milliseconds are required for the bouncing to settle. For large switches, the bouncing can last for 50 ms. (Sometimes, several bouncy switches are placed in parallel to increase the probability of at least one switch making contact.)

1.8 Speed of Switching

Why is the speed of switching a factor in breakdown? [Bell, '71; Paul, '92(b)]

Initially it may be surprising that the speed of the switch is a factor in breakdown. The following discussion should provide the necessary illumination. Rather than sketching the voltage across a switch vs. the contact separation, *d*, it is more helpful to sketch the switch voltage vs. time, *t*. The contact separation for a switch usually does not remain at a static fixed value but varies as the switch is opened or closed. Assuming that the speed of the switch is constant, the voltage vs. time graph for breakdown is as shown in Figure 1.12. This graph of the required voltage for arcing and gas discharge is for an opening switch. The time, *t*, and the distance, *d*, between the contacts are related by $t = d/s$, where *s* is the constant speed of the opening. When *t* is small, *d* is small, and the electric field is intense. Arcing is the major breakdown process for small distances. For larger times, *d* is larger, the electric field is less intense, and gas discharge is the more likely breakdown process. Of course, once breakdown occurs for a gas discharge and sufficient current is available to sustain it, the voltage across the switch drops from

$$V_B = 320 + 7 \times 10^6 d \quad \text{to} \quad V_G = 280 + 1{,}000d$$

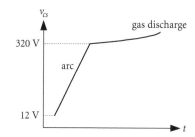

FIGURE 1.12 Voltage across a switch vs. time as the contact separation increases.

For large times and distances, the electric field guideline of 3 MV/m is often used instead of these voltage expressions. For gas discharge at these larger distances, the voltage must often be greater than 320 V and the electric field greater than 3 MV/m.

The slope of the arc portion of the voltage vs. time sketch is a factor in the design of arc-suppression circuitry. If the voltage across the switch were plotted vs. distance, the slope of the arc portion would be equal to 50 MV/m or more depending on the contact material and its state (e.g., cleanliness):[6]

$$\frac{\partial v_{cs}}{\partial d} = E_A \approx 50 \text{ MV/m} \tag{1.20}$$

The slope of the arc region for the voltage vs. time sketch has units of V/sec. To "convert" the 50 MV/m to V/sec, the speed of the switch is required. Typical switching speeds range from 1 to 50 cm/sec. Therefore,

$$\frac{dv_{cs}}{dt} = \frac{\partial v_{cs}}{\partial d}\frac{\partial d}{\partial t} \approx E_A s \tag{1.21}$$

A constant speed, $s = d/t$, was assumed. If $s = 5$ cm/sec and $E_A = 50$ MV/m, then

$$\frac{dv_{cs}}{dt} \approx (50 \times 10^6)\,0.05 = 2.5 \times 10^6 \text{ V/s}$$

To prevent arcing, the initial rise in voltage across the switch should not exceed this value. Another way of writing this relationship is

$$E_A > \frac{v_{cs}}{d} = \frac{v_{cs}}{st} \quad \text{or} \quad t > \frac{v_{cs}}{sE_A} = \frac{v_{cs}}{2.5 \times 10^6} \tag{1.22}$$

This result is a function of the switching speed. When the time is greater than v_{cs}/sE_A, the distance between the contacts is likely too great for arcing. Therefore, the breakdown is a function of the switching speed because the speed determines the distance between the contacts. Obviously, the time after which arcing is less likely to occur decreases as the switching speed increases. These results apply for arcing. A similar relationship for gas discharge breakdown is

$$320 + 7 \times 10^6 d > v_{cs} \quad \text{or} \quad t > \frac{v_{cs} - 320}{7 \times 10^6 s} \tag{1.23}$$

This equation is for switch voltages greater than 330 V. For times satisfying this inequality, gas discharge breakdown is unlikely. With all of these discussions, there are additional complicating factors. For example, one factor is the ability of the breakdown processes (e.g., thermal hysteresis) to respond to the changing voltage and current. This response time, which is around 0.01 to 1 μsec, can be important for faster switching speeds.

1.9   Suppressing the Breakdown

How can the likelihood of breakdown be reduced when switching dc circuits? What is the best time for the closing and opening of a switch in an ac circuit? [Paul, '92(b); Bell, '71; Ott; Holm]

There are many factors that affect breakdown when switching such as the voltage, current, speed of the switching, contact material, humidity, contact bounce, parasitic inductance of the leads, load and source impedance, and age of the contacts. Since many of these factors are often not known, typical snubbing (or antibreakdown) circuits are used to reduce the likelihood of breakdown. Initially, the values

---

[6]The partial differentiation symbol, ∂, is used instead of the ordinary differentiation symbol, *d*, to help avoid mistaking it with the distance variable *d*.

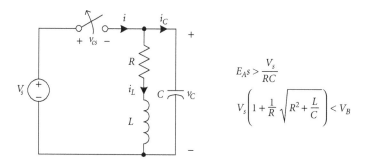

FIGURE 1.13   *RLC* circuit and the inequalities that should be satisfied to reduce the likelihood of breakdown across the opening switch.

for the components in the circuitry are obtained via design equations, then the component values are adjusted through experimental trial and error.

To limit or prevent breakdown, the voltage across the switch, the current through the switch, or both must be controlled. By limiting the voltage to less than the various breakdown voltages (e.g., V_B), breakdown will not be initiated. By limiting the current, breakdown will not be sustained if it is initiated. The voltage and current can be limited by adding to or modifying the circuitry connected to the switch.

The restrictions on the size of the components in the circuit shown in Figure 1.13, to help avoid breakdown, will be derived. Then, a common protection network will be analyzed. Before the switch is opened, it is assumed that the dc voltage has been applied to the circuit for a long time. Consequently, the inductor appears like a short circuit and the capacitor appears like an open circuit. The initial current through the resistor and inductor is therefore

$$I_o = \frac{V_s}{R} = i_L(0)$$

Immediately after the switch is opened, the current through the capacitor is equal to this initial current since the capacitor is in series with the inductor (and the current cannot change instantaneously through an inductor):

$$i_C = C\frac{dv_C}{dt} = -I_o \quad \Rightarrow \quad \frac{dv_C}{dt} = -\frac{I_o}{C} = -\frac{V_s}{RC} \quad @t = 0^+ \tag{1.24}$$

Since the voltage across the switch is the difference between the dc supply voltage V_s and v_C, the magnitude of the initial *change* in the voltage across the switch is also I_o/C (the derivative of V_s is zero since it is a constant dc source). For the switch not to arc, the magnitude of this initial rise in the voltage across the switch must be less than V_s/RC:

$$\left|\frac{dv_{cs}}{dt}\right| \approx E_A s > \frac{V_s}{RC} \tag{1.25}$$

where s is the speed of the switch and E_A is the field strength required for arcing. Equation (1.25) is the first of the inequalities given.

The second breakdown mechanism that should be avoided is gas discharge. To prevent a gas discharge, the peak voltage across the switch must be less than V_G. After the switch is opened, the maximum value that the voltage across the capacitor can ever obtain is when the losses are set to zero, which occurs when $R = 0$. The total energy in the circuit is the sum of the capacitive and inductive energies:

$$E_T = \frac{1}{2}Cv_C^2 + \frac{1}{2}Li_L^2 = \frac{1}{2}CV_s^2 + \frac{1}{2}LI_o^2 = \frac{1}{2}CV_s^2 + \frac{1}{2}L\left(\frac{V_s}{R}\right)^2 \tag{1.26}$$

This total electrical energy remains constant with time if the losses are zero. The peak value of the voltage across the capacitor occurs when all of this initial energy is contained in the capacitor. In other words, the peak voltage occurs when $i_L = 0$:

$$\frac{1}{2}Cv_{C,max}^2 + \frac{1}{2}L(0)^2 = \frac{1}{2}CV_s^2 + \frac{1}{2}L\left(\frac{V_s}{R}\right)^2 \Rightarrow v_{C,max} = V_s\sqrt{1+\frac{L}{CR^2}} = \frac{V_s}{R}\sqrt{R^2+\frac{L}{C}} \qquad (1.27)$$

The upper bound for the voltage across the capacitor is given by this expression. This voltage is greater than V_s. Another equation sometimes seen for this upper bound voltage across the capacitor is

$$v_{C,max2} = \frac{V_s}{R}\sqrt{\frac{L}{C}} \qquad (1.28)$$

This result is obtained by assuming that the initial energy in the capacitor is small compared to the initial energy in the inductor. If C represents a small parasitic capacitance and R represents a small ohmic resistance, then this would be a reasonable approximation for the upper bound on the peak voltage. The actual voltage across the capacitor will oscillate with an amplitude given by this upper-bound expression when there are no losses. With losses, the amplitude will decay with time. The *upper* bound for the voltage *across the switch* is then

$$v_{cs} = V_s + \frac{V_s}{R}\sqrt{R^2+\frac{L}{C}} = V_s\left(1+\frac{1}{R}\sqrt{R^2+\frac{L}{C}}\right) \qquad (1.29)$$

(The sign of the voltage across the capacitor can be positive or negative.) Equation (1.29) is the second of the inequalities given. The contribution of the supply voltage can be neglected when

$$V_s \ll \frac{V_s}{R}\sqrt{R^2+\frac{L}{C}} \qquad (1.30)$$

This inequality is frequently satisfied, especially if C represents a parasitic capacitance. To summarize, to prevent both arcing and gas discharge, both of the following inequalities should be satisfied:

$$E_A s > \frac{V_s}{RC} \qquad (1.31)$$

$$V_s\left(1+\frac{1}{R}\sqrt{R^2+\frac{L}{C}}\right) < V_B \qquad (1.32)$$

These bounds indicate that a small supply voltage and inductance and a large capacitance (and larger resistance) are desirable to help prevent breakdown. If the RL circuit represents a fixed load and C its parasitic capacitance, placing a large capacitor in parallel with the load should reduce the likelihood of breakdown. One disadvantage of placing a capacitor in parallel with the load, however, is the current surge that occurs when the switch is being *closed*. When the inductance or resistance leading to the switch or when the source impedance is included in the model, it is clear that the voltage across the capacitor immediately before the switch is closed is not necessarily equal to the supply voltage, V_s. Because the voltage cannot change instantaneously across the capacitor, a current, limited by the inductance and resistance of the wiring leading to the switch, surges through the switch. (This current surge is similar to the initial current between two batteries of different voltages when connected in parallel.) This current surge can damage or reduce the lifetime of the switch.

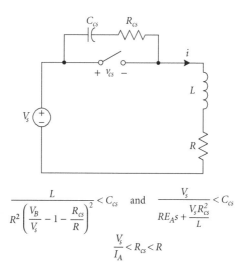

$$\frac{L}{R^2\left(\dfrac{V_B}{V_s}-1-\dfrac{R_{cs}}{R}\right)^2}<C_{cs} \quad \text{and} \quad \frac{V_s}{RE_As+\dfrac{V_sR_{cs}^2}{L}}<C_{cs}$$

$$\frac{V_s}{I_A}<R_{cs}<R$$

FIGURE 1.14 *RC protection network connected across a switch, and the corresponding design equations.*

An add-on circuit commonly used to reduce the likelihood of breakdown and contact damage for *RL* loads is a series *RC* circuit placed in parallel with the switch. This circuit, shown in Figure 1.14, was devised around the mid 1800's. In this book, it is referred to as the *RC* protection network. The parasitic capacitance across the load is not included in this model. The stated design expressions will be derived in the remaining portion of this section.

When the switch has been closed for a long time (compared to the time constants L/R and $R_{cs}C_{cs}$), the voltage source is in series with the load, and the initial current through the inductor is again V_s/R. In this circuit containing the *RC* protection network, the current through the inductor does not have to change from V_s/R to zero over a short period of time after the switch is opened: the current can pass through the $R_{cs}C_{cs}$ circuit. This will reduce the kickback voltage across the inductor and the voltage across the switch. After the switch has been opened a long time, the voltage across the switch will be equal to V_s. Referring to Figure 1.15, this voltage is easily verified using Kirchoff's voltage law:

$$-V_s+V_s+0R_{cs}+0R=0$$

The capacitor forces the current through the load to be zero after a long time since the capacitor appears like an open circuit after a long time to a dc source. The value of the capacitance is important

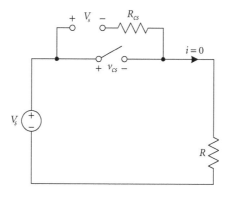

FIGURE 1.15 Circuit in Figure 1.14 after the switch has been opened for a long time.

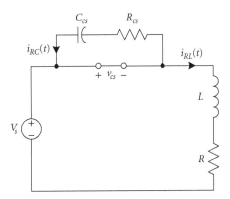

FIGURE 1.16 Circuit in Figure 1.15 immediately after the switch has been closed.

in reducing breakdown. After the switch is closed, the current through the switch is the sum of the normal load current

$$i_{RL}(t) = \frac{V_s}{R}\left(1 - e^{-\frac{R}{L}t}\right) \quad \Rightarrow i_{RL}(0) = 0, i_{RL}(\infty) = \frac{V_s}{R}$$

and the current from the discharging capacitor, C_{cs}:

$$i_{RC}(t) = \frac{V_s}{R_{cs}}e^{-\frac{t}{R_{cs}C_{cs}}} \quad \Rightarrow i_{RC}(0) = \frac{V_s}{R_{cs}}, i_{RC}(\infty) = 0$$

These variables are defined in Figure 1.16. The resistor R_{cs} limits the current through the switch from the discharging capacitor. A range of acceptable values for R_{cs} is obtained by realizing that the maximum current through the switch immediately after it is closed should be less than the minimum arcing current, I_A:

$$\frac{V_s}{R_{cs}} < I_A \quad \Rightarrow \frac{V_s}{I_A} < R_{cs}$$

The other bound for R_{cs} is obtained by determining the voltage across the switch immediately after it is opened. After the switch has been closed for a long time, the current through the *RL* load and $R_{cs}C_{cs}$ circuit immediately after opening the switch is

$$i_{RL} = -i_{RC} = \frac{V_s}{R}$$

This current is shown in Figure 1.17. The voltage across the switch is the sum of the voltage across the capacitor C_{cs} and resistor R_{cs}:

$$v_{cs} = 0 + \frac{V_s}{R}R_{cs}$$

The voltage across the capacitor immediately before (and after) the switch is opened is zero since its charge was dissipated through the resistor R_{cs}. The voltage across the switch immediately after it is opened is often selected to be less than the supply voltage:

$$V_s > \frac{V_s}{R}R_{cs} \quad \Rightarrow R > R_{cs}$$

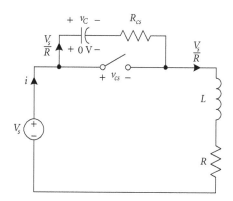

FIGURE 1.17 Circuit immediately after the switch is opened for the circuit in Figure 1.16 (after the switch has been closed for a long time).

Combining the previous results, provides a lower and an upper bound for R_{cs}:

$$\frac{V_s}{I_A} < R_{cs} < R \tag{1.33}$$

A guideline seen, based on these inequalities, is that R_{cs} should be approximately between the load resistance and 1 Ω/V of the source voltage. This guideline is assuming an arcing current of 1 A and a sufficiently large load resistor so that the upper limit is R.

The range on the protection network capacitance C_{cs} is obtained from the arc and gas discharge conditions. For the arc condition, the initial rise of the voltage across the switch must be limited by the rise in voltage across the protection network's capacitance and resistance. The voltage across the switch after it is opened is

$$v_{cs} = v_C + iR_{cs} \tag{1.34}$$

and the change in switch voltage immediately afterward is

$$\frac{dv_{cs}}{dt} = \frac{dv_C}{dt} + \frac{di}{dt} R_{cs} = \frac{I_o}{C_{cs}} + \frac{di}{dt} R_{cs} = \frac{V_s}{RC_{cs}} + \frac{di}{dt} R_{cs} \tag{1.35}$$

where I_o is the initial current through the capacitor (also, see (1.24)). The current, i, is a function of the elements in the series *RLC* circuit ($R + R_{cs}$, L, C_{cs}). The form of the current equation changes depending on whether the circuit is overdamped, critically damped, or underdamped. Rather than determining the current for each of these cases, the change in the current immediately after opening the switch will be obtained by using Kirchoff's voltage law:

$$-V_s + 0 + \frac{V_s}{R} R_{cs} + L \frac{di}{dt} + \frac{V_s}{R} R = 0 \quad \Rightarrow \quad \frac{di}{dt} = -\frac{V_s R_{cs}}{RL} \tag{1.36}$$

Therefore, immediately after opening the switch,

$$\frac{dv_{cs}}{dt} = \frac{V_s}{RC_{cs}} - \frac{V_s R_{cs}}{RL} R_{cs} \tag{1.37}$$

Frequently, the upper bound for this change in voltage across the switch is taken as

$$\frac{dv_{cs}}{dt} \approx \frac{V_s}{RC_{cs}} \quad \text{if} \quad \frac{V_s}{RC_{cs}} \gg \frac{V_s R_{cs}}{RL} R_{cs} \quad \text{or} \quad \frac{L}{C_{cs}} \gg R_{cs}^2 \tag{1.38}$$

To prevent arcing, the commonly seen expression is

$$E_A s > \frac{V_s}{RC_{cs}} \quad \text{or} \quad C_{cs} > \frac{V_s}{RE_A s} \tag{1.39}$$

The more general expression that does not involve the limitations on the values of the components is

$$E_A s > \frac{V_s}{RC_{cs}} - \frac{V_s R_{cs}^2}{RL} \quad \text{or} \quad C_{cs} > \frac{V_s}{RE_A s + \dfrac{V_s R_{cs}^2}{L}} \tag{1.40}$$

For the gas discharge condition, the voltage across the switch should not exceed V_B. The upper bound for the voltage across the capacitor for a source-free *RLC* circuit was shown to be

$$V_s \sqrt{1 + \frac{L}{C_{cs} R^2}}$$

when the initial voltage across the capacitor is equal to V_s and the initial current through the inductor is equal to V_s/R. After the switch is opened in this circuit, however, the circuit is not source free: it contains the V_s dc source. Furthermore, the initial voltage across the capacitor is 0 V not V_s. In this case, the total energy in the circuit, if the dc source is turned off, is

$$E_T = \frac{1}{2} C_{cs} v_C^2 + \frac{1}{2} L i_L^2 = \frac{1}{2} C_{cs}(0) + \frac{1}{2} L I_o^2 = \frac{1}{2} L \left(\frac{V_s}{R}\right)^2 \tag{1.41}$$

The peak value of the voltage across the capacitor occurs when all of this initial energy is contained in the capacitor:

$$\frac{1}{2} C_{cs} v_{C,max}^2 + \frac{1}{2} L(0)^2 = \frac{1}{2} L \left(\frac{V_s}{R}\right)^2 \quad \Rightarrow v_{C,max} = \frac{V_s}{R} \sqrt{\frac{L}{C_{cs}}} \tag{1.42}$$

The actual peak value across the capacitor is the sum of Equation (1.42) and the dc supply voltage (the average value of the oscillation is V_s):

$$v_{C,max} = V_s + \frac{V_s}{R} \sqrt{\frac{L}{C_{cs}}} = V_s \left(1 + \frac{1}{R} \sqrt{\frac{L}{C_{cs}}}\right) \tag{1.43}$$

This upper bound is obtained by setting the losses equal to zero. The voltage across the switch in this protection network is the sum of the voltage across the capacitor C_{cs} and resistor R_{cs}:

$$v_{cs} = v_C + i R_{cs}$$

The peak voltage occurred when the *oscillation* current was equal to zero (i.e., at the time that the total electrical energy is in the capacitor). The voltage drop across the resistor R_{cs}, slightly after the switch opens, due to the dc supply is

$$I_{dc} R_{cs} = \frac{V_s}{R} R_{cs}$$

Therefore, an upper bound for the voltage across the switch is

$$V_s \left(1 + \frac{1}{R} \sqrt{\frac{L}{C_{cs}}} + \frac{R_{cs}}{R} \right) < V_B$$

Solving for the capacitance,

$$\frac{L}{R^2 \left(\dfrac{V_B}{V_s} - 1 - \dfrac{R_{cs}}{R} \right)^2} < C_{cs} \tag{1.44}$$

Collecting all of the previous expressions, R_{cs} and C_{cs} must satisfy the following inequalities:

$$\frac{L}{R^2 \left(\dfrac{V_B}{V_s} - 1 - \dfrac{R_{cs}}{R} \right)^2} < C_{cs} \quad \text{and} \quad \frac{V_s}{RE_A s + \dfrac{V_s R_{cs}^2}{L}} < C_{cs} \tag{1.45}$$

$$\frac{V_s}{I_A} < R_{cs} < R \tag{1.46}$$

Finally, the complete set of equations has been derived!

In ac circuits, a switch should be closed when the voltage across the switch is small or at its minimum. This will help reduce the likelihood of breakdown. When opening a switch, the current through it should be small or at its minimum. This will help reduce inductive kickback or reduce the likelihood of sustaining a breakdown. (For a dc circuit, there is no natural "weak phase" since the current does not pass periodically through zero as with ac circuits.) The *RC* protection network discussed may not be suitable for ac circuits because of the finite reactance of the capacitance across the switch. Because of this finite reactance, when the switch is opened, the load is still partially connected to the source. As will be discussed later, the *RC* network can be used across *the load* for ac circuits. Electronic switches are available that close when the voltage across the switch is zero and open when the current through the switch is zero.

1.10 Switch Network Example

A 100 V dc supply is connected via a switch to an *RL* load where $R = 400 \ \Omega$ and $L = 10$ mH. The parasitic capacitance across the load is 100 pF. If the switch is expected to breakdown, determine the values for a series *RC* protection network to be placed across the switch to prevent this breakdown. How fast can the switch be opened and closed? What is the minimum value for the capacitance to avoid oscillation? If the maximum voltage across the switch should be much less than 100 V, what can be done? [Ott]

To prevent breakdown for this load, both the arcing and gas discharge conditions should be satisfied:

$$E_A s > \frac{V_s}{RC}, \quad V_s \left(1 + \frac{1}{R} \sqrt{R^2 + \frac{L}{C}} \right) < V_B$$

If the switching speed, s, is assumed to be 0.05 m/sec, the arcing field, E_A, equal to 100 MV/m, and the breakdown potential, V_B, equal to 320 V, then

$$E_A s = 5 \times 10^6 \underset{?}{>} \frac{V_s}{RC} = 2.5 \times 10^9$$

$$V_s \left(1 + \frac{1}{R} \sqrt{R^2 + \frac{L}{C}} \right) \approx 2.6 \times 10^3 \underset{?}{<} V_B = 320$$

Since both of these inequalities are obviously not satisfied, the switch should break down. For small switch gaps, arcing can occur since 100 V is greater than 12 V: the electric field is sufficiently intense to emit electrons from the contacts. For large gaps, even though the supply voltage, 100 V, is less than the breakdown voltage of 320 V, the inductance and capacitance in the system can cause the peak voltage to exceed 320 V across the switch. Again, the gas discharge equation is an upper bound for the switch voltage and was derived by assuming an oscillatory no-loss solution. To determine the true peak value of the voltage across the switch, the actual voltage across the capacitor must be determined. Since the Q of the circuit without a protection network is greater than 0.5

$$Q = \frac{1}{R} \sqrt{\frac{L}{C}} = 25$$

the system is definitely underdamped (oscillatory). The general expression for the voltage across the capacitor for an underdamped series *RLC* circuit is (assuming that the final voltage across the capacitor is zero)

$$v_C(t) = A e^{-\alpha t} \cos(\omega_d t) + B e^{-\alpha t} \sin(\omega_d t)$$

where

$$\alpha = \frac{R}{2L} = 2 \times 10^4 \text{ s}^{-1}, \quad \omega_d = \sqrt{\omega_o^2 - \alpha^2} = \sqrt{\frac{1}{LC} - \left(\frac{R}{2L} \right)^2} \approx 10^6 \text{ rad/s}$$

The initial voltage across the capacitor is $V_s = 100 \text{ V} = v_C(0)$ and the initial current through the inductor (and capacitor) is

$$I_o = \frac{V_s}{R} = 0.25 \text{ A} = C \frac{dv_C}{dt}$$

Using this information, the voltage expression becomes

$$v_C(t) = V_s e^{-\alpha t} \cos(\omega_d t) + V_s \left(\frac{\alpha - \dfrac{1}{RC}}{\omega_d} \right) e^{-\alpha t} \sin(\omega_d t) \tag{1.47}$$

$$\approx 100 e^{-\alpha t} \cos(\omega_d t) - 2,500 e^{-\alpha t} \sin(\omega_d t) \text{ V}$$

From the plot of this voltage given in Mathcad 1.2, the minimum value is about −2.4 kV. The voltage across the switch is then about $100 + 2,400 = 2.5$ kV. This voltage is very close to the value obtained from

$\alpha := 2 \cdot 10^4 \qquad \omega_d := 10 \cdot 10^5$

$t := 0.10^{-9} .. 4 \cdot 10^{-5}$

$v_C(t) := 100 \cdot e^{-\alpha \cdot t} \cdot \cos(\omega_d \cdot t) - 2500 \cdot e^{-\alpha \cdot t} \cdot \sin(\omega_d \cdot t)$

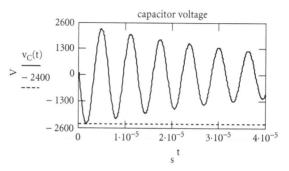

MATHCAD 1.2   Voltage given in Equation (1.47).

the upper-bound expression, (1.27). As the losses increase (i.e., larger α), the difference between the actual maximum voltage and the upper-bound estimate will be more significant.

To prevent breakdown, the resistance of a series $R_{cs}C_{cs}$ circuit placed in parallel with the switch must satisfy the following inequalities:

$$\frac{V_s}{I_A} \approx 333 < R_{cs} < R = 400$$

A minimum arcing current, I_A, of 300 mA was assumed. The lower bound on the resistance helps prevent an arc from being sustained by limiting the current through the switch, and the upper bound on the resistance limits the voltage across the switch immediately after the switch is opened. Selecting $R_{cs} = 350 \; \Omega$, the limits on the capacitance, C_{cs}, can be obtained:

$$\frac{L}{R^2 \left(\dfrac{V_B}{V_s} - 1 - \dfrac{R_{cs}}{R} \right)^2} \approx 0.04 \times 10^{-6} < C_{cs} \quad \text{and} \quad \frac{V_s}{RE_A s + \dfrac{V_s R_{cs}^2}{L}} \approx 0.03 \times 10^{-6} < C_{cs}$$

A value of $C_{cs} = 0.047 \; \mu F$ will satisfy both of these inequalities. The effect of the parasitic capacitance of 100 pF was assumed negligible compared to C_{cs}. The first capacitance inequality is to prevent gas breakdown, and the second capacitance inequality is to prevent arcing. This capacitor C_{cs} should be rated for a voltage much greater than the supply voltage (the maximum voltage across the capacitor is 2.5 kV). Typical capacitance values are 0.1 to 2 μF.

This $R_{cs}C_{cs}$ protection network is popular since it is simple and inexpensive, and its response time can be controlled. During the closing of the switch, the time constant of the $R_{cs}C_{cs}$ circuit

$$\tau_{RC} = R_{cs}C_{cs} \tag{1.48}$$

determines how long it takes the capacitor to discharge. After about $5 \tau_{RC} \approx 82 \; \mu\text{sec}$, the capacitor should be mostly discharged. The time constant of the load

$$\tau_{RL} = \frac{L}{R} \tag{1.49}$$

also has an influence over the response. After the switch is opened, the time constant of the decaying oscillation is

$$\frac{1}{\alpha} = \frac{2L}{R_{tot}} = \frac{2\,(10\times10^{-3})}{350+400} \approx 27\ \mu\text{sec}$$

where R_{tot} is the total series resistance, $R + R_{cs}$. After about $5/\alpha \approx 130\ \mu\text{sec}$, the amplitude of the oscillation will be small. Since typical switch contacts speeds range from 1 to 50 cm/sec, over distances of 100 to 400 μm, 200 μsec is a reasonable estimate for a minimal switching time. A protection network with a large resistance, compared to the inductance, will provide a small "release time."

To prevent the circuit from oscillating after the switch is opened, the Q of the series RLC circuit must be less than 0.5:

$$Q = \frac{1}{R_{tot}}\sqrt{\frac{L}{C_{cs}}} < \frac{1}{2} \quad \text{or} \quad C_{cs} > \frac{4L}{R_{tot}^2} \tag{1.50}$$

Again, the resistance, R_{tot}, is the total series resistance, $R + R_{cs}$. Therefore,

$$C_{cs} > 0.07\ \mu\text{F}$$

A value of $C_{cs} = 0.1\ \mu\text{F}$ will keep the circuit overdamped and nonoscillatory. Oscillations are a source of electrical noise.

The maximum voltage across the switch with the protection network is obtained by determining the voltage across the switch immediately after it is opened. The current through the RL load and $R_{cs}C_{cs}$ circuit is V_s/R, and the voltage across the switch is the sum of the voltages initially across the capacitor C_{cs} ($= 0$ V) and resistor R_{cs}:

$$v_{cs} = 0 + \frac{V_s}{R}R_{cs} = \frac{100}{400}(350) \approx 88\ \text{V}$$

If R_{cs} were reduced below 330 Ω, the current through the switch would exceed the minimum arcing current. If the voltage across the switch is to be significantly reduced (e.g., below 12 V), then the more complex protection network shown in Figure 1.18, involving a diode, can be used. (The diode is assumed ideal in the following analysis.) When the switch is opened, the capacitor charges to V_s (as

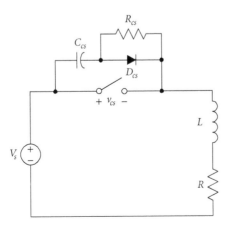

FIGURE 1.18   *RCD* switch protection network for an *RL* load.

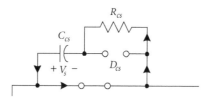

FIGURE 1.19   Protection network in Figure 1.18 immediately after the switch is closed.

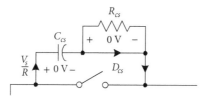

FIGURE 1.20   Protection network in Figure 1.19 (after the capacitor is completely discharged) immediately after the switch is opened.

before) but now through the forward-biased diode, D_{cs}. When the switch is closed, the capacitor discharges through the resistor, R_{cs}, as with the standard $R_{cs}C_{cs}$ circuit, since the diode is reversed biased. As shown in Figure 1.19, the discharge current is through the switch and is limited by R_{cs}. After the capacitor completely discharges, the voltage across it is 0 V. When the switch is opened, the diode is forward biased and the resistor is short circuited as shown in Figure 1.20. The voltage across the switch is now zero. The breakdown voltage of the diode must be greater than the supply voltage, and it must be capable of handling more than the load current, V_s/R. The upper limit on the resistance can be eliminated:

$$\frac{V_s}{I_A} < R_{cs} \qquad (1.51)$$

The resistance can be selected to limit the current to one-tenth or less of the arcing current:

$$10\left(\frac{V_s}{I_A}\right) < R_{cs} \qquad (1.52)$$

Obviously, this *RCD* protection network cannot be used for ac circuits, and it is more costly than the popular *RC* network.

1.11   Arc Suppression with Resistive Loads

When is an arc-suppression network needed for loads that are mainly resistive?

If the circuitry connected to a switch is almost entirely resistive, the maximum voltage across the switch is about equal to the supply voltage. Unlike inductive loads, the current can change instantaneously across a resistor, and there is no inductive kickback voltage. A tungsten lamp is an example of a mostly resistive load. The resistance in this case is temperature dependent: its initial cold resistance is much less than its final hot resistance (e.g., a cold resistance equal to one-tenth of its hot resistance).

Whether an arc or glow discharge is initiated is dependent on the supply voltage. If the supply voltage is less than the voltage required to arc, which is about 12 V, then an arc cannot be initiated. If the voltage is greater than 12 V, then an arc can be initiated and sustained if the current through the switch is greater than the minimum arcing current I_A:

$$\frac{V_s}{R} > I_A$$

where V_s is the supply voltage and R is the load resistance. If this current is less than the minimum arcing current, I_A, then an arc cannot be sustained. An RC protection network can be used to limit the current. If the supply voltage is less than 320 V, then a glow discharge cannot be initiated. If the supply voltage is greater than 320 V, then a glow discharge can be initiated and sustained if the current is greater than I_G, which is less than I_A. Again, an RC protection network can be used to limit the rate of voltage increase across the switch.

Note that parasitic inductances and capacitances should be considered even when a load is resistive. The inductance of the wires leading to the switch and the capacitance across a load can sometimes be sufficient to cause the actual voltage across the switch to be much greater than the supply voltage.

1.12 Arc Suppression with Capacitive Loads

When is an arc-suppression network needed for loads that are mainly capacitive? [Khalifa]

Capacitive loads, such as a capacitor bank or collection of cables, do not provide an inductive kickback voltage, but they can produce in-rush currents. For dc or ac circuits, the voltage across a capacitive load is the same immediately before or after the switch opens. When the load is in parallel with the supply, the voltage across the load is the same as the supply voltage. If the supply voltage is sinusoidal, when a switch is *closed*, the voltage of the supply at that instant may not be equal to the voltage across the capacitor. Because of this voltage difference across the switch, which can be twice the amplitude of the supply voltage, the current through the switch can be large. Only the impedance of the wiring to the switch and load limits this in-rush current.

For ideal dc circuits with an ideal capacitive load, the voltage across the capacitor does not change if there is no discharge path. Therefore, the potential difference across the switch is zero when it is closed, which is the same as when the switch was opened. Real capacitors, however, have dielectric loss. This loss is modeled via a large resistor in parallel with the capacitor. After the switch is opened, a fraction of the charge on the capacitor dissipates, and the voltage across it decreases. When the switch is closed, there is a potential difference across the switch. This potential difference and the resultant in-rush current is a function of the rate of this discharge.

To limit this in-rush current, a resistor or an inductor in series with the capacitor is necessary. An inductor resists changes in current but could allow the circuit to oscillate. A resistor limits the current but also limits the current during normal operating conditions. If the expected in-rush current is much greater than the normal operating current of the load, a nonlinear current-dependent resistor can be used. The resistance should be large for larger in-rush currents while small for smaller operating currents.

Another interesting but troubling aspect of capacitive loads is revealed by briefly studying the circuit shown in Figure 1.21. Working in the frequency domain, the voltage across the capacitors *before* the switch is opened is

$$V_{ys} = V_{xs} = A \frac{\dfrac{1}{j\omega\,(C_x + C_y)}}{\dfrac{1}{j\omega\,(C_x + C_y)} + j\omega L_y} = \frac{A}{1 - \omega^2 L_y (C_x + C_y)}$$

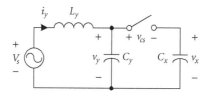

FIGURE 1.21 Switching two parallel capacitors.

where A is the amplitude of the sinusoidal source. The current through the inductor is

$$I_{ys} = \frac{A}{\dfrac{1}{j\omega (C_x + C_y)} + j\omega L_y} = \frac{jA\omega (C_x + C_y)}{1 - \omega^2 L_y (C_x + C_y)}$$

The corresponding voltage and current in the time domain are

$$v_y = v_x = \frac{A}{1 - \omega^2 L_y (C_x + C_y)} \cos(\omega t), \quad i_y = \frac{A\omega (C_x + C_y)}{1 - \omega^2 L_y (C_x + C_y)} \cos(\omega t + 90°)$$

After the switch is opened, the steady-state voltage across C_y is

$$v_y = \frac{A}{1 - \omega^2 L_y C_y} \cos(\omega t) \tag{1.53}$$

If the frequency of the source is much less than the resonant frequency of the circuit, before and after the switch is opened,

$$\omega \ll \omega_o = \frac{1}{\sqrt{L_y (C_x + C_y)}} < \frac{1}{\sqrt{L_y C_y}}$$

then the equations for the steady-state voltage and current before the switch is opened reduce to

$$v_y = v_x \approx A\cos(\omega t), \quad i_y \approx A\omega (C_x + C_y)\cos(\omega t + 90°)$$

and the voltage across the capacitor C_y is $v_y \approx A\cos(\omega t)$. Finally, if the switch is opened during a zero-current crossing (e.g., $\omega t = 0$),

$$v_y = v_x \approx A, \quad i_y \approx 0$$

Therefore, at the zero-current crossing, the voltage across both capacitors is A at lower frequencies. Since there is no discharge path modeled, the voltage across the load capacitor, C_x, remains at this value. The voltage across capacitor C_y changes with the supply voltage. At a half cycle later ($\omega t = 180°$), the voltage across the capacitor changes to $A\cos(180°) = -A$. Therefore, at a half cycle later, the voltage magnitude across the switch is twice the supply voltage or $A - (-A) = 2A$.

Equation (1.53) indicates that the magnitude of the voltage across C_y can exceed A at (or even around) the resonant frequency. If an arc develops across the switch, the voltage across the load capacitor, C_x, can be charged from C_y. The arc, a low-voltage and high-current breakdown, is a low-resistance path through the switch. The voltage across the switch can now exceed $2A$. Repeated restriking of the arc can occur as the switch is opened or closed.

1.13 Arc Suppression with Inductive Loads

Besides the *RC* protection network, what other arc-suppression networks can be used for loads that are mainly inductive? [Ott; Holm; National; Auger]

To suppress switching arcs for inductive loads, various networks or snubbing circuits can be placed directly across the inductive load. In severe cases, a network can be placed across both the load and switch. Note that although a load may appear only resistive, parasitic inductances (e.g., wiring) could be sufficiently large to cause a voltage surge or kickback when the current is interrupted.

The problem with an energy-storage inductive load is its tendency to resist sudden changes in current. Therefore, if a switch in series with an inductor breaks the current path during a nonzero current crossing, the inductor strongly resists this sudden current change by producing a large kickback voltage. This voltage increases the likelihood of switch arcing. The following suppression networks allow the current from the inductor to take another path, avoiding this voltage surge.

For an inductive load, the most obvious means of providing a current path after the switch is opened is to place a resistor in parallel with the load. This method works for both dc and ac circuits. Of course, the resistor will absorb power during normal circuit operation, load the supply, and affect the response or drop-out time of the voltage across the inductive load. Referring to Figure 1.22, the new time constant of the load after the switch is opened is

$$\tau = \frac{L}{R_{eq}} = \frac{L}{R+R_x} \tag{1.54}$$

After three to five time constants, the voltage across the load is small compared with its initial value. As R_x increases, the time constant and time required for the voltage across the load to drop to some specified level decreases. Although the drop-out time decreases as R_x increases, as will be seen, the maximum voltage across the switch increases.[7] The voltage across the snubbing resistor, R_x, immediately after the switch is thrown open is

$$v_x = i_x R_x = -\frac{V_s}{R} R_x \tag{1.55}$$

and the voltage across the switch is

$$v_{cs} = V_s - v_x = V_s \left(1 + \frac{R_x}{R} \right) \tag{1.56}$$

If the snubber resistor is set equal to the load resistance, $R_x = R$, then the voltage across the switch is twice the supply voltage. For this dc circuit, the steady-state power absorbed by the load would be equal

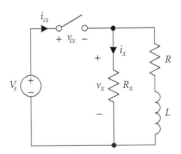

FIGURE 1.22 Resistor in shunt with an *RL* load to provide a path for the inductor current.

---

[7]"Faster" or smaller drop-out times will have more high-frequency energy and a greater potential to interfere.

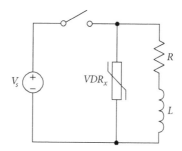

FIGURE 1.23 Varistor in shunt with an *RL* load to reduce the kickback voltage.

to the power absorbed by R_x. For this reason, the snubber resistor is often selected as several times the value of R. Also note that the steady-state current through the switch with the snubber resistor, before the switch is opened, is greater than the current without it:

$$i_{cs} = \frac{V_s}{\dfrac{RR_x}{R+R_x}} > \frac{V_s}{R} \qquad (1.57)$$

With these protection networks, there is a tradeoff between the maximum magnitude of the voltage across the load after the switch is opened and the rate of decrease in the load current and energy dissipation in the load.

A second circuit that is similar to that in Figure 1.22 is obtained by replacing the resistor R_x with a varistor. A varistor is a nonlinear resistor: its resistance is high when the magnitude of the voltage across it is low and its resistance is low when the magnitude of the voltage across it is high. Referring to Figure 1.23, before the switch is opened, the voltage across the varistor is equal to the supply voltage. (The varistor should be selected so that its resistance is high when the switch is closed.) After the switch is opened, the current from the inductor must pass through the high resistance of the varistor, and the resultant voltage across the varistor is now high. Since the voltage across the varistor is high, its resistance drops and the voltage across the switch remains low. The inductive kickback voltage is thus avoided. (Voltage ratings from 10 V to 1 kV are available for varistors with current ratings beyond 1 kA. The lifetime of the varistor should also be considered.) Because of the variable resistance, the losses are less than with a standard fixed resistance. This circuit can be used for both dc and ac circuits.

Previously, a series *RC* suppression network was placed across the switch. It can also be placed directly across the load as shown in Figure 1.24. The values for the components are obtained in the same manner

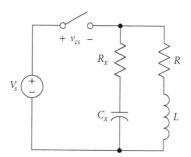

FIGURE 1.24 *RC* protection network placed across the *RL* load instead of across the switch.

as when the network is placed across the switch. The voltage across the capacitor can be much greater than the supply voltage. An advantage of placing the *RC* network across the load is that the area circumscribed by the closed path of the current after opening the switch is less than when placing the network across the switch. This smaller loop area will reduce electromagnetic emissions. The circuit shown in Figure 1.24 works for both dc and ac circuits. For a dc supply, the power absorbed by this circuit in steady state is zero when the switch is closed since the capacitor appears like an open circuit. After the switch is opened, the current from the inductor passes through the *RC* network. The drop-out time is a function of the damping coefficient of this *RLC* circuit. If this circuit is underdamped, then the time constant for the exponential decay of the oscillation is

$$\tau = \frac{1}{\alpha} = \frac{2L}{R_{eq}} = \frac{2L}{R+R_x} \tag{1.58}$$

where α is the damping coefficient. If $R_x = R$, then $\tau = L/R$, which is identical to the time constant of an *RL* circuit. There is another advantage of placing the *RC* network across the load instead of across the switch. If the capacitor is short circuited, the switch will remain operational.

For dc circuits, a "freewheeling" diode in series with a current-limiting resistor can be placed across the inductive load as shown in Figure 1.25. When the switch is closed, the diode is reversed biased and no current passes through the diode. When the switch is opened, the current from the inductor forward biases the diode, and a closed path is provided for the inductor current, preventing the kickback voltage across the inductor. The diode must be capable of handling the high current levels carried by the load, and the reverse breakdown voltage rating of the diode must be greater than the supply voltage. As with the previous circuits, the time constant of the network is affected by the addition of the resistor. This time constant is, when the diode is forward biased,

$$\tau = \frac{L}{R+R_x} \tag{1.59}$$

The forward resistance of the diode was assumed negligible compared to R and R_x. The voltage across the switch immediately after the switch is opened is

$$v_{cs} = V_s + R_x \frac{V_s}{R} = V_s \left(1 + \frac{R_x}{R} \right) \tag{1.60}$$

The voltage drop across the diode was neglected in this equation because it is small compared to the voltages of interest. If $R_x = 0$, the voltage across the switch is equal to the voltage of the supply. As R_x increases, the maximum voltage across the switch increases, but the drop-out time decreases. This protection circuit does not absorb power when the switch is closed. Again, this single diode snubber circuit will not work for loads such as ac motors or drives.

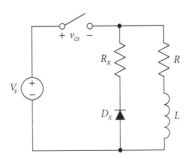

FIGURE 1.25 Protection network containing a "freewheeling" diode and resistor.

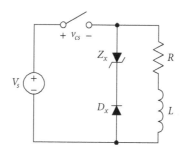

FIGURE 1.26 Protection network containing both a regular and zener diode.

There are several variations of the network shown in Figure 1.25. The resistor can be replaced with a zener diode as shown in Figure 1.26. A zener diode is a diode that is designed to work in a predictable manner when it is reverse biased. The zener diode will turn on when the negative voltage across it is greater in magnitude than its reverse-biased breakdown voltage. The breakdown voltage of the zener is selected to be less than the breakdown voltage of the switch (minus the voltage drop of the regular diode and supply voltage) but as large as possible so that the energy of the load is quickly dissipated: beyond the breakdown knee of the zener's curve, the power dissipated by the zener is approximately the product of the current through it multiplied by its breakdown voltage. After the switch is opened, both the zener diode, Z_x, and regular diode, D_x, are turned on. The initial current from the inductive load must pass through the diodes since the inductor is initially acting like a current supply. The power, voltage, and current ratings of the zener must be properly selected since the current-voltage product of the zener is large. (The current-voltage product for the regular diode is much less since the forward-bias voltage across it is small, which is approximately 0.7 V.) The regular diode's reverse breakdown voltage must be greater than the supply voltage. Eventually, the voltage across the zener will drop below its breakdown voltage, and the zener will turn off. The current through the zener when it is operated below its breakdown voltage is small. This circuit only works for dc circuits. By using this two-diode circuit across the load with an RC network across the switch, it can be shown that the size of the C across the switch can be reduced (relative to an RC network without the diode circuit).

The resistor can also be replaced with a varistor as shown in Figure 1.27. With the diode in series with the varistor, power is only dissipated in the varistor after the switch is opened. Without the diode, power is dissipated when the switch is opened or closed. Again, this protection network is only for dc circuits. Finally, the resistor can be replaced with a capacitor as shown in Figure 1.28. This network (for an ideal capacitor and diode) does not absorb power when the switch is opened or closed. For a real diode, the voltage across the ideal capacitor will be equal to the source voltage after the switch has been closed for a long time (a small but nonzero current exists in a reversed-biased diode). For a period of time after the switch is opened, the inductor current will forward bias the diode. Then, depending on the voltage across the diode and its reverse breakdown voltage, the current may stop. As long as the voltage is sufficient to

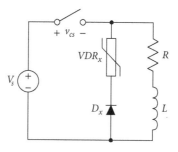

FIGURE 1.27 Protection network containing both a diode and varistor.

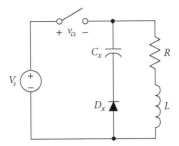

FIGURE 1.28 Protection network that absorbs little power.

break down the diode, a current will exist through it. As soon as the voltage across the diode drops below its breakdown voltage, the current will cease (and a voltage will remain across the capacitor). For the current not to oscillate or reverse direction,

$$C_x > \frac{4L}{R^2} \tag{1.61}$$

(The forward resistance of the diode is assumed much less than R.) The time constant, which is a measure of the drop-out time, is

$$\tau = \begin{cases} \dfrac{2L}{R} & \text{if } C_x = \dfrac{4L}{R^2} \text{ (critically damped)} \\[2mm] \dfrac{L}{R} & \text{if } C_x \gg \dfrac{4L}{R^2} \text{ (overdamped)} \end{cases} \tag{1.62}$$

Obviously, the breakdown voltage of the diode should be greater than the supply voltage; otherwise, the diode will conduct when the switch is first closed. This circuit protection network is for dc sources.

Protection networks are available for ac circuits and inductive loads. One network is shown in Figure 1.29. With ac circuits, the current passes through zero, which helps prevent a stable arc from developing across the contacts. (With dc, there is no weak phase where the current passes to zero.) The lifetime of the contacts can be increased, however, by using the network shown in Figure 1.29. The back-to-back zener diodes can break down for currents in either direction. The breakdown voltage should be greater than the supply voltage but less than the switch breakdown voltage. The energy dissipated by a zener diode increases with the diode's breakdown voltage. Another ac protection network is shown in Figure 1.30. Current is present in the protection network when the switch is closed. This current is limited by the reactance of the capacitor C_x and the resistance, R_x. After the switch is opened, the current from

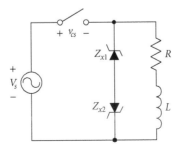

FIGURE 1.29 Protection network for an ac source.

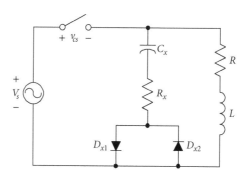

FIGURE 1.30   Protection network for an ac source with Q control.

the load passes through the appropriate diode depending on the direction of the current. The resistor, R_x, performs several functions. It dissipates energy, thereby decreasing the drop-out time, and it provides greater control of the Q of the circuit. The capacitor also affects the Q of the circuit. For the circuit not to oscillate after the switch is opened,

$$C_x > \frac{4L}{(R+R_x)^2} \qquad (1.63)$$

Sometimes designers believe that selecting $R = R_x$ and setting the two time constants equal to each other

$$R_x C_x = \frac{L}{R} \;\; \Rightarrow C_x = \frac{L}{RR_x} = \frac{L}{R^2} \qquad (1.64)$$

will optimize the drop-out time. In this case, $Q = 0.5$, which corresponds to critical damping. If the circuit does oscillate, the set of parallel diodes will allow current to pass in either direction. (When this network is used across the switch, the impedance of the network should be considered since it will allow current to bypass the opened switch.) Although not discussed in this book, there are other methods of terminating an arc for ac circuits that involve cooling, air evacuation, air blowouts, and multiple contacts.

1.14   Sparking at Very Low Voltages?

Can a switch spark at voltages less than about 12 V?

Generally, for large gaps (greater than about 100 μm) in standard conditions, the electric field must exceed about 3 MV/m, while for small gaps (less than about 100 μm), the voltage must exceed about 320 V. A 12 V difference across a 100 μm gap can only generate a uniform electric field of 0.12 MV/m. Obviously, voltages less than 12 V will generate uniform fields less than 0.12 MV/m at 100 μm. For nonuniform fields, the requirements are different, but this low voltage is still not sufficient to produce sparking, corona, arcing, or breakdown.

For very small gaps, field emission can take place if the electric field is very large (around 50 MV/m for realistic contaminated surfaces). The gap distance to generate a 50 MV/m uniform field with 12 V is 0.24 μm! It would be very difficult to determine whether any form of breakdown is occurring at these small distances since the electrode surfaces would have to be extremely smooth, flat, and stable.[8] Even if the large gap criterion of 3 MV/m were used, the distance between parallel flat electrodes would be only 4 μm for 12 V. With the naked eye, a few μm of separation appears like the electrodes are in contact. If

---

[8]Nearby vehicle movement would affect the measurement unless special precautions are taken.

a supply voltage is placed across the electrodes, the current through the electrodes can be monitored to determine if contact is made (or some type of breakdown is occurring).

It is sometimes claimed that sparking is seen for supply voltages less than 12 V. When a switch opens, a 12 V or less *supply* voltage can generate a very large voltage and electric field across a gap, especially if the load is inductive or wire length is long. Recall that inductance resists changes in current. When the current is interrupted in a circuit containing inductance, the inductance "reacts" by producing a large kickback voltage. Therefore, even a low supply voltage such as 2 V can generate a spark across an opening switch. However, to produce a *sustained* arc, sufficient current must be available to the gap and 12 V or more must exist across the gap. If 12 V and the minimum arcing current are not available in steady state, the arc will be extinguished. Even if one or both of the electrodes are sharp with high curvature, thereby producing a strong nonuniform field, about 12 V or more is required across the gap.

Sometimes, it appears in the closing of a switch that voltages less than 12 V can generate a spark across the switch. For example, if a large capacitor is first charged to 9 V and then a screwdriver is placed across the terminals of the capacitor, a spark can be observed. Although it is difficult to detect with the naked eye because of the small distances involved and the unsteadiness of the hand, the screwdriver is first making contact with the terminal and then bouncing off the contact. The spark is observed in the opening of the switch (or the screwdriver) not the closing. The capacitor is responding to the sudden short circuit across its terminals by generating a current surge. The inductances of the screwdriver and real capacitor respond to this sudden current surge by generating a large kickback voltage.

When an insulating surface between two electrodes is "dirty," arcing can occur for voltages less than about 300 V. The "dirt" or other material along the surface (e.g., moisture or water film) can decrease the surface resistance of the insulator. The surface resistance can vary along the surface or length of an insulator. Sufficient current can pass between the electrodes to heat the surface and initiate an arc. Arcing below 300 V or 30 kV/cm can also occur in a hot gas atmosphere (e.g., in a fire).

1.15   Switch Corrosion and Erosion

Often with age, switches corrode and erode. Discuss why this occurs and what can be done to reduce this degradation. Tabulate the materials used for contacts and provide their major properties. [Bosich; Kussy; Slade; Schweitzer; Benner; Browne; Dummer; Riggs]

Many factors including wear, corrosion, and arcing affect the lifetime of a switch. The hoi polloi probably know that switches deteriorate due to wear. However, the deterioration of a switch due to corrosion and arcing is probably not understood by most engineers.

Corrosion can be easily understood at an "executive level." Corrosion is the deterioration of a metal as a result of the loss of electrons. It has been known for a long time that charge flows between two different metals when in an electrolytic environment (e.g., wet or conductive). The conductive environment between the electrodes can be seawater, soil, or even lemon juice! Imagine that two electrodes of different materials are immersed in an electrolytic liquid. This "cell" is shown in Figure 1.31. There will be a potential difference between the two electrodes, labeled cathode and anode in the figure, referred to as battery action. The ammeter will measure a charge flow between the two electrodes. For one particular dry cell battery, where the anode electrode is composed of zinc (the circular outer case) and the cathode electrode is composed of carbon (the rod in the center of the zinc case), approximately 1.55 V will exist between the electrodes when the battery is new. The greater the voltage across the electrodes, the greater the electrical driving force.

Because there is typically some confusion about the direction of the current for these electrochemical cells, a few remarks concerning terminology are appropriate. First, the anode is defined as "the electrode from which electrons leave the cell," while the cathode is defined as "the electrode from which electrons enter the cell." Also, for conductors, the conventional engineering direction for current (as shown in Figure 1.31) is in the opposite direction to the electron flow. The confusion occurs when an electromotive force (EMF) or a voltage reference is assigned to the terminals of the

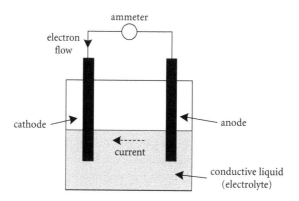

FIGURE 1.31   Measurement of the current between two electrodes in a cell.

cell. In the given figure, the cell is a voltaic cell or supplying current. The cathode is the positive terminal and the anode is the negative terminal, which is easily seen by imagining that the cell is a battery or voltage source. This labeling is in agreement with what is pontificated by the instructor in the first few weeks of an elementary circuits course: "current leaves the positive terminal." Therefore, for a cell used as a voltage source, the terms cathode and positive electrode are used interchangeably, and the terms anode and negative electrode are used interchangeably. Now, imagine that the ammeter in the figure is replaced by an external voltage source, where the positive terminal (of the external source) is connected to the anode and the negative terminal (of the external source) is connected to the cathode. For this same cell, which is referred to as an electrolytic cell, the anode is the positive electrode and the cathode is the negative terminal. (The assignment of the positive and negative electrodes is reversed.) The external source is "charging" or forcing current into the anode.

The magnitude of the voltage difference is dependent on various factors including the relative positions of the two electrode materials in the electrochemical series. For example, the "distance" between copper and aluminum in the series is large and, hence, the voltage expected between the copper and aluminum electrodes with an electrolyte present is relatively large. The more active metal in this series (i.e., the metal with the more negative electromotive force) will lose or give up electrons to the electrolyte. This active electrode, which loses electrons, is referred to as the anode. Oxidation and corrosion occur at the anode. If, for example, magnesium and platinum are the two electrodes, magnesium will act as the anode and corrode while platinum will act as the cathode. If a galvanized steel support structure is in contact with a copper grounding strap under the ground (or in snow or flood water), the zinc coating on the steel will corrode and act as a sacrificial anode. Eventually the zinc coating will be lost, leaving the steel bare to the corrupting influence of the soil.

Corrosion is the deterioration of a material (usually a metal) due to a reaction with its environment. The difference in the EMFs of the metals, surface conditions (and imperfections) of the metals, history of the metals, area of the metals (affecting the current density), distance between the metals, temperature of the metals and their surroundings, flow rate of the surrounding electrolyte, and resistivity of the electrolyte all determine the rate of this corrosion. The potential difference between metals will also vary with time. When two different metals are involved (and an electrolyte is present such as moisture), it is referred to as galvanic corrosion. The two metals can be in direct contact or electrically connected via some conducting medium. It is interesting to note that both gold and silver have nonnegative EMFs and are found in their natural state. Iron and aluminum, however, both have negative EMFs and are mined as oxides (i.e., strong oxidation of the metal has occurred).

Corrosion (or electrolytic corrosion) can also occur between two pieces of metal of the same material or even on a single piece of metal of the same material. If a single piece of unprotected steel, for example, is left outside, it will corrode or rust. This corrosion can take place if there is an electrolyte composition, a temperature, or a surface state variation along the metal. When the electrolyte or charge (e.g., ground current in the soil) is flowing past the surface of the metal, corrosion can also occur.

Corrosion has been studied for many years. Although specialized sources should be consulted for detailed information, the following guidelines should be helpful in guiding the nonspecialist in the slowing down of corrosion:

1. Decrease the distance between the metals on the electrochemical table or galvanic series.
2. Insulate, electrically, the metals from each other.
3. Ensure that the area of the anode is not small compared to the area of the cathode (sacrificial electrodes are an exception to this guideline).
4. Coat the metals (e.g., paint the metals or plate them with an intermediate metal on the electrochemical table).
5. Increase the physical distance between the metals.
6. Add inhibiting chemicals to the electrolyte between the metals.
7. Reduce the moisture content of the surrounding environment.
8. Install nearby sacrificial bare electrodes.

Short arcs (lower voltages across a small gap) can produce erosion of the contacts. This *erosion* should not be confused with corrosion. With each spark or breakdown, carbon (or other materials) will build up on the contacts of the switch. This carbon buildup decreases the effective, surface contact area, increases the impedance of the switch when in its closed position, and increases the power dissipated in the switch. However, often this carbon buildup can be partially removed if a sufficiently large arcing voltage is applied (or self-wiping contacts are used). This cleaning action is probably the only benefit of switch arcing. A range of arc levels can burn off the buildup developed on the contacts. With arcing, the metal is evaporated at spots or transferred to the other electrode. As expected, the rate of erosion is a function of the energy in the arc and other factors such as the melting point of the electrodes.

Pitting is a type of erosion that is localized or confined to a point or small area. It can produce small holes or cavities in the metal. It often occurs when the anode area (the active corroding electrode) is small compared to the cathode area, resulting in a large current density at the anode.

Whiskers are threadlike formations that can form between metal surfaces such as between traces on a circuit board. They can grow slowly eventually forming a short-circuit or arcing path. Coatings are available to slow or eliminate this growth. Gold and palladium resist whisker production.

On the surface of most metals, poor conducting oxides will form. Over a long period of time, the oxide can become so thick that the switch becomes useless. These oxides (and sulfides) are a source of electrical noise when the switch or contact must break through the film or layer before electrical contact is made. Silver plating of copper will stop the copper oxide growth, but for *power* contacts, the silver can burn off. Also silver has a greater tendency to "cold" weld, since silver has a lower melting point than copper. A sulfidation film on silver increases the contact resistance but can be burned off through controlled intentional arcing. Silver is used, when gold plated or flashed, for low-power switches or switches that are used infrequently. Gold may also be used to reduce the corrosion, however, gold cannot handle large currents because of its low melting point. Gold, because it is soft, provides for a solid contact area when the contacts are closed. Switching contacts can be rubbed to wear off the oxide. One simple method of rubbing off or cleaning the oxide buildup on copper ac power plugs is to pull out and then reinsert the plug.

The corrosion and erosion of switch contacts reduces the lifetime of the switch and increases the impedance between the contacts when in the closed position. Both corrosion and arcing between the contacts should normally be avoided to increase the life of the switch. Table 1.1 can be used as a guide in the selection of contact material for switches, relays, and other devices. Unfortunately, there is no one universal contact material that can be used for all switching applications.

TABLE 1.1   Typical Properties and Applications of Common Contact Materials

Contact Material	Properties	Applications
Aluminum	Thin but tough nonconducting oxide, should not be used if arcing is expected, severe erosion possible, welding can occur	
Copper	Oxide layer but easily removed, high conductivity, any arc that forms should not remain on the same spot, more resistant to welding than silver	Large voltage, lower current opening switch in nonoxidizing atmosphere with oil, short-life wiping action switches that remove the oxide buildup
Copper with beryllium	Beryllium content decreases the conductivity but increases the strength of the copper	Contact springs
Copper with cadmium	Cadmium content decreases the conductivity but increases strength of the copper	Backing material for silver-based contacts
Copper with chromium	High strength and high conductivity	Backing material for silver-based contacts, light-weight contact arms
Copper with tungsten	Resistance to arc erosion but eventual erosion, more susceptible to oxidation than copper	Oil-immersed circuit breakers
Gold	Does not tarnish or oxidize, very soft, used as a coating, welding and sticking can occur, fast wearing, chemically inert	Light-duty plug and sockets
Gold with copper and silver	Good resistance to tarnishing, stable contact resistance	Light-duty sliding contacts against rhodium, palladium-silver, or silver
Gold with platinum and silver	Platinum substitute but inferior in electrical wear	Low mechanical-wear telephone relays
Gold with silver	Platinum substitute but inferior in electrical wear, gold-flashed over silver reduces sulfidation	Relays that remain idle for long periods of time, connectors
Iridium	Hardener for platinum, difficult to work with	
Mercury	Liquid, pool of mercury in a tube when tilted forms electrical contact, sealed, no moving parts, low currents	Very long operating life, static mercury wetted relays, bounceless switches
Molybdenum	Similar to tungsten but greater oxidation	Electrodes in mercury switches, resistance welding electrodes
Nickel	Thin hard oxide, used as hard substrate	
Palladium	Platinum substitute, good resistance to tarnishing, oxidation, and corrosion, poor electrical conductivity	Reed switches
Palladium with copper	Similar to palladium-silver but less material transfer, better conductivity than palladium	Current in-rush switches, potentiometers, motor brushes
Palladium with ruthenium	Hard, low material transfer	Repetitive make-and-break switches, flasher relays
Palladium with silver	Resistant to tarnishing at normal temperatures	Make-and-brake and sliding contacts, edge connectors

TABLE 1.1   Typical Properties and Applications of Common Contact Materials (Continued)

Contact Material	Properties	Applications
Platinum	Nearly tarnish and oxidation free but soft	Light duty, low-current and low-voltage switch, spark plug tips, small fuses, relays with long idle times
Platinum with iridium	Iridium content increase hardness and wear resistance	High mechanical-wear relays, severe mechanical-wear dc thermostats, automobile flashers
Platinum with iridium and ruthenium	Very resistant to wear, does not tarnish at normal temperatures	High mechanical-wear brush and slip rings
Platinum with ruthenium	Ruthenium content increase hardness, substitute for platinum-iridium, begins to tarnish for higher ruthenium percentages	Severe mechanical wear dc thermostats
Rhodium	Does not tarnish, excellent resistance to oxidation, hard, should not be used if arcing expected, used as coating	Light-duty make-and-brake and sliding contacts
Ruthenium	Hard and similar to rhodium but lower internal stress, hardener for platinum and palladium	
Silver (fine)	Commonly used general purpose contact material, highest electrical conductivity but soft, forms sulfide tarnish but still clean, silver generates a very thin oxide layer and therefore not appropriate for very light electrical duty and low-level audio circuits, low and constant contact resistance, welding a problem at high currents and voltages, material loss and pitting, electrical migration	Moderate current and voltage circuit breakers, low-current dc thermostats, miniature low-voltage circuit breakers, small relays, fuses
Silver with cadmium oxide	Cadmium oxide increases contact resistance but increases hardness, superior in tarnish resistance compared to silver with copper, widely used contact material for even high currents, tendency to extinguish arcs, sulfide formation possible	Slow acting, arcing dc thermostats, switches, relays, miniature circuit breakers, motors
Silver with copper	Copper content increases hardness but increases tendency to tarnish and increases contact resistance	Commutators for micromotors, sliding switches with large forces
Silver with copper and nickel	Copper and nickel content increases resistance to welding, handle high in-rush currents	Capacitive loads, incandescent lamps
Silver with gold	Gold content increases resistance to tarnishing	
Silver with graphite	Lubrication properties provided by graphite, severe arc erosion at medium currents, no welding if graphite content between 2 to 5%	Light-duty sliding contacts against silver and silver-nickel, low-voltage circuit breakers
Silver with iron		Surge current dc thermostats
Silver with molybdenum	Similar to silver with tungsten but lower conductivity and slightly harder	

(Continued)

TABLE 1.1   Typical Properties and Applications of Common Contact Materials (Continued)

Contact Material	Properties	Applications
Silver with nickel	Nickel content increases resistance to arc erosion and hardness, increased lifetime compared to fine silver	Lower current contactors,[a] high-current and high-voltage circuit breakers, low-current thermostats, lower current ac and dc relays
Silver with palladium	Palladium content increases resistance to tarnishing and mechanical wear but much lower conductivity	Light contact pressure ac thermostats, connectors, micromotor brushes
Silver with platinum		Low inductive-current ac thermostats
Silver with tungsten	Susceptible to welding for higher silver concentrations, contact resistance changes with time, arcs can cause erosion	High inductive-current starters, switches for residential lamps
Silver with tungsten and carbide	Carbide content increases life and hardness compared to silver with tungsten but decreases the conductivity	
Tin	Soft but hard oxide forms	
Tungsten	Very hard, resistant to material transfer-wear low, oxide layer troublesome at low voltages and contact pressures, high limiting arcing voltage, welding and sticking not a problem, high contact resistance	High-voltage reed relays, magneto ignition, furnace igniters, automotive horns
Tungsten with silver	Similar to tungsten with copper but preferable in air, silver increases conductivity	

[a]Contactors are relays with heavy-duty contacts.

1.16   Maximum Electric Field and Breakdown Table

Provide a table of the maximum electric field between two electrodes that exhibit symmetry. Then, provide a listing of the breakdown strength of various gases, liquids, and solids.

The source of electric fields is charge or time-varying magnetic fields, which are due to the movement of charge. The maximum electric field expressions given in Table 1.2 were derived using Poisson's and Laplace's equations, which assume electrostatic conditions or no charge movement. However, a very good estimate of the potential and electric field distributions between two electrodes, including the maximum electric field, can be obtained for electroquasistatic situations using Poisson's and Laplace's equations.

The expressions in Table 1.2 can be obtained from Laplace's equation

$$\nabla^2 \Phi = 0 \tag{1.65}$$

using the classical boundary condition approach, where Φ is the potential distribution function.[9] If the region between the electrodes is not homogeneous or contains multiple dielectrics, Laplace's equation is used to determine the potential function, Φ, in each homogeneous region. Once the potential function is determined in the various regions, the electric field is determined from the potential function:

$$\vec{E} = -\nabla \Phi \tag{1.66}$$

[9]A number of authors prefer to use the variable Φ for the potential or voltage function while the variable V for the actual potential or voltage difference between two objects or between an object and a "ground" reference.

TABLE 1.2 Maximum Electric Field for Simple Electrode Configurations

Electrode Configuration	Maximum Magnitude of the Electric Field
Two Close, Parallel Plane Conductors 	$\dfrac{V}{d}$ everywhere away from the ends
Two Close, Parallel Plane Conductors with Parallel Dielectrics 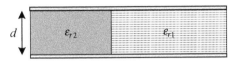	$\dfrac{V}{d}$ everywhere in both regions away from the ends
Two Close, Parallel Plane Conductors with Layered Dielectrics 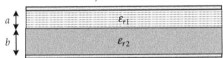	$V\dfrac{\varepsilon_{r1}}{\varepsilon_{r2}a+\varepsilon_{r1}b}$ if $\varepsilon_{r1}>\varepsilon_{r2}$ in region ε_{r2} $V\dfrac{\varepsilon_{r2}}{\varepsilon_{r2}a+\varepsilon_{r1}b}$ if $\varepsilon_{r2}>\varepsilon_{r1}$ in region ε_{r1} and away from the ends
Two Slanted Conductors 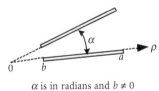 α is in radians and $b\neq 0$	$\dfrac{V}{\alpha b}$ near $\rho=b$ and away from the edges
Two Slanted Conductors with Layered Dielectrics 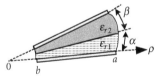 α and β are in radians and $b\neq 0$	$\dfrac{V}{b}\dfrac{\varepsilon_{r1}}{\varepsilon_{r2}\alpha+\varepsilon_{r1}\beta}$ if $\varepsilon_{r1}>\varepsilon_{r2}$ in region ε_{r2} near $\rho=b$ and away from the edges $\dfrac{V}{b}\dfrac{\varepsilon_{r2}}{\varepsilon_{r2}\alpha+\varepsilon_{r1}\beta}$ if $\varepsilon_{r2}>\varepsilon_{r1}$ in region ε_{r1} near $\rho=b$ and away from the edges
Two Concentric Cylindrical Conductors 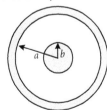	$\dfrac{V}{b\ln\left(\dfrac{a}{b}\right)}$ near $\rho=b$

(Continued)

TABLE 1.2   Maximum Electric Field for Simple Electrode Configurations (Continued)

Electrode Configuration	Maximum Magnitude of the Electric Field
Two Partial, Close, Concentric Cylindrical Conductors 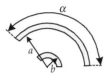	$\dfrac{V}{b\ln\left(\dfrac{a}{b}\right)}$ near $\rho = b$ and away from the edges
Two Concentric Cylindrical Conductors with Two Parallel Dielectrics 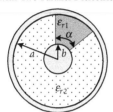	$\dfrac{V}{b\ln\left(\dfrac{a}{b}\right)}$ near $\rho = b$ in both regions
Two Concentric Cylindrical Conductors with Layered Dielectrics 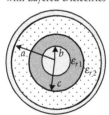	$\dfrac{V}{c}\dfrac{\varepsilon_{r1}}{\varepsilon_{r2}\ln\left(\dfrac{c}{b}\right)+\varepsilon_{r1}\ln\left(\dfrac{a}{c}\right)}$ if $c\varepsilon_{r2} < b\varepsilon_{r1}$ in region ε_{r2} near $\rho = c$ $\dfrac{V}{b}\dfrac{\varepsilon_{r2}}{\varepsilon_{r2}\ln\left(\dfrac{c}{b}\right)+\varepsilon_{r1}\ln\left(\dfrac{a}{c}\right)}$ if $c\varepsilon_{r2} > b\varepsilon_{r1}$ in region ε_{r1} near $\rho = b$ $\dfrac{V}{b\ln\left(\dfrac{c}{b}\right)+c\ln\left(\dfrac{a}{c}\right)}$ if $c\varepsilon_{r2} = b\varepsilon_{r1}$ two maximums: one in region ε_{r1} near $\rho = b$ AND one in region ε_{r2} near $\rho = c$
Two Concentric Spherical Conductors 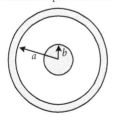	$V\dfrac{a}{b(a-b)}$ near $r = b$
Two Partial, Close, Concentric Spherical Conductors  α is in radians	$V\dfrac{a}{b(a-b)}$ near $r = b$ and away from the edges

TABLE 1.2   Maximum Electric Field for Simple Electrode Configurations (Continued)

Electrode Configuration	Maximum Magnitude of the Electric Field
Two Concentric Spherical Conductors with Two Parallel Dielectrics 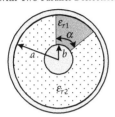 α is in radians of the wedge dielectric (in the θ direction)	$V \dfrac{a}{b(a-b)}$ near $r = b$ in both regions
Two Concentric Spherical Conductors with Layered Dielectrics 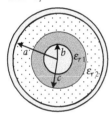	$\dfrac{V}{c} \dfrac{\varepsilon_{r1}}{\varepsilon_{r1}\left(1-\dfrac{c}{a}\right)+\varepsilon_{r2}\left(\dfrac{c}{b}-1\right)}$ if $c^2\varepsilon_{r2} < b^2\varepsilon_{r1}$ in region ε_{r2} near $r = c$ $\dfrac{V}{b} \dfrac{\varepsilon_{r2}}{\varepsilon_{r1}\left(\dfrac{b}{c}-\dfrac{b}{a}\right)+\varepsilon_{r2}\left(1-\dfrac{b}{c}\right)}$ if $c^2\varepsilon_{r2} > b^2\varepsilon_{r1}$ in region ε_{r1} near $r = b$ $\dfrac{Vac}{abc + ac^2 - ab^2 - c^3}$ if $c^2\varepsilon_{r2} = b^2\varepsilon_{r1}$ two maximums: one in region ε_{r1} near $r = b$ AND one in region ε_{r2} near $r = c$
Two Conical Conductors 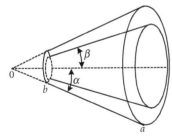 α and β are in radians and $b \neq 0$	$\dfrac{V}{b \sin\beta \ln\left[\dfrac{\tan\left(\dfrac{\alpha}{2}\right)}{\tan\left(\dfrac{\beta}{2}\right)}\right]}$ near $r = b$ at the inner cone at $\theta = \beta$
Two Identical Spheres [Attwood; Khalifa] 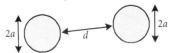	$\dfrac{V}{d}\left[\dfrac{\dfrac{d}{a}+1+\sqrt{\left(\dfrac{d}{a}+1\right)^2+8}}{4}\right]$ $\approx \dfrac{V}{2a}$ if $d \gg a$ near the two points on the spheres closest to each other

TABLE 1.2   Maximum Electric Field for Simple Electrode Configurations (Continued)

Electrode Configuration	Maximum Magnitude of the Electric Field
Spherical Conductor and Large Flat Conductor 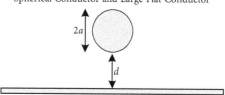	$$\frac{V}{d}\left[\frac{\frac{2d}{a}+1+\sqrt{\left(\frac{2d}{a}+1\right)^2+8}}{4}\right] \approx \frac{V}{d}\left(0.94\frac{d}{a}+0.8\right)$$ near the point on the sphere closest to the flat plane
Two Distant, Parallel, Unequal Cylindrical Conductors [Moon, '61] 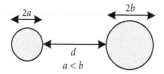	$$\frac{V}{2a\ln\left(\dfrac{d}{\sqrt{ab}}\right)} \quad \text{if } d \gg b$$ near the side of the smaller cylinder closest to the larger cylinder
Two Parallel, Identical Cylindrical Conductors [Khalifa; Watt] 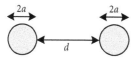	$$\frac{V}{2a\sqrt{\dfrac{\frac{2a+d}{2a}-1}{\frac{2a+d}{2a}+1}}\ln\left[\frac{2a+d}{2a}+\sqrt{\left(\frac{2a+d}{2a}\right)^2-1}\right]}$$ $$\approx \frac{V}{2a\ln\left(\dfrac{2a+d}{a}\right)} \quad \text{if } \frac{d}{a} \gg 4$$ near both sides of the cylinders closest to each other
Cylindrical Conductor above Large Flat Conductor [Khalifa; Watt] 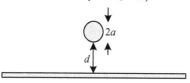	$$\frac{V}{a\sqrt{\dfrac{\frac{a+d}{a}-1}{\frac{a+d}{a}+1}}\ln\left[\frac{a+d}{a}+\sqrt{\left(\frac{a+d}{a}\right)^2-1}\right]}$$ $$\approx \frac{V}{a\ln\left(\dfrac{2d}{a}\right)} \quad \text{if } d \gg a$$ $$\approx \frac{V}{d}\left(\frac{d}{4a}+1\right) \quad \text{if } d < 20a$$ near the side of the cylinder closest to the flat plane
Dielectric-Coated Cylindrical Conductor above Large Flat Conductor [Harper, '72] 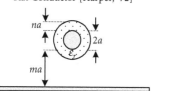	$$\frac{V}{a}\frac{2p\varepsilon_r^2}{p^2-(m-n)^2}$$ $$\left\{\left(\varepsilon_r^2+1\right)\ln\left[\frac{p+(m-n)}{p-(m-n)}\right]\right.$$ $$\left.+\ln\left[\frac{p+m+(\varepsilon_r-1)n}{p-m-(\varepsilon_r-1)n}\right]\right\}$$ where $p=\sqrt{[m+1+(\varepsilon_r-1)n]^2-1}$ this is the maximum field in the air near the side of the coating closest to the flat plane

TABLE 1.2   Maximum Electric Field for Simple Electrode Configurations (Continued)

Electrode Configuration	Maximum Magnitude of the Electric Field
Cylindrical Conductor between Two Large Flat Conductors [Haus] 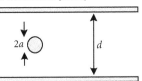 the cylindrical conductor is charge neutral and floating and the voltage is applied across the flat plates	$2\dfrac{V}{d}$ if $d \gg a$ near both sides of the cylinder closest to the two flat planes
Semicylinder on One of Two Large Flat Conductors [Bouwers] 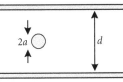	$2\dfrac{V}{d}$ if $d \gg a$ near the side of the semicylinder closest to the upper flat plane
Cylindrical Insulator between Two Large Flat Conductors [Haus; Zahn] 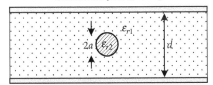 the voltage is applied across the flat plates	$\dfrac{V}{d}\dfrac{2\varepsilon_{r1}}{\varepsilon_{r1}+\varepsilon_{r2}}$ if $d \gg a,\, \varepsilon_{r1} > \varepsilon_{r2}$ everywhere inside the cylinder $\dfrac{V}{d}\dfrac{2\varepsilon_{r2}}{\varepsilon_{r1}+\varepsilon_{r2}}$ if $d \gg a,\, \varepsilon_{r2} > \varepsilon_{r1}$ near both sides of the outer surface of the cylinder closest to the two flat planes
Spherical Conductor between Two Large Flat Conductors [Haus; Moon, '61] 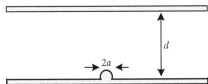 the spherical conductor is charge neutral and floating and the voltage is applied across the flat plates	$3\dfrac{V}{d}$ if $d \gg a$ near both points on the sphere closest to the two flat planes
Hemisphere on One of Two Large Flat Conductors [Moon, '61; Bouwers] 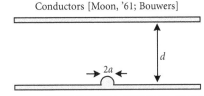	$3\dfrac{V}{d}$ if $d \gg a$ near the point of the hemisphere closest to the upper flat plane

(Continued)

TABLE 1.2   Maximum Electric Field for Simple Electrode Configurations (Continued)

Electrode Configuration	Maximum Magnitude of the Electric Field
Spherical Insulator between Two Large Flat Conductors [Jefimenko] 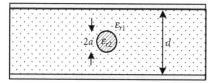 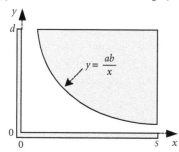 the voltage is applied across the flat plates	$\dfrac{V}{d}\dfrac{3\varepsilon_{r1}}{2\varepsilon_{r1}+\varepsilon_{r2}}$ if $d \gg a$, $\varepsilon_{r1}>\varepsilon_{r2}$ everywhere inside the sphere $\dfrac{V}{d}\dfrac{3\varepsilon_{r2}}{2\varepsilon_{r1}+\varepsilon_{r2}}$ if $d \gg a$, $\varepsilon_{r2}>\varepsilon_{r1}$ near both points on the outer surface of the sphere closest to the two flat planes

| Hyperbolic Conductor in Corner Trough [Zahn] 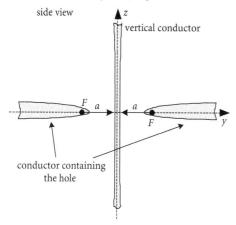 | $V\sqrt{\left(\dfrac{s}{ab}\right)^2+\dfrac{1}{s^2}}$ if $s>d$

 between the electrodes at $x=s$

 $V\sqrt{\left(\dfrac{d}{ab}\right)^2+\dfrac{1}{d^2}}$ if $d>s$

 between the electrodes at $y=d$

 $\dfrac{Vs\sqrt{2}}{ab}$ if $d=s$

 between the electrodes at $y=d$ and $x=s$ |

Vertical Hyperboloid Conductor through Hyperboloid-Shaped Circular Hole

[Moon, '61]

side view

z

vertical conductor

F   a   a

F

y

conductor containing
the hole

$$\frac{2V}{a\ln\left[\dfrac{\cot\left(\dfrac{\alpha}{2}\right)}{\cot\left(\dfrac{\beta}{2}\right)}\right]}\sin(2\alpha)$$

near the surface of the vertical conductor at $z=0$

$$\frac{2V}{a\ln\left[\dfrac{\cot\left(\dfrac{\alpha}{2}\right)}{\cot\left(\dfrac{\beta}{2}\right)}\right]}\sin(2\beta)$$

near the surface of the conductor
containing the hole at $z=0$

$$\left(\frac{x}{a\sin\alpha}\right)^2+\left(\frac{y}{a\sin\alpha}\right)^2-\left(\frac{z}{a\cos\alpha}\right)^2=1$$

describes surface of vertical conductor where
α is almost 0 and almost π, as $\alpha \rightarrow 0$ and π
the vertical conductor becomes more slender

TABLE 1.2 Maximum Electric Field for Simple Electrode Configurations (Continued)

Electrode Configuration	Maximum Magnitude of the Electric Field

$$\left(\frac{x}{a\sin\beta}\right)^2 + \left(\frac{y}{a\sin\beta}\right)^2 - \left(\frac{z}{a\cos\beta}\right)^2 = 1$$

describes surface of conductor containing the hole where β is almost $\pi/2^\pm$, as $\beta \to \pi/2$ the conductor containing the hole becomes flatter note that a is not equal to the radius of the hole unless $\beta = \pi/2$

Hyperboloid Conductors above Large Flat Conductor [Weber]

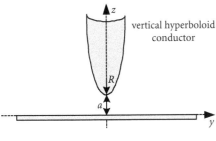

vertical hyperboloid conductor

$$\left(\frac{z}{a}\right)^2 - \left(\frac{x}{b}\right)^2 - \left(\frac{y}{b}\right)^2 = 1$$

describes surface of hyperboloid conductor sharpness of electrode increases as b decreases or a increases

$$\text{focus} = f = \sqrt{a^2 + b^2}$$

R = radius of curvature at the tip of the paraboloid conductor

$$\frac{V}{f\left(\frac{b}{f}\right)^2 \tanh^{-1}\left(\frac{a}{f}\right)}$$

near the apex or tip of the vertical conductor at $(0,0,a)$

$$\frac{Va}{b^2\ln\left(\frac{2a}{b}\right)} = \frac{2V}{R\ln\left(\frac{4a}{R}\right)} \quad \text{if } a \gg b$$

near the apex or tip of the vertical conductor at $(0,0,a)$

Two Hyperboloid Conductors [Weber]

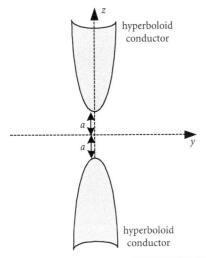

hyperboloid conductor

hyperboloid conductor

$$\frac{V}{2f\left(\frac{b}{f}\right)^2 \tanh^{-1}\left(\frac{a}{f}\right)}$$

near the apex or tip of either conductor at $(0,0,\pm a)$

$$\frac{Va}{2b^2\ln\left(\frac{2a}{b}\right)} \quad \text{if } a \gg b$$

near the apex or tip of either conductor at $(0,0,\pm a)$

(Continued)

TABLE 1.2   Maximum Electric Field for Simple Electrode Configurations (Continued)

Electrode Configuration	Maximum Magnitude of the Electric Field

$$\left(\frac{z}{a}\right)^2 - \left(\frac{x}{b}\right)^2 - \left(\frac{y}{b}\right)^2 = 1$$

describes the surfaces of the hyperboloid
conductors, sharpness of electrodes increases
as b decreases or a increases

$$\text{foci at } f = \sqrt{a^2 + b^2}$$

Oblate Semispheroidal on One of Two
Large Flat Conductors [Moon, '61]

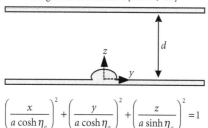

$$\left(\frac{x}{a\cosh\eta_o}\right)^2 + \left(\frac{y}{a\cosh\eta_o}\right)^2 + \left(\frac{z}{a\sinh\eta_o}\right)^2 = 1$$

describes surface of the oblate spheroidal
as $\eta_o \to 0$ surface approaches flat disk
  of radius a
as $\eta_o \to \infty$ surface approaches sphere
  of radius a

$$\frac{V}{d}\left[1 + \frac{\sinh\eta_o \cot^{-1}(\sinh\eta_o) - \tanh^2\eta_o}{1 - \sinh\eta_o \cot^{-1}(\sinh\eta_o)}\right]$$

near the point on the semispheroid
closest to the upper flat plane

Two Paraboloids of Revolution [Moon, '61; Weber]

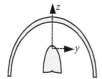

$$x^2 + y^2 = a^2(a^2 - 2z)$$

describes surface of the lower paraboloid of
  revolution
as $a \to 0$ surface approaches thin vertical
  conductor

$$x^2 + y^2 = b^2(b^2 - 2z)$$

describes surface of the upper paraboloid of
  revolution
as $b \to \infty$ surface approaches flat conductor

$$\frac{V}{a^2 \ln\left(\frac{b}{a}\right)}$$

near the point on the lower conductor
closest to the upper conductor

TABLE 1.2   Maximum Electric Field for Simple Electrode Configurations (Continued)

Electrode Configuration	Maximum Magnitude of the Electric Field
Paraboloid of Revolution below Large Flat Conductor [Weber; Khalifia] $x^2 + y^2 = a^2(a^2 - 2z)$ describes surface of the paraboloid of revolution R = radius of curvature at the tip of the paraboloid conductor	$\dfrac{2V}{R \ln\left(\dfrac{2d}{R}\right)}$ if $d \gg R$ near the apex or tip of the parabolic conductor

Voltage V is applied between the given electrodes or conductors.

Again, Equation (1.66) is an electrostatic-based definition, but it can be used for many situations where the electrical energy is mainly due to the electric fields. The del operator, ∇, is a vector operator involving differentiation. It transforms a scalar function, Φ, into a vector function, $\vec{E}$. In the Cartesian, cylindrical, and spherical coordinate systems, respectively,

$$\vec{E} = -\nabla\Phi = -\frac{\partial\Phi}{\partial x}\hat{a}_x - \frac{\partial\Phi}{\partial y}\hat{a}_y - \frac{\partial\Phi}{\partial z}\hat{a}_z \tag{1.67}$$

$$\vec{E} = -\nabla\Phi = -\frac{\partial\Phi}{\partial\rho}\hat{a}_\rho - \frac{1}{\rho}\frac{\partial\Phi}{\partial\phi}\hat{a}_\phi - \frac{\partial\Phi}{\partial z}\hat{a}_z \tag{1.68}$$

$$\vec{E} = -\nabla\Phi = -\frac{\partial\Phi}{\partial r}\hat{a}_r - \frac{1}{r}\frac{\partial\Phi}{\partial\theta}\hat{a}_\theta - \frac{1}{r\sin\theta}\frac{\partial\Phi}{\partial\phi}\hat{a}_\phi \tag{1.69}$$

As is seen from these expressions, the electric field is a function of the derivative or change in the potential in each of the three orthogonal directions in the coordinate system. When the potential changes quickly in a particular direction, the electric field in that particular direction is large. When the potential changes slowly in a particular direction, the electric field in that particular direction is small. The medium between electrodes will often first break down where the electric field is the largest.

Another interesting property of the gradient operation on the potential function is that it generates vectors that are perpendicular to the constant potential contours: if the voltage contours or equipotentials are plotted, the electric field lines are normal or perpendicular to these equipotentials. The electric fields are directed from the most positive electrode to the least positive electrode. (If a small positive charge is placed in the field, this charge will move in the direction of the electric field, away from the most positive electrode and toward the least positive electrode.)

In the design of many products, it is desirable to be able to predict where the electric field is expected to be the greatest. Sometimes, the shape of the electrode can be changed to reduce the intensity of the field or the material between the electrodes changed to decrease the probability of breakdown. As is apparent by glancing over Table 1.2, the electric field is often the greatest where the electrodes are the sharpest (i.e., smallest radii of curvature) and most closely spaced.

If the medium is homogeneous or the same everywhere, the electric field for a given electrode configuration is not affected by the medium's dielectric constant. However, all materials have a maximum electric field rating. This rating is the maximum electric field that a material can handle before it begins to break down (or change in some noticeable and important manner). This applies to solid, liquid, and gaseous mediums. Sometimes, when the field is too intense for a medium, the "weaker" medium can be replaced with a "stronger" medium. Sometimes, a material with a higher electric field strength and different dielectric constant can be inserted or introduced in only those regions where the electric field is the most troublesome, reducing costs and simplifying structural requirements. However, this must be done carefully. Although not a strict requirement, the new material generally should not disturb the electric field lines between the electrodes. This implies that the surface of the new insulative material (if its dielectric constant is different from the original medium) should conform to the electric field lines; in other words, the surface of the new medium should not crossover any electric field lines. When the surface follows the equipotential contours, the electric field distribution is affected by the introduction of the new material of different dielectric constant. There is, however, a potential disadvantage in introducing solid insulative material that follow the shape of the electric field lines: surface currents and lower-than-normal breakdown voltage levels can result. If the material is not well protected from its environment, dirt, grease, and other impurities can build up along its surface. These impurities can dramatically decrease the effectiveness of the insulative properties of the medium. The conductivity of the insulative material along its surface, referred to as the surface conductivity, can be much greater than in its bulk. When the electric field is tangential to the new material, the resultant charge flow (i.e., current) is directly proportional to this tangential field. The term creepage is used to describe the current along insulating surfaces. The creepage path is the length that the current "travels" between the electrodes. By increasing the creepage path, the probability of breakdown is reduced. (Insulators sometimes have a wavy shape, in part, to increase this path length.)

As seen in several of the expressions in Table 1.2, the introduction of a dielectric material between electrodes may actually increase the electric field strength in certain regions. Also, when a dielectric material, such as an insulating oil for a transformer, contains air pockets, gas bubbles, or voids, corona may occur in these air regions. There may be no visible signs of this corona, but it may produce electrical noise and damage the material. In some cases, such as with coaxial cables, two dielectric layers can be used to increase the working voltage of the cable.

It is true that analytical expressions for the maximum electric field between many real electrode configurations are not available. However, many of these electrode configurations can be approximated, possibly in only a small local region, using the expressions given in Table 1.2. The electric field has units of V/m when the applied voltage across the electrodes is given in volts (V). If not otherwise shown, the material between and around the electrodes is assumed homogeneous. Although the $\vec{D} = \varepsilon_r \varepsilon_o \vec{E}$ field is affected by the relative permittivity or dielectric constant of the medium, the electric field is not a function of the permittivity for a homogeneous environment (for a given applied voltage). Besides the surface charge on the electrodes, there are no point, line, surface, or volume charges between the electrodes. For those two-dimensional fields, the lengths of the electrodes are assumed long compared to any of the cross-sectional dimensions. The formulas apply to either dc or slowly varying ac applied voltages. The label "close" as applied to two conductors usually implies that the distance between the two conductors is small compared to the other dimensions of the electrodes. The label "large" as applied to a conductor(s) usually implies that the width and length of the conductor are large compared to the dimensions of the "other" conductor of interest.

The breakdown strength or gradient of a material, also referred to as the dielectric or electric strength, is a function of many parameters. For example, it is a function of the: (1) frequency and duration of the applied signal, (2) temperature, (3) condition of the surface including level and type of contaminates, moisture content, orientation, thickness, and method of preparation, and (4) uniformity of applied field.[10] For these and many other reasons, the breakdown strengths given in Table 1.3 should only be used as a rough guideline. In any application, the actual material should be tested in the expected environment. There are several standard methods of determining the dielectric strength of an insulator, partially depending on whether it is a liquid, solid, or gas. For solids, the thickness of the sample can also vary

---

[10]The purity of a liquid can have a dramatic effect on the breakdown strength.

TABLE 1.3   Breakdown Strength of Solids, Gases, and Liquids

Substance	E_{bk} (MV/m)
2,2-Dimethylbutane	130
2,3-Dimethylbutane	140
2-Methylpentane	150
Acetal	16–20
Acrylic	14–30
Acrylonitrile-butadiene-styrene (ABS)	12–19
Acrylonitrile-butadiene-styrene polycarbonate	17
Air	3
Air (compressed to 20 atm)	19
Alkyd	12–16
Alumina (aluminum oxide)	8.3–15 (246 mils)
	100–800 (thin film)
Aluminum nitride	15 (246 mils)
Amber	91 (125 mils)
Ammonia (gas)	3
Aniline-formaldehyde (Dilectene)	32 (68 mils)
Barium titanate (ceramic)	2–12
	35 (thin film)
Benzene	163
Beryllium oxide	12 (125 mils)
Bismuth oxide (Bi_2O_3)	70 (thin film)
Boron nitride	370 (246 mils)
Calcium titanate (ceramic)	3.9
Carbon dioxide (gas)	2.4
Carbon tetrachloride (gas)	19
Cellulose acetate	8
Cellulose acetate-butyrate, plasticized	9.8–16 (125 mils)
Cellulose propionate	12
Chlorinated polyether	16 (125 mils)
Chlorinated polyether	16
Chlorine (gas)	2.6
Chlorotrifluoro-ethylene	18
Cordierite	5.1 (246 mils)
Cresylic acid-formaldehyde, 50% α-cellulose	40 (33 mils)
Decafluorobutane (gas)	9.2
Decane	190
Delrin AF	16 (125 mils)
Diallyl phthalate	14–16
Epoxy	16–40
Ethyl cellulose	60
Ethylbenzene	226
FEP-Teflon	71 (20 mils) @ 60 Hz
	260 (1 mils) @ 60 Hz
Fluoroelastomer	20
Fluoropolymer	8
Freon 11 (gas)	11
Freon 12 (gas)	7.3
Freon 13 (gas)	4.3
Freon 14 (gas)	3.0
Freon 21 (gas)	4.0
Freon 22 (gas)	4.2
Glass (soda lime)	450
Glass cloth (yellow varnished)	47 (10 mils)
Glassine (paper)	16–20

(Continued)

TABLE 1.3   Breakdown Strength of Solids, Gases, and Liquids (Continued)

Substance	E_{bk} (MV/m)
GR-S (rubber)	34 (40 mils)
Helium (gas)	0.45
Heptane	170
Hexafluoroethane (gas)	5.5
Hexane	160
Hydrogen (gas)	2.0
Hydrogen sulfide (gas)	2.7
i-Butylbenzene	220
i-Propylbenzene	240
Iso-Octane	140
Lava, grade A	3.9 (246 mils)
Lead titanate	2–12
	230 (thin film)
Liquid crystal polymer	33 (125 mils)
Macor (glass ceramic)	40 (246 mils)
Magnesium difluoride	200 (thin film)
Magnesium oxide	200
Melamine formaid	6–18
Melamine formaldehyde	11
Melamine resin	9.4–11 (125 mils)
Melamine, glass fiber	9.6–12
Melamine, α-cellulose	12–16
Monobromotrifluoromethane (gas)	4.8
Monochloromonofluoromethane (gas)	3.1
Monomeric styrene (organic liquid)	12 (100 mils)
Mullite	9.8 (246 mils)
Mylar polyster	300 (1 mils) @ 60 Hz
	550 (1 mils) @ dc
n-Butylbenzene	280
Neoprene (rubber)	12 (125 mils)
Nitrogen (gas)	3.3
Nonane	180
n-Propylbenzene	250
Nylon	16–26
Nylon 101	12–16 (80 mils)
Nylon 11	26–30
Nylon 12	16–30
Nylon 6	17–30
Nylon 6/6	16–24
Nylon, polyamide	13
Octane	180
Octofluoropropane (gas)	6.6
Oil, capacitor	12
Oil, caster	14
Oil, pipe cable	12
Oil, polybutene pipe cable/capacitor	14
Oil, transformer	13
Oil, transil	12 (100 mils)
Oxygen (gas)	2.6
Paper, royalgrey	8 (125 mils)
Parylene	220–280 (1 mils)
Pentane	140
Perchlorylfluoride (gas)	7.4
Perfluorocyclobutane (gas)	8.4
Phenol-formaldehyde (bakelite)	11–12 (125 mils)

TABLE 1.3   Breakdown Strength of Solids, Gases, and Liquids (Continued)

Substance	E_{bk} (MV/m)
Phenolic	9–15
Phenoxy	16
Polyamide (paper)	18 (2 mils)
	32 (15 mils)
Polyaryl ether	17 (125 mils)
	20
Polyarylate	15
Polybutylene terephthalate	16
Polybutylene terephthalate-polycarbonate	18
Polycarbonate	15–19
Polyester	14–18
Polyester (unclad, flexible)	200 (film)
Polyether sulfone	16
Polyethylene-high density (plastic)	18–20
Polyethylene-low density (plastic)	18–28
Polyethylene-medium density (plastic)	20–30
Polyethylene-terephthalate	21
Polyimide	14–22
Polyimide (FEP coated)	140 (5 mils)
Polyisobutylene	24 (10 mils)
Polymethyl methacrylate	20
Polyolefin	28–38
Polyoxymethylene	20
Polyphenylene oxide	16–20
Polyphenylene sulfide	18
Polypropylene	18–32
Polystyrene	20–23
Polysulfone	16–17
Polytetrafluoroethylene (Teflon)	39–79
Polyurethane	19
Polyvinyl fluoride	140 (1 mils)
Polyvinyl formal	34 (34 mils)
Polyvinyl-chloride (PVC)	15–20
Polyvinylidene fluoride	10 (125 mils)
Porcelain (mullite)	9.4–16
Press/paper board	8–12 (94–125 mils)
Pyrex (Corning 7740)	13–14
Quartz (silicon dioxide)	25–40 (246 mils)
Rag paper	13
Rubber (depolymerized-flowable)	15
Rubber (EPDM hydrocarbon)	31
Ruby mica	150–220 (40 mils)
Sapphire crystal	15–50 (246 mils)
Silicon dioxide	150–1,000 (thin film)
Silicon monoxide	200 (thin film)
Silicon nitride (hot pressed)	18 (246 mils)
Silicone	12–16
Silicone fluid	14
Silicone oil	30–40
Steatite	8.9–9.3 (246 mils)
	20
Strontium titanate (ceramic)	3.9
Styrene-acrylonitrile	12
Styrene-maleic anhydride	19
Sulfur dioxide (gas)	0.9

(Continued)

TABLE 1.3   Breakdown Strength of Solids, Gases, and Liquids (Continued)

Substance	E_{bk} (MV/m)
Sulfur hexafluoride (gas)	7.9
Tantalum oxide (Ta$_2$O$_5$)	100–600 (thin film)
Tetradecane	200
Tetrafluoro-ethylene	17
Thorium dioxide	100 (film)
Titanium dioxide	4–8.4
	100 (thin film)
Toluene	200
Trichloromethane (gas)	13
Urea resin	8.7–12 (125 mils)
Urea-formaldehyde, α-cellulose	12–16
Urethane elastomer	18–20 (125 mils)
Vinylidene-vinyl chloride copolymer	12 (125 mils)
Wax	12
Wood-flour filled phenolic	7.9–16 (125 mils)
Zinc blende	100 (thin film)
Zirconia	9–9.4 (246 mils)
	400 (thin film)

but 32 to 125 mils is common (unless the sample's intended use is as a film). For uniform electric fields, increasing the pressure generally increases the breakdown strength for gases. The data in Table 1.3 assumes standard atmospheric conditions at 25° C. Also, although this table provides the breakdown strength in MV/m, many references provide the breakdown strength in V/mils or vpm. To convert V/mils to MV/m, multiply the V/mils quantity by 3.94×10^{-2}. Thickness is provided in mils, which is common. One mils is equal to 0.001 of an inch. To convert from mils to meter, multiply by 2.54×10^{-5}. A film is usually considered less than a few mils thick and, for thin films, less than 1/25 of a mils. Again, the breakdown values provided should only be used as a rough guideline. [Howard W. Sams; ASM, '89; Dorf; Plastics; Saums; Maissel; Adamczewski; Thomson; product data sheets]

1.17   Minimum Corona Voltage

List the empirically derived expressions for the onset of visual corona under normal atmospheric conditions for (1) a circular conductor above a ground plane, (2) two parallel circular conductors, (3) two concentric cylindrical conductors, and (4) two identical spherical conductors. Plot the variation of the critical voltage vs. spacing and conductor radius for two parallel circular conductors. [Peek; Watt; Grigsby; Saums; Weber; Taylor, '94]

Corona, although a low-energy discharge, is a source of electrical noise (with cutoff frequencies around 10 to 50 MHz), audible noise, power loss, and surface degradation. It often occurs near sharp points, edges, and other areas with a small radius of curvature.

The expressions given in this section are for normal atmospheric conditions. The equations given are for 60 Hz ac.[11] For higher frequencies, the expressions should be multiplied by a frequency factor that varies slowly with frequency and is less than one. At 20 kHz, for example, this multiplier is about 0.9. The roughness of the surface will also affect these corona onset voltages. A surface irregularity multiplier, which is also less than one, if known, can be used to improve these results. For high-voltage overhead power lines, the surface condition of the conductors often changes with the weather and with age. For dry, aged conductors, this surface multiplier is typically 0.8 to 0.9, while for conductors in poor weather conditions, the surface multiplier is typically 0.5 to 0.7.

---

[11]As a side note of possible interest, for high-voltage transmission lines in foul weather, the electrical and audible noise generated by dc-based coronas is less than ac-based coronas.

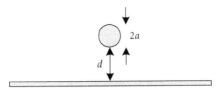

FIGURE 1.32 Long circular conductor parallel to a large flat ground plane.

For all of the following configurations, the electric field is nonuniform between the electrodes. However, if the distance between the electrodes is small compared to the radii of curvature of both electrodes, then the field directly between the electrodes is approximately uniform.

Corona occurs where the electric field is nonuniform. For the long circular conductor parallel to a large flat ground plane shown in Figure 1.32, the critical electric field strength for the onset of corona is

$$E_c = 3.0 \varsigma \left(1 + \frac{0.030}{\sqrt{\varsigma a}} \right) \times 10^6 \text{ V/m} \quad \text{where } \varsigma = \frac{273}{273 + T} e^{-\frac{3.45 \times 10^{-2} h}{273 + T}} \tag{1.70}$$

where T is the temperature in degrees Centigrade, h is the height above sea level in meters, and ς is the relative air density factor. Notice that this critical electric field is greater than 3 MV/m, the standard result used for breakdown of air in a uniform electric field. Using the expression for the maximum electric field for this configuration listed in Table 1.2, the critical voltage is

$$V_c = 3.0 a \varsigma \left(1 + \frac{0.030}{\sqrt{\varsigma a}} \right) \sqrt{\frac{\frac{a+d}{a} - 1}{\frac{a+d}{a} + 1}} \ln \left[\frac{a+d}{a} + \sqrt{\left(\frac{a+d}{a} \right)^2 - 1} \right] \times 10^6 \text{ V}$$

$$\approx 3.0 a \varsigma \left(1 + \frac{0.030}{\sqrt{\varsigma a}} \right) \ln \left(\frac{2d}{a} \right) \times 10^6 \text{ V} \quad \text{if } d \gg a \quad \text{where } \varsigma = \frac{273}{273 + T} e^{-\frac{3.45 \times 10^{-2} h}{273 + T}} \tag{1.71}$$

For corona to occur for a circular conductor parallel to a ground plane, $d/a > 3$ (or greater). Otherwise, if the voltage difference across the two conductors is sufficiently large, a sparkover will occur instead of corona. For very small conductor radii (e.g., $a = 0.05$ mm or less and $d = 12.7$ mm), the critical voltage obtained from these expressions can be too small.

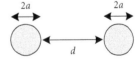

FIGURE 1.33 Two long, parallel circular conductors.

For the two long, parallel circular conductors shown in Figure 1.33, the critical electric field strength for the onset of corona is also given by (1.70). Using the expression for the maximum electric field for this configuration, the critical voltage is

$$V_c = 6.0 \varsigma a \left(1 + \frac{0.030}{\sqrt{\varsigma a}} \right) \sqrt{\frac{\frac{2a+d}{2a} - 1}{\frac{2a+d}{2a} + 1}} \ln \left[\frac{2a+d}{2a} + \sqrt{\left(\frac{2a+d}{2a} \right)^2 - 1} \right] \times 10^6 \text{ V}$$

$$\approx 6.0 \varsigma a \left(1 + \frac{0.030}{\sqrt{\varsigma a}} \right) \ln \left(\frac{d}{a} \right) \times 10^6 \text{ V} \quad \text{if } d \gg a \quad \text{where } \varsigma = \frac{273}{273 + T} e^{-\frac{3.45 \times 10^{-2} h}{273 + T}} \tag{1.72}$$

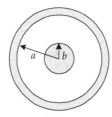

FIGURE 1.34 Two long, concentric cylindrical conductors.

For corona to occur for two parallel conductors, $d/a > 6$ (or greater). Otherwise, if the voltage difference across the two conductors is sufficiently large, a sparkover could occur instead of corona.

For the two long, concentric cylindrical conductors shown in Figure 1.34, the critical electric field in air for the onset of corona is

$$E_c = 3.08\varsigma\left(1 + \frac{0.034}{\sqrt{\varsigma b}}\right) \times 10^6 \text{ V/m} \quad \text{where } \varsigma = \frac{273}{273+T}e^{-\frac{3.45\times10^{-2}h}{273+T}} \tag{1.73}$$

where T is the temperature in degrees Centigrade, h is the height above sea level in meters, and ς is the relative air density factor. Using the expression for the maximum electric field for this configuration, the critical voltage is

$$V_c = 3.08\varsigma b\left(1 + \frac{0.034}{\sqrt{\varsigma b}}\right)\ln\left(\frac{a}{b}\right) \times 10^6 \text{ V} \quad \text{where } \varsigma = \frac{273}{273+T}e^{-\frac{3.45\times10^{-2}h}{273+T}} \tag{1.74}$$

For corona to occur (along the smaller inner conductor) for two concentric cylindrical conductors, $a/b > e^1 \approx 2.7$ (or greater). Otherwise, if the voltage difference across the two conductors is sufficiently large, a sparkover could occur instead of corona.

For the two identical spheres shown in Figure 1.35, the critical electric field in air for normal atmospheric conditions for the onset of corona is

$$E_c = 2.72\left(1 + \frac{0.054}{\sqrt{a}}\right) \times 10^6 \text{ V/m} \tag{1.75}$$

Using the expression for the maximum electric field for this configuration, the critical voltage is

$$V_c = 11\left(1 + \frac{0.054}{\sqrt{a}}\right)\frac{d}{\frac{d}{a}+1+\sqrt{\left(\frac{d}{a}+1\right)^2+8}} \times 10^6 \text{ V} \tag{1.76}$$

$$V_c \approx 5.4a\left(1 + \frac{0.054}{\sqrt{a}}\right) \times 10^6 \text{ V} \quad \text{if } d \gg a$$

For corona to occur for two identical spheres, $d/a > 2$ (or greater), but sparkover can be difficult to avoid when $d/a < 8$.

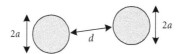

FIGURE 1.35 Two identical spheres.

$$T := 25 \qquad h := 500 \qquad d_m := 0.01$$

$$d := d_m, \frac{d_m}{0.9} \ .. \ 10 \qquad \zeta(h, T) := \frac{273}{273 + T} \cdot e^{\frac{-3.45 \cdot 10^{-2} \cdot h}{273 + T}} \qquad \zeta(h, T) = 0.865$$

$$V_c(a, d) := 6.0 \cdot \zeta(h, T) \cdot a \cdot \left(1 + \frac{0.03}{\sqrt{\zeta(h, T) \cdot a}}\right) \cdot \sqrt{\frac{\frac{2 \cdot a + d}{2 \cdot a} - 1}{\frac{2 \cdot a + d}{2 \cdot a} + 1}} \cdot \ln\left[\frac{2 \cdot a + d}{2 \cdot a} + \sqrt{\left(\frac{2 \cdot a + d}{2 \cdot a}\right)^2 - 1}\right] \cdot 10^6$$

Critical Voltage for Visual Corona

MATHCAD 1.3 Critical voltage vs. distance for two parallel conductors.

The plots in Mathcad 1.3 clearly show that as the distance between two parallel conductors increases, the critical voltage for corona to occur increases. As the spacing increases, for a fixed voltage, the maximum electric field decreases. Also shown is the variation in the critical voltage with conductor radius. As the conductor radius decreases, for a fixed distance, the critical voltage decreases.

There are various methods of reducing or eliminating corona. The most common is avoiding or eliminating sharp edges on the conductors. Surprisingly, the thread on a bolt can be a possible location for corona. For high-voltage terminals, such as seen on some high-voltage resistor dividers, the terminals have a spherical or thick rounded-edge disk shape to avoid corona. The electric field between electrodes can also be reduced by changing the position of nearby objects and electrodes and by using guard or stress rings. In high-voltage transmission lines, conductors of similar potential and phase can be bundled or grouped together to reduce the electric field between them. To help prevent corona (when the surrounding medium is nonhomogeneous), materials with high dielectric constants should be avoided. The introduction of higher dielectric constant materials can increase the electric field in neighboring regions. Another method of preventing corona is to increase the gas pressure between the electrodes or replace the air between the electrodes with SF_6 gas. Sometimes an entire device is surrounded by a high-breakdown liquid, such as transformer oil. Of course, this method will probably increase the cost since a special chamber or enclosure is often required.

1.18   Voltage Rating of Coax

Starting with Gauss' law, derive the expressions for the voltage and electric field between the conductors of a coaxial structure. Do smaller coaxial cables have lower voltage ratings? [Crowley, '86]

One of Maxwell's equations states that the divergence of the $\vec{D}$ field is equal to the volume charge density, ρ_V:

$$\nabla \cdot \vec{D} = \rho_V \tag{1.77}$$

Since $\vec{D} = \varepsilon_r \varepsilon_o \vec{E}$, this expression states (roughly) that charge is a source of electric field. Substituting the expression for $\vec{D}$ into (1.77) and assuming that the relative permittivity, ε_r, is constant or not varying with position,

$$\varepsilon_r \varepsilon_o \nabla \cdot \vec{E} = \rho_V \tag{1.78}$$

For electrostatic or almost-electrostatic conditions, the electric field is related to the voltage or potential function via the gradient operator:

$$\vec{E} = -\nabla \Phi \tag{1.79}$$

Substituting this relationship into (1.78),

$$\varepsilon_r \varepsilon_o \nabla \cdot \left(-\nabla \Phi\right) = -\varepsilon_r \varepsilon_o \nabla^2 \Phi = \rho_V$$

an expression that relates the charge density with the potential function is obtained:

$$\nabla^2 \Phi = -\frac{\rho_V}{\varepsilon_r \varepsilon_o} \tag{1.80}$$

Equation (1.80) is known as Poisson's equation. When the volume charge density, ρ_V, is zero, $\nabla \cdot \vec{D} = 0$,[12] and Poisson's equation reduces to Laplace's equation:

$$\nabla^2 \Phi = 0 \tag{1.81}$$

This expression, which involves second-order partial derivatives, can be easily solved when Φ varies in only one direction (and in some other situations when a great deal of symmetry is present), without resorting to any special techniques. The Laplacian operator, or ∇^2, is a scalar operator since $\nabla^2 = \vec{\nabla} \cdot \vec{\nabla}$ is the dot product of two vectors. In the Cartesian, cylindrical, and spherical coordinate systems, respectively, the Laplacian of Φ is defined as

$$\nabla^2 \Phi = \frac{\partial^2 \Phi}{\partial x^2} + \frac{\partial^2 \Phi}{\partial y^2} + \frac{\partial^2 \Phi}{\partial z^2} \tag{1.82}$$

$$\nabla^2 \Phi = \frac{1}{\rho} \frac{\partial}{\partial \rho}\left(\rho \frac{\partial \Phi}{\partial \rho}\right) + \frac{1}{\rho^2} \frac{\partial^2 \Phi}{\partial \phi^2} + \frac{\partial^2 \Phi}{\partial z^2} \tag{1.83}$$

$$\nabla^2 \Phi = \frac{1}{r^2} \frac{\partial}{\partial r}\left(r^2 \frac{\partial \Phi}{\partial r}\right) + \frac{1}{r^2 \sin\theta} \frac{\partial}{\partial \theta}\left(\sin\theta \frac{\partial \Phi}{\partial \theta}\right) + \frac{1}{r^2 \sin^2\theta} \frac{\partial^2 \Phi}{\partial \phi^2} \tag{1.84}$$

---

[12]When the divergence of a field is equal to zero, the field is said to be solenoidal.

Each of these expressions involves second-order partial derivatives. The coordinate system that is selected for a particular problem is mainly a function of the similarity of the electrode surfaces to one of the planes or surfaces easily generated in that particular coordinate system.

For a number of highly symmetric electrodes with no space charge between them, the potential will vary with only one coordinate variable if the in-between region has a constant permittivity. For concentric cylindrical conductors, the potential only varies in the ρ direction. This assumes, of course, that the conductors are very long in the z direction and that fringing or end effects are ignored. For this coaxial arrangement, the cylindrical version of Laplace's equation reduces to

$$\nabla^2 \Phi = \frac{1}{\rho}\frac{d}{d\rho}\left(\rho \frac{d\Phi}{d\rho}\right) = 0 \tag{1.85}$$

Again, this assumes that the relative permittivity or dielectric constant between the inner and outer conductors is constant (i.e., not a function of any of the variables ρ, ϕ, or z) and that no charge exists between the electrodes. Multiplying both sides of (1.85) by ρ and then integrating both sides with respect to ρ,

$$\frac{d}{d\rho}\left(\rho \frac{d\Phi}{d\rho}\right) = 0 \quad \Rightarrow \rho \frac{d\Phi}{d\rho} = A \tag{1.86}$$

where A is a constant of integration. Next, dividing both sides of Equation (1.86) by ρ and again integrating,

$$\frac{d\Phi}{d\rho} = \frac{A}{\rho} \quad \Rightarrow \Phi = A \ln\rho + B$$

where B is another constant of integration. This potential function, Φ, contains two constants of integration since Laplace's equation is a second-order differential equation. The constants of integration are determined by using boundary conditions. For the coaxial cable shown in Figure 1.36, these boundary conditions are quite obvious since the potential at the inner and outer conducting electrodes is known. Assuming that the outer electrode located at $\rho = a$ is at 0 volts and the inner electrode located at $\rho = b$ is at V volts, the two equations required to solve for these two unknowns are

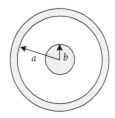

FIGURE 1.36 Coaxial structure.

$$\Phi(\rho = a) = A \ln a + B = 0, \quad \Phi(\rho = b) = A \ln b + B = V$$

Solving for A and B, the equation for the potential function between the two electrodes is

$$\Phi = \frac{V}{\ln\left(\frac{b}{a}\right)}\ln\rho - \frac{V}{\ln\left(\frac{b}{a}\right)}\ln a = V\frac{\ln\left(\frac{\rho}{a}\right)}{\ln\left(\frac{b}{a}\right)}$$

The electric field between the electrodes is obtained from this potential function:

$$\vec{E} = -\nabla\Phi = -\frac{d\Phi}{d\rho}\hat{a}_\rho = -\frac{V}{\rho\ln\left(\frac{b}{a}\right)}\hat{a}_\rho = \frac{V}{\rho\ln\left(\frac{a}{b}\right)}\hat{a}_\rho \tag{1.87}$$

As expected, the electric field is directed radially from the inner electrode to the outer electrode (when $V > 0$).

The maximum value of the electric field occurs at the inner electrode at $\rho = b$. The curvature of the inner electrode is greater than the curvature of the outer electrode. Breakdown is first expected along the inner conductor, assuming the medium between the electrodes is homogeneous with a constant dielectric constant. If the maximum breakdown field of the medium is given as E_{bk} and the corresponding voltage as V_{bk}, then at the inner electrode

$$E_{bk} = \frac{V_{bk}}{b \ln\left(\dfrac{a}{b}\right)} \quad \Rightarrow V_{bk} = E_{bk} b \ln\left(\frac{a}{b}\right) = E_{bk} a \frac{\ln\left(\dfrac{a}{b}\right)}{\dfrac{a}{b}} \tag{1.88}$$

The maximum value of this breakdown voltage can be obtained in the normal manner using the power of calculus:

$$\frac{dV_{bk}}{db} = E_{bk} \ln\left(\frac{a}{b}\right) + E_{bk} b \frac{-\dfrac{a}{b^2}}{\dfrac{a}{b}} = E_{bk} \ln\left(\frac{a}{b}\right) - E_{bk} = 0 \quad \Rightarrow b = ae^{-1} \approx 0.37a \tag{1.89}$$

Therefore, for a fixed outer radius (a) and breakdown strength (E_{bk}), selecting the inner radius equal to about 0.37a will provide the highest voltage rating for the coaxial line. The location of this maximum value is clearly seen in Mathcad 1.4 where the breakdown voltage (divided by aE_{bk}) is plotted vs. b/a.

For this optimum b/a ratio, the characteristic impedance of an air-filled coaxial structure is near that of the commonly used 50 and 75 Ω lines:

$$Z_o = 60 \ln\left(\frac{a}{ae^{-1}}\right) = 60\ \Omega \tag{1.90}$$

There are other factors that determine the optimum value for b/a. Obviously, as the distance between the inner and outer electrodes increases, the maximum electric field strength between them must decrease,

$$E_{bk} := 1 \quad a := 1$$
$$b := 0.01, 0.011.. \ 1$$
$$V_{bk}(b) := a \cdot E_{bk} \cdot \frac{\ln\left(\dfrac{a}{b}\right)}{\left(\dfrac{a}{b}\right)}$$

Max Breakdown Voltage vs. Ratio of Radii

MATHCAD 1.4   Voltage breakdown strength (divided by aE_{bk}) vs. the ratio of the inner to outer radius for a coaxial cable.

for a given applied voltage. So, larger coaxial cables with larger distances between the inner and outer conductors have greater maximum voltage ratings than smaller coaxial cables with smaller conductor spacings. For a fixed outer radius, however, decreasing the radius of the inner electrode so that the spacing between the electrodes is large will produce a thin, high-curvature inner electrode. To reduce the probability of corona or partial discharges, high-curvature sharp electrodes should be avoided. Thus, the optimum value for b ($\approx 0.37a$) is a compromise between a small curvature inner electrode and large spacing between the inner and outer electrodes. Therefore, smaller coaxial cables generally have smaller voltage ratings than larger coaxial cables. (They also have smaller current ratings since the cross-sectional area of the electrodes is probably smaller.) Generally, for various reasons, including the degradation of the insulation, breakdown in a cable is avoided. However, in some special applications,[13] the cable is designed with an inner radius smaller than the given optimum value. If the breakdown voltage is reached, a less harmful partial discharge occurs near the smaller-than-optimum inner radius, rather than a more complete rupture of the insulator or full breakdown between the electrodes. The partial discharge will produce charges around the inner electrode that tend to stabilize the discharge.

The expressions derived in this section for the voltage and electric field between two concentric, cylindrical electrodes assumed zero net charge between the electrodes. Electric fields determined when there is zero net volume charge between the electrodes are referred to as Laplacian fields. However, whenever a discharge occurs, charge is generated. This charge distorts the electric field and can be an important factor in the breakdown process. If the charge distribution or some rough approximation to it is known at some time during a discharge, Poisson's equation can be used to determine the electric field distribution (if the field can be assumed electroquasistatic) at this time.

1.19   Solutions to Poisson's Equation

Tabulate the solutions to Poisson's equation in the three major coordinate systems for various volume charge densities. Assume that the potential only varies with one variable.

In a region with a constant relative permittivity or dielectric constant, ε_r, where a volume charge density, ρ_V (in C/m³), is present, Poisson's equation can be used to determine the potential distribution, Φ, in this region:

$$\nabla^2 \Phi = -\frac{\rho_V}{\varepsilon_r \varepsilon_o} \quad \text{where } \varepsilon_o = 8.854 \times 10^{-12} \text{ F/m} \tag{1.91}$$

This equation can be used for electrostatic or electroquasistatic systems. If external charged electrodes are present, these charged electrodes and the volume charge are both sources of electric field and, thus, both have an impact on the potential distribution. In the Cartesian, cylindrical, and spherical coordinate systems, respectively, Poisson's equation is

$$\frac{\partial^2 \Phi}{\partial x^2} + \frac{\partial^2 \Phi}{\partial y^2} + \frac{\partial^2 \Phi}{\partial z^2} = -\frac{\rho_V}{\varepsilon_r \varepsilon_o} \tag{1.92}$$

$$\frac{1}{\rho}\frac{\partial}{\partial \rho}\left(\rho \frac{\partial \Phi}{\partial \rho}\right) + \frac{1}{\rho^2}\frac{\partial^2 \Phi}{\partial \phi^2} + \frac{\partial^2 \Phi}{\partial z^2} = -\frac{\rho_V}{\varepsilon_r \varepsilon_o} \tag{1.93}$$

$$\frac{1}{r^2}\frac{\partial}{\partial r}\left(r^2 \frac{\partial \Phi}{\partial r}\right) + \frac{1}{r^2 \sin\theta}\frac{\partial}{\partial \theta}\left(\sin\theta \frac{\partial \Phi}{\partial \theta}\right) + \frac{1}{r^2 \sin^2\theta}\frac{\partial^2 \Phi}{\partial \phi^2} = -\frac{\rho_V}{\varepsilon_r \varepsilon_o} \tag{1.94}$$

---

[13]For example, it may be necessary to use occasionally larger-than-designed voltages on an older underwater power cable to a developing island.

If the potential, Φ, varies with more than one variable, the solution of these partial differential equations is often beyond the scope of many introductory courses in electromagnetics. If the potential only varies with one variable, then the solution can be readily obtained for several volume charge distributions by merely integrating. *However*, there must be a great deal of symmetry in the charge distribution so that the electric field produced by the volume charge is in the same direction as the electric field produced by the electrodes. As seen in Table 1.4 through Table 1.10, this symmetry condition severely limits the number of simple analytical solutions that can be obtained in this manner. For example, even a constant

TABLE 1.4 Volume Charge between Two Wide Flat Conductors

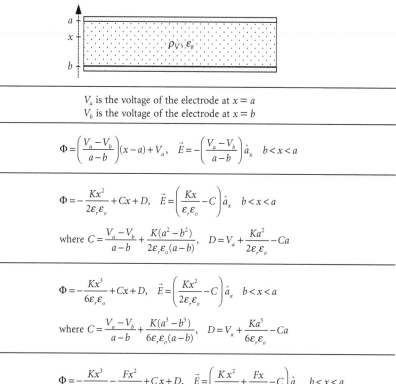

V_a is the voltage of the electrode at $x = a$
V_b is the voltage of the electrode at $x = b$

$\rho_V = 0$	$\Phi = \left(\dfrac{V_a - V_b}{a - b}\right)(x - a) + V_a, \quad \vec{E} = -\left(\dfrac{V_a - V_b}{a - b}\right)\hat{a}_x \quad b < x < a$

$\rho_V = K$

$$\Phi = -\frac{Kx^2}{2\varepsilon_r\varepsilon_o} + Cx + D, \quad \vec{E} = \left(\frac{Kx}{\varepsilon_r\varepsilon_o} - C\right)\hat{a}_x \quad b < x < a$$

$$\text{where } C = \frac{V_a - V_b}{a - b} + \frac{K(a^2 - b^2)}{2\varepsilon_r\varepsilon_o(a - b)}, \quad D = V_a + \frac{Ka^2}{2\varepsilon_r\varepsilon_o} - Ca$$

$\rho_V = Kx$

$$\Phi = -\frac{Kx^3}{6\varepsilon_r\varepsilon_o} + Cx + D, \quad \vec{E} = \left(\frac{Kx^2}{2\varepsilon_r\varepsilon_o} - C\right)\hat{a}_x \quad b < x < a$$

$$\text{where } C = \frac{V_a - V_b}{a - b} + \frac{K(a^3 - b^3)}{6\varepsilon_r\varepsilon_o(a - b)}, \quad D = V_a + \frac{Ka^3}{6\varepsilon_r\varepsilon_o} - Ca$$

$\rho_V = Kx + F$

$$\Phi = -\frac{Kx^3}{6\varepsilon_r\varepsilon_o} - \frac{Fx^2}{2\varepsilon_r\varepsilon_o} + Cx + D, \quad \vec{E} = \left(\frac{Kx^2}{2\varepsilon_r\varepsilon_o} + \frac{Fx}{\varepsilon_r\varepsilon_o} - C\right)\hat{a}_x \quad b < x < a$$

$$\text{where } C = \frac{V_a - V_b}{a - b} + \frac{K(a^3 - b^3)}{6\varepsilon_r\varepsilon_o(a - b)} + \frac{F(a^2 - b^2)}{2\varepsilon_r\varepsilon_o(a - b)}, \quad D = V_a + \frac{Ka^3}{6\varepsilon_r\varepsilon_o} + \frac{Fa^2}{2\varepsilon_r\varepsilon_o} - Ca$$

$\rho_V = Ke^{Fx}$

$$\Phi = -\frac{K}{F^2\varepsilon_r\varepsilon_o}e^{Fx} + Cx + D, \quad \vec{E} = \left(\frac{K}{F\varepsilon_r\varepsilon_o}e^{Fx} - C\right)\hat{a}_x \quad b < x < a$$

$$\text{where } C = \frac{V_a - V_b}{a - b} + \frac{K(e^{Fa} - e^{Fb})}{F^2\varepsilon_r\varepsilon_o(a - b)}, \quad D = V_a + \frac{Ke^{Fa}}{F^2\varepsilon_r\varepsilon_o} - Ca$$

$\rho_V = K\cos(Fx + \vartheta)$

$$\Phi = \frac{K}{F^2\varepsilon_r\varepsilon_o}\cos(Fx + \vartheta) + Cx + D, \quad \vec{E} = \left[\frac{K}{F\varepsilon_r\varepsilon_o}\sin(Fx + \vartheta) - C\right]\hat{a}_x \quad b < x < a$$

$$\text{where } C = \frac{V_a - V_b}{a - b} + \frac{K[\cos(Fb + \vartheta) - \cos(Fa + \vartheta)]}{F^2\varepsilon_r\varepsilon_o(a - b)}, \quad D = V_a - \frac{K\cos(Fa + \vartheta)}{F^2\varepsilon_r\varepsilon_o} - Ca$$

TABLE 1.5   Volume Charge between Two Concentric Cylindrical Conductors

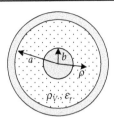

V_a is the voltage of the electrode at $\rho = a > b$

V_b is the voltage of the electrode at $\rho = b > 0$

$\rho_V = 0$	$\Phi = \left[\dfrac{V_a - V_b}{\ln\left(\dfrac{a}{b}\right)}\right]\ln\left(\dfrac{\rho}{a}\right) + V_a, \quad \vec{E} = -\left[\dfrac{V_a - V_b}{\rho\ln\left(\dfrac{a}{b}\right)}\right]\hat{a}_\rho \quad b < \rho < a$
$\rho_V = K$	$\Phi = -\dfrac{K\rho^2}{4\varepsilon_r\varepsilon_o} + C\ln\rho + D, \quad \vec{E} = \left(\dfrac{K\rho}{2\varepsilon_r\varepsilon_o} - \dfrac{C}{\rho}\right)\hat{a}_\rho \quad b < \rho < a$ where $C = \dfrac{V_a - V_b}{\ln\left(\dfrac{a}{b}\right)} + \dfrac{K(a^2 - b^2)}{4\varepsilon_r\varepsilon_o\ln\left(\dfrac{a}{b}\right)}, \quad D = V_a + \dfrac{Ka^2}{4\varepsilon_r\varepsilon_o} - C\ln a$
$\rho_V = K\rho$	$\Phi = -\dfrac{K\rho^3}{9\varepsilon_r\varepsilon_o} + C\ln\rho + D, \quad \vec{E} = \left(\dfrac{K\rho^2}{3\varepsilon_r\varepsilon_o} - \dfrac{C}{\rho}\right)\hat{a}_\rho \quad b < \rho < a$ where $C = \dfrac{V_a - V_b}{\ln\left(\dfrac{a}{b}\right)} + \dfrac{K(a^3 - b^3)}{9\varepsilon_r\varepsilon_o\ln\left(\dfrac{a}{b}\right)}, \quad D = V_a + \dfrac{Ka^3}{9\varepsilon_r\varepsilon_o} - C\ln a$
$\rho_V = K\rho + F$	$\Phi = -\dfrac{K\rho^3}{9\varepsilon_r\varepsilon_o} - \dfrac{F\rho^2}{4\varepsilon_r\varepsilon_o} + C\ln\rho + D, \quad \vec{E} = \left(\dfrac{K\rho^2}{3\varepsilon_r\varepsilon_o} + \dfrac{F\rho}{2\varepsilon_r\varepsilon_o} - \dfrac{C}{\rho}\right)\hat{a}_\rho \quad b < \rho < a$ where $C = \dfrac{V_a - V_b}{\ln\left(\dfrac{a}{b}\right)} + \dfrac{K(a^3 - b^3)}{9\varepsilon_r\varepsilon_o\ln\left(\dfrac{a}{b}\right)} + \dfrac{F(a^2 - b^2)}{4\varepsilon_r\varepsilon_o\ln\left(\dfrac{a}{b}\right)}, \quad D = V_a + \dfrac{Ka^3}{9\varepsilon_r\varepsilon_o} + \dfrac{Fa^2}{4\varepsilon_r\varepsilon_o} - C\ln a$
$\rho_V = \dfrac{K}{\rho}$	$\Phi = -\dfrac{K\rho}{\varepsilon_r\varepsilon_o} + C\ln\rho + D, \quad \vec{E} = \left(\dfrac{K}{\varepsilon_r\varepsilon_o} - \dfrac{C}{\rho}\right)\hat{a}_\rho \quad b < \rho < a$ where $C = \dfrac{V_a - V_b}{\ln\left(\dfrac{a}{b}\right)} + \dfrac{K(a - b)}{\varepsilon_r\varepsilon_o\ln\left(\dfrac{a}{b}\right)}, \quad D = V_a + \dfrac{Ka}{\varepsilon_r\varepsilon_o} - C\ln a$

volume charge density between two slanted conductors in the cylindrical coordinate system will introduce additional field components preventing a one-dimensional solution for the potential.

To illustrate the approach, one example will be given. Assume that the volume charge distribution between two very long, cylindrical concentric electrodes is given by

$$\rho_V = K\rho \tag{1.95}$$

TABLE 1.6   Volume Charge within a Cylindrical Conductor

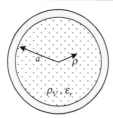

V_a is the voltage of the electrode at $\rho = a > 0$	
$\rho_V = 0$	$\Phi = V_a, \quad \vec{E} = 0 \quad \rho < a$
$\rho_V = K$	$\Phi = \dfrac{K(a^2 - \rho^2)}{4\varepsilon_r \varepsilon_o} + V_a, \quad \vec{E} = \dfrac{K\rho}{2\varepsilon_r \varepsilon_o}\hat{a}_\rho \quad \rho < a$
$\rho_V = K\rho$	$\Phi = \dfrac{K(a^3 - \rho^3)}{9\varepsilon_r \varepsilon_o} + V_a, \quad \vec{E} = \dfrac{K\rho^2}{3\varepsilon_r \varepsilon_o}\hat{a}_\rho \quad \rho < a$
$\rho_V = K\rho + F$	$\Phi = \dfrac{K(a^3 - \rho^3)}{9\varepsilon_r \varepsilon_o} + \dfrac{F(a^2 - \rho^2)}{4\varepsilon_r \varepsilon_o} + V_a, \quad \vec{E} = \left(\dfrac{K\rho^2}{3\varepsilon_r \varepsilon_o} + \dfrac{F\rho}{2\varepsilon_r \varepsilon_o}\right)\hat{a}_\rho \quad \rho < a$

TABLE 1.7   Zero Volume Charge between Two Wide Slanted Conductors

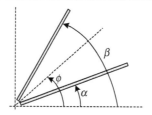

V_α is the voltage of the electrode at $\phi = \alpha$ V_β is the voltage of the electrode at $\phi = \beta$	
$\rho_V = 0$	$\Phi = \left(\dfrac{V_\alpha - V_\beta}{\alpha - \beta}\right)(\phi - \alpha) + V_\alpha, \quad \vec{E} = -\dfrac{1}{\rho}\left(\dfrac{V_\alpha - V_\beta}{\alpha - \beta}\right)\hat{a}_\phi \quad \alpha < \phi < \beta, \ \rho > 0$

where K is a constant and ρ is the distance from the center of the inner conductor as shown in Figure 1.37. The volume charge density increases with ρ. The voltage of the outer electrode is V_a, while the voltage of the inner electrode is V_b. The outer radius of the inner electrode is b and the inner radius of the outer electrode is a. The potential, Φ, does not vary along the length of the cylinder, z (if not too close to the cylinder's ends). Also, the potential does not vary at a given distance

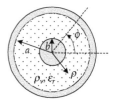

FIGURE 1.37   Volume charge density between two cylindrical conductors.

from the inner electrode in the ϕ direction. Because of the symmetry, Poisson's equation in the cylindrical

TABLE 1.8   Volume Charge between Two Concentric Spherical Conductors

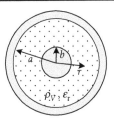

V_a is the voltage of the electrode at $r = a > b$
V_b is the voltage of the electrode at $r = b > 0$

$\rho_V = 0$	$\Phi = (V_a - V_b)\left(\dfrac{ab}{b-a}\right)\left(\dfrac{1}{r} - \dfrac{1}{a}\right) + V_a, \quad \vec{E} = (V_a - V_b)\left(\dfrac{ab}{b-a}\right)\dfrac{1}{r^2}\hat{a}_r \quad b < r < a$

$$\rho_V = K$$

$$\Phi = -\frac{Kr^2}{6\varepsilon_r\varepsilon_o} - \frac{C}{r} + D, \quad \vec{E} = \left(\frac{Kr}{3\varepsilon_r\varepsilon_o} - \frac{C}{r^2}\right)\hat{a}_r \quad b < r < a$$

$$\text{where } C = (V_a - V_b)\left(\frac{ab}{a-b}\right) + \frac{Kab(a+b)}{6\varepsilon_r\varepsilon_o}, \quad D = V_a + \frac{Ka^2}{6\varepsilon_r\varepsilon_o} + \frac{C}{a}$$

$$\rho_V = Kr$$

$$\Phi = -\frac{Kr^3}{12\varepsilon_r\varepsilon_o} - \frac{C}{r} + D, \quad \vec{E} = \left(\frac{Kr^2}{4\varepsilon_r\varepsilon_o} - \frac{C}{r^2}\right)\hat{a}_r \quad b < r < a$$

$$\text{where } C = (V_a - V_b)\left(\frac{ab}{a-b}\right) + \frac{Kab(a^2+ab+b^2)}{12\varepsilon_r\varepsilon_o}, \quad D = V_a + \frac{Ka^3}{12\varepsilon_r\varepsilon_o} + \frac{C}{a}$$

$$\rho_V = Kr + F$$

$$\Phi = -\frac{Kr^3}{12\varepsilon_r\varepsilon_o} - \frac{Fr^2}{6\varepsilon_r\varepsilon_o} - \frac{C}{r} + D, \quad \vec{E} = \left(\frac{Kr^2}{4\varepsilon_r\varepsilon_o} + \frac{Fr}{3\varepsilon_r\varepsilon_o} - \frac{C}{r^2}\right)\hat{a}_r \quad b < r < a$$

$$\text{where } C = (V_a - V_b)\left(\frac{ab}{a-b}\right) + \frac{Kab(a^2+ab+b^2)}{12\varepsilon_r\varepsilon_o} + \frac{Fab(a+b)}{6\varepsilon_r\varepsilon_o}, \quad D = V_a + \frac{Ka^3}{12\varepsilon_r\varepsilon_o} + \frac{Fa^2}{6\varepsilon_r\varepsilon_o} + \frac{C}{a}$$

$$\rho_V = \frac{K}{r}$$

$$\Phi = -\frac{Kr}{2\varepsilon_r\varepsilon_o} - \frac{C}{r} + D, \quad \vec{E} = \left(\frac{K}{2\varepsilon_r\varepsilon_o} - \frac{C}{r^2}\right)\hat{a}_r \quad b < r < a$$

$$\text{where } C = (V_a - V_b)\left(\frac{ab}{a-b}\right) + \frac{Kab}{2\varepsilon_r\varepsilon_o}, \quad D = V_a + \frac{Ka}{2\varepsilon_r\varepsilon_o} + \frac{C}{a}$$

$$\rho_V = Ke^{Fr}$$

$$\Phi = \frac{Ke^{Fr}}{F^2\varepsilon_r\varepsilon_o}\left(\frac{2}{Fr} - 1\right) - \frac{C}{r} + D, \quad \vec{E} = \left[\frac{2Ke^{Fr}}{F^3\varepsilon_r\varepsilon_o r^2} - \frac{Ke^{Fr}}{F\varepsilon_r\varepsilon_o}\left(\frac{2}{Fr} - 1\right) - \frac{C}{r^2}\right]\hat{a}_r \quad b < r < a$$

$$\text{where } C = (V_a - V_b)\left(\frac{ab}{a-b}\right) - \frac{Ke^{Fa}}{F^2\varepsilon_r\varepsilon_o}\left(\frac{2}{Fa} - 1\right)\left(\frac{ab}{a-b}\right) + \frac{Ke^{Fb}}{F^2\varepsilon_r\varepsilon_o}\left(\frac{2}{Fb} - 1\right)\left(\frac{ab}{a-b}\right)$$

$$D = V_a - \frac{Ke^{Fa}}{F^2\varepsilon_r\varepsilon_o}\left(\frac{2}{Fa} - 1\right) + \frac{C}{a}$$

TABLE 1.9   Volume Charge within a Spherical Conductor

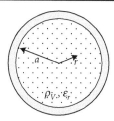

	V_a is the voltage of the electrode at $r = a > 0$
$\rho_V = 0$	$\Phi = V_a, \quad \vec{E} = 0 \quad r < a$
$\rho_V = K$	$\Phi = \dfrac{K(a^2 - r^2)}{6\varepsilon_r\varepsilon_o} + V_a, \quad \vec{E} = \dfrac{Kr}{3\varepsilon_r\varepsilon_o}\hat{a}_r \quad r < a$
$\rho_V = Kr$	$\Phi = \dfrac{K(a^3 - r^3)}{12\varepsilon_r\varepsilon_o} + V_a, \quad \vec{E} = \dfrac{Kr^2}{4\varepsilon_r\varepsilon_o}\hat{a}_r \quad r < a$
$\rho_V = Kr + F$	$\Phi = \dfrac{K(a^3 - r^3)}{12\varepsilon_r\varepsilon_o} + \dfrac{F(a^2 - r^2)}{6\varepsilon_r\varepsilon_o} + V_a, \quad \vec{E} = \left(\dfrac{Kr^2}{4\varepsilon_r\varepsilon_o} + \dfrac{Fr}{3\varepsilon_r\varepsilon_o}\right)\hat{a}_r \quad r < a$
$\rho_V = \dfrac{K}{r}$	$\Phi = \dfrac{K(a - r)}{2\varepsilon_r\varepsilon_o} + V_a, \quad \vec{E} = \dfrac{K}{2\varepsilon_r\varepsilon_o}\hat{a}_r \quad r < a$

TABLE 1.10   Zero Volume Charge between Two Cone Conductors

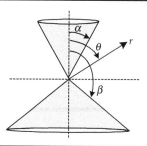

V_α is the voltage of the conical electrode at $\theta = \alpha$
V_β is the voltage of the conical electrode at $\theta = \beta$

$\rho_V = 0$	$\Phi = (V_\beta - V_\alpha)\dfrac{\ln\left[\dfrac{\tan\left(\dfrac{\theta}{2}\right)}{\tan\left(\dfrac{\alpha}{2}\right)}\right]}{\ln\left[\dfrac{\tan\left(\dfrac{\beta}{2}\right)}{\tan\left(\dfrac{\alpha}{2}\right)}\right]} + V_a, \quad \vec{E} = -\left\{\dfrac{V_\beta - V_\alpha}{r\sin(\theta)\ln\left[\dfrac{\tan\left(\dfrac{\beta}{2}\right)}{\tan\left(\dfrac{\alpha}{2}\right)}\right]}\right\}\hat{a}_\theta \quad \alpha < \theta < \beta, \; r > 0$

coordinate system reduces to

$$\frac{1}{\rho}\frac{d}{d\rho}\left(\rho\frac{d\Phi}{d\rho}\right) = -\frac{\rho_V}{\varepsilon_r\varepsilon_o}$$

The partial derivatives become normal derivatives since the potential only varies with ρ. Next, (1.95) is inserted into this expression and both sides multiplied by ρ:

$$\frac{d}{d\rho}\left(\rho\frac{d\Phi}{d\rho}\right) = -\frac{K\rho^2}{\varepsilon_r\varepsilon_o}$$

By multiplying both sides of the expression by ρ, it is assumed that the final expression is not valid at $\rho = 0$. Next, both sides are integrated with respect to ρ:

$$\rho\frac{d\Phi}{d\rho} = -\frac{K\rho^3}{3\varepsilon_r\varepsilon_o} + C$$

where C is a constant of integration. It is common, but usually incorrect, to substitute the expression for the charge density after performing all of the given steps. Unless the charge density is constant or does not vary with position, this is not correct since the integration is with respect to ρ. Dividing both sides of the last result by ρ,

$$\frac{d\Phi}{d\rho} = -\frac{K\rho^2}{3\varepsilon_r\varepsilon_o} + \frac{C}{\rho} \tag{1.96}$$

Again, integrating both sides with respect to ρ,

$$\Phi = -\frac{K\rho^3}{9\varepsilon_r\varepsilon_o} + C\ln\rho + D \tag{1.97}$$

where D is another constant of integration. Two constants of integration are expected since the original differential equation was of second order. The constants of integration are determined by applying boundary conditions. In this example, the boundaries are merely the electrodes. Recall that the voltage of the outer electrode is V_a, while the voltage of the inner electrode is V_b. Substituting these boundary conditions into Equation (1.97),

$$V_a = -\frac{Ka^3}{9\varepsilon_r\varepsilon_o} + C\ln a + D, \quad V_b = -\frac{Kb^3}{9\varepsilon_r\varepsilon_o} + C\ln b + D$$

There are two equations and two unknowns. Solving for C by subtracting one equation from the other:

$$V_a - V_b = -\frac{Ka^3}{9\varepsilon_r\varepsilon_o} + C\ln a + D - \left(-\frac{Kb^3}{9\varepsilon_r\varepsilon_o} + C\ln b + D\right) = -\frac{Ka^3}{9\varepsilon_r\varepsilon_o} + \frac{Kb^3}{9\varepsilon_r\varepsilon_o} + C\ln\left(\frac{a}{b}\right)$$

$$\Rightarrow C = \frac{V_a - V_b}{\ln\left(\frac{a}{b}\right)} + \frac{K(a^3 - b^3)}{9\varepsilon_r\varepsilon_o\ln\left(\frac{a}{b}\right)}$$

Once C is determined, the constant D can be obtained from either of the two expressions:

$$D = V_a + \frac{Ka^3}{9\varepsilon_r\varepsilon_o} - C\ln a = V_b + \frac{Kb^3}{9\varepsilon_r\varepsilon_o} - C\ln b$$

The electric field expression is obtained from the potential expression. Again, since Φ only varies in the ρ direction, the electric field expression reduces to

$$\vec{E} = -\nabla\Phi = -\frac{\partial\Phi}{\partial\rho}\hat{a}_\rho - \frac{1}{\rho}\frac{\partial\Phi}{\partial\phi}\hat{a}_\phi - \frac{\partial\Phi}{\partial z}\hat{a}_z = -\frac{d\Phi}{d\rho}\hat{a}_\rho = -\frac{d}{d\rho}\left(-\frac{K\rho^3}{9\varepsilon_r\varepsilon_o} + C\ln\rho + D\right)\hat{a}_\rho$$

$$= \left(\frac{K\rho^2}{3\varepsilon_r\varepsilon_o} - \frac{C}{\rho}\right)\hat{a}_\rho$$

It was not necessary to perform this differentiation since the derivative is already given in Equation (1.96).

Students often automatically assume that the magnitude of the electric field is just the derivative of the function (for one-dimensional fields). However, referring to Equations (1.68) and (1.69), it is clear that the gradient of a function sometimes involves more than a derivative. For example, if the potential only varies with θ in the spherical coordinate system, the electric field is not just the negative derivative with respect to θ but

$$\vec{E} = -\nabla\Phi = -\frac{1}{r}\frac{d\Phi}{d\theta}\hat{a}_\theta$$

Notice the $1/r$ multiplier.

In obtaining the general expression for the potential in the previous example, expressions were multiplied and divided by ρ. This assumes that $\rho \neq 0$. When the potential or electric field expression is desired at $\rho = 0$ or $r = 0$, the general solution can be obtained in a nonrigorous manner, in some cases, by merely dropping those parts of the solution where the potential "blows up." Equation (1.97), for example, reduces to

$$\Phi = -\frac{K\rho^3}{9\varepsilon_r\varepsilon_o} + D$$

Now, only one boundary condition is required to solve for the one constant, D. To obtain the solution of a partial differential equation, almost any method is legitimate, including guessing. Although dropping certain terms that "blow up" is mathematically sloppy, Φ can be checked by verifying that the solution satisfies the boundary conditions and is a solution of Poisson's equation, Equation (1.91).

In Table 1.4 through Table 1.10, the SI system of units is assumed: the voltage of the electrodes is in V, potential function is in V, electric field is in V/m, and charge density is in C/m³. In these tables, K, F, and ϑ are constant with respect to position. Obviously, the conductors or electrodes are assumed perfectly conducting so that the potential is constant along their surfaces and the electric field is normal to their surfaces. Fringing fields are also neglected; otherwise, the potential would be a function of more than one variable. Also, the charge density is assumed stationary and known.

1.20   Arcing in a Silo

While transporting grain into a silo, the grain becomes tribocharged. Assuming that the silo has an inner radius of a, determine the maximum electric field in the silo. State all assumptions. Can this field be reduced by including a ground wire down the length of the silo? [Crowley, '86]

The charged grain in the silo is a source of electric field. If this electric field is sufficiently strong, a discharge can occur. Under the right conditions, this discharge, which is an ignition source, can result in an explosion. In the following analysis, it is assumed that the charge distribution is uniformly distributed in a very long cylindrical-shaped silo. However, other distributions could be proposed that can generate larger electric fields. Using Table 1.6, for a uniform volume charge distribution equal to K, the potential distribution and electric field inside a very long cylindrical structure are given by the expressions

$$\Phi = \frac{K(a^2-\rho^2)}{4\varepsilon_r\varepsilon_o}, \quad \vec{E} = \frac{K\rho}{2\varepsilon_r\varepsilon_o}\hat{a}_\rho \quad \rho < a$$

The 0 V reference is at a distance $a = \rho$ from the center of the silo (i.e., at the silo's inner surface). The electric field is perpendicular to the sides of the silo. *If* the silo were constructed of a very good metal, then surface charge would exist along the inner surface of the silo and there would be a discontinuity in the normal electric field component at the surface. The maximum electric field is at the silo's inner surface (for $K > 0$):

$$\left| E_{max} \right| = \frac{Ka}{2\varepsilon_r \varepsilon_o} \tag{1.98}$$

With the high degree of symmetry present, this result can also be obtained using Gauss' law. Even if the silo's surface were not a conductor, this would be the maximum electric field. This maximum field is a strong indicator of whether a discharge will occur. At the center of the silo, the electric field is zero.

When a grounded wire or rod of radius b is inserted down the length of the center of the silo, the expressions given in Table 1.5 may be used:

$$\Phi = -\frac{K\rho^2}{4\varepsilon_r \varepsilon_o} + C \ln \rho + D, \quad \vec{E} = \left(\frac{K\rho}{2\varepsilon_r \varepsilon_o} - \frac{C}{\rho} \right) \hat{a}_\rho \quad b < \rho < a$$

$$\text{where} \quad C = \frac{K(a^2 - b^2)}{4\varepsilon_r \varepsilon_o \ln \left(\frac{a}{b} \right)}, \quad D = \frac{Ka^2}{4\varepsilon_r \varepsilon_o} - C \ln a$$

In this equation, the potential is equal to zero at both the silo's inner surface and at the wire's outer surface. Checking these two boundary conditions,

$$\Phi\left(\rho = a \right) = -\frac{Ka^2}{4\varepsilon_r \varepsilon_o} + \left[\frac{K(a^2 - b^2)}{4\varepsilon_r \varepsilon_o \ln \left(\frac{a}{b} \right)} \right] \ln a + \frac{Ka^2}{4\varepsilon_r \varepsilon_o} - \left[\frac{K(a^2 - b^2)}{4\varepsilon_r \varepsilon_o \ln \left(\frac{a}{b} \right)} \right] \ln a = 0$$

$$\Phi(\rho = b) = -\frac{Kb^2}{4\varepsilon_r \varepsilon_o} + \left[\frac{K(a^2 - b^2)}{4\varepsilon_r \varepsilon_o \ln \left(\frac{a}{b} \right)} \right] \ln b + \frac{Ka^2}{4\varepsilon_r \varepsilon_o} - \left[\frac{K(a^2 - b^2)}{4\varepsilon_r \varepsilon_o \ln \left(\frac{a}{b} \right)} \right] \ln a = 0$$

(The property $\ln (a/b) = -\ln (b/a)$ was used.) By connecting (external to the silo) the inner wire to the silo's surface (assuming that it is metallic), their potentials would be the same. Substituting the expression for C into the electric field equation,

$$\vec{E} = \left[\frac{K\rho}{2\varepsilon_r \varepsilon_o} - \frac{K(a^2 - b^2)}{4\varepsilon_r \varepsilon_o \rho \ln \left(\frac{a}{b} \right)} \right] \hat{a}_\rho \quad b < \rho < a \tag{1.99}$$

Plots of the *magnitude* of the field, for a 6 m diameter charged silo, vs. ρ with a #1/0 AWG (4.13 mm radius) center wire and without a center wire are given in Mathcad 1.5. The large field strength near the grounding wire is apparent. Without a grounding wire, the field is the smallest at the center of the silo. A minimum in the field *magnitude* is present between the wire and silo. The sign of the field between

$$\varepsilon_r := 1 \qquad \varepsilon_o := 8.854 \cdot 10^{-12} \qquad K := 10^{-5} \qquad a := 3 \qquad b := 4.13 \cdot 10^{-3}$$

$$\rho := b, \frac{b}{0.99} \, .. \, a$$

$$E_{wire}(\rho, b) := \frac{K \cdot \rho}{2 \cdot \varepsilon_r \cdot \varepsilon_o} - \frac{K \cdot (a^2 - b^2)}{4 \cdot \varepsilon_r \cdot \varepsilon_o \cdot \rho \cdot \ln\left(\frac{a}{b}\right)} \qquad\qquad E_{without}(\rho) := \frac{K \cdot \rho}{2 \cdot \varepsilon_r \cdot \varepsilon_o}$$

MATHCAD 1.5   Magnitude of the electric field vs. radial distance with and without a ground wire.

the wire and the silo must change since the potential difference between the wire and silo is zero. This can be quickly shown. The potential difference between the two radial locations b and a is zero:

$$V = -\int_b^a \vec{E} \cdot d\vec{L} = -\int_b^a E_\rho \, d\rho = 0$$

Since $a \neq b$, for this integral to be equal to zero, E_ρ must be both positive and negative. (For the area under a nonzero analytic function to be zero over a nonzero width, the function must have both positive and negative "area.")

To determine whether the field has a maximum or minimum somewhere between the inner ground wire and outer silo, the derivative of Equation (1.99) is first set equal to zero:

$$\frac{d\vec{E}}{d\rho} = \left[\frac{K}{2\varepsilon_r\varepsilon_o} + \frac{K(a^2 - b^2)}{4\varepsilon_r\varepsilon_o\rho^2 \ln\left(\frac{a}{b}\right)} \right] \hat{a}_\rho = 0$$

By solving for ρ, the position of any maximums or minimums is obtained:

$$\rho^2 = -\frac{(a^2 - b^2)}{2\ln\left(\frac{a}{b}\right)} \tag{1.100}$$

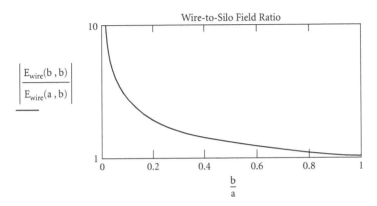

$\varepsilon_r := 1 \qquad \varepsilon_o := 8.854 \cdot 10^{-12} \qquad K := 10^{-5} \qquad a := 3$

$$b := \frac{a}{100}, \frac{a}{99} \, .. \, a$$

$$E_{wire}(\rho, b) := \frac{K \cdot \rho}{2 \cdot \varepsilon_r \cdot \varepsilon_o} - \frac{K \cdot (a^2 - b^2)}{4 \cdot \varepsilon_r \cdot \varepsilon_o \cdot \rho \cdot \ln\left(\frac{a}{b}\right)}$$

Wire-to-Silo Field Ratio

$\left| \dfrac{E_{wire}(b, b)}{E_{wire}(a, b)} \right|$

$\dfrac{b}{a}$

MATHCAD 1.6 Ratio of the magnitude of the electric field at the outer surface of the inner wire to the inner surface of the silo vs. the ratio of their radii.

Since $a > b$, there are no real values for ρ, independent of the sign and value of the volume charge, K. Therefore, for a uniform charge distribution, there are no maximums or minimums of the electric field between the wire and silo. However, as was shown in Mathcad 1.5, the *magnitude* of the field can have a minimum (where the sign of the electric field changes) between the wire and silo.

Since there are no maximums or minimums between the wire and silo, the largest magnitude of the electric field must exist at $\rho = a$ or $\rho = b$. By examining the electric field expression, it is easily seen that the largest positive value of the electric field still occurs at $\rho = a$ when $K > 0$. (In (1.99), the first term is at its maximum at $\rho = a$ and the second term is at its minimum at $\rho = a$.) However, the largest magnitude of the electric field occurs along the surface of the inner wire. At the inner surface of the silo and outer surface of the wire, the fields are, respectively,

$$\vec{E}(\rho = a) = \left[\frac{Ka}{2\varepsilon_r\varepsilon_o} - \frac{K(a^2 - b^2)}{4\varepsilon_r\varepsilon_o a \ln\left(\frac{a}{b}\right)} \right] \hat{a}_\rho, \quad \vec{E}(\rho = b) = \left[\frac{Kb}{2\varepsilon_r\varepsilon_o} - \frac{K(a^2 - b^2)}{4\varepsilon_r\varepsilon_o b \ln\left(\frac{a}{b}\right)} \right] \hat{a}_\rho \qquad (1.101)$$

The ratio of the magnitude of these fields is not a function of K. It is easily shown numerically that the field at the wire's surface is greater than the field at the silo's surface. A plot of this ratio for one specific set of parameters is given in Mathcad 1.6. The ratio of the maximum field magnitude with the grounded wire to the maximum field magnitude without the wire is

$$\frac{\left| E_{with,max} \right|}{\left| E_{without,max} \right|} = \frac{\left| \dfrac{Kb}{2\varepsilon_r\varepsilon_o} - \dfrac{K(a^2 - b^2)}{4\varepsilon_r\varepsilon_o b \ln\left(\frac{a}{b}\right)} \right|}{\left| \dfrac{Ka}{2\varepsilon_r\varepsilon_o} \right|} = \frac{\dfrac{a^2 - b^2}{4\varepsilon_r\varepsilon_o b \ln\left(\frac{a}{b}\right)} - \dfrac{b}{2\varepsilon_r\varepsilon_o}}{\dfrac{a}{2\varepsilon_r\varepsilon_o}} = \frac{a^2 - b^2}{2ab \ln\left(\frac{a}{b}\right)} - \frac{b}{a} \qquad (1.102)$$

Notice that this expression is independent of the volume charge density, K. Furthermore, for larger values of b, the maximum electric field can actually decrease with the addition of a grounding "wire." That is, this ratio can be less than one. A very good numerical approximation for when this ratio is less than one is given by

$$b > \frac{a}{3.513}$$ (1.103)

As a check, when $a = 3.513b$,

$$\frac{|E_{with,max}|}{|E_{without,max}|} = \frac{(3.513b)^2 - b^2}{2\,(3.513b)\,b \ln\left(\dfrac{3.513b}{b}\right)} - \frac{b}{3.513b} \approx 1.000$$

For very large wire diameters, a grounded center "wire" can actually reduce the maximum electric field inside a long charged silo. Again, this assumes a constant volume charge with a center wire at the same potential as the surface of the silo. A specific example with and without a grounding wire is given in Mathcad 1.7 when $b > a/3.513$. Also, the variation in the maximum field magnitude (at $\rho = b$) is shown for various silo diameters in Mathcad 1.8. The maximum field magnitude is plotted vs. the wire radius

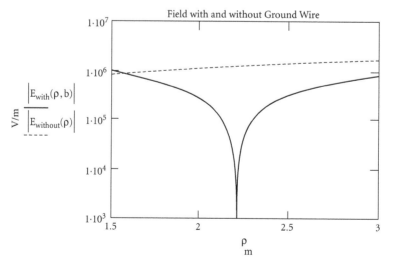

$$\varepsilon_r := 1 \qquad \varepsilon_o := 8.854 \cdot 10^{-12} \qquad K := 10^{-5} \qquad a := 3 \qquad b := 1.5$$

$$\rho := b, \frac{b}{0.999} \, .. \, a$$

$$E_{with}(\rho, b) := \frac{K \cdot \rho}{2 \cdot \varepsilon_r \cdot \varepsilon_o} - \frac{K \cdot (a^2 - b^2)}{4 \cdot \varepsilon_r \cdot \varepsilon_o \cdot \rho \cdot \ln\left(\dfrac{a}{b}\right)} \qquad\qquad E_{without}(\rho) := \frac{K \cdot \rho}{2 \cdot \varepsilon_r \cdot \varepsilon_o}$$

$$b_c := \frac{a}{3.513} \qquad E_{with}(b_c, b_c) = -1.694 \times 10^6 \qquad E_{without}(a) = 1.694 \times 10^6$$

MATHCAD 1.7 Magnitude of the electric field vs. distance with and without a ground wire.

$\varepsilon_r := 1 \qquad \varepsilon_o := 8.854 \cdot 10^{-12} \qquad K := 10^{-5}$

$b := 0.01, 0.011 .. \ 10$

$$E_{with}(a,b) := \frac{K \cdot b}{2 \cdot \varepsilon_r \cdot \varepsilon_o} - \frac{K \cdot (a^2 - b^2)}{4 \cdot \varepsilon_r \cdot \varepsilon_o \cdot b \cdot \ln\left(\dfrac{a}{b}\right)} \qquad b_c(a) := \frac{a}{3.513}$$

Max Field vs. Wire Radius

$\left|E_{with}(1,b)\right|$ (V/m) vs. $\dfrac{b}{b_c(1)}, \dfrac{b}{b_c(3)}, \dfrac{b}{b_c(9)}$

MATHCAD 1.8 Maximum field magnitude (at the inner wire) vs. the ratio of the wire radius to the critical value for three silo diameters.

divided by the critical value $a/3.513$. The sensitivity of the maximum field about this critical value can be obtained from this plot.

When a voltage is applied to the center wire, relative to the silo's surface, the analysis is more complicated. The volume charge in the silo is still a source of electric field. In addition, the wire and silo electrodes that are now at different potentials also generate an electric field. Assuming the 0 V reference is at the silo's surface, the equation in Table 1.5 for concentric cylinders is again used:

$$\vec{E} = \left[\frac{K\rho}{2\varepsilon_r\varepsilon_o} + \frac{V_b}{\rho \ln\left(\dfrac{a}{b}\right)} - \frac{K(a^2 - b^2)}{4\varepsilon_r\varepsilon_o\, \rho \ln\left(\dfrac{a}{b}\right)} \right] \hat{a}_\rho \quad b < \rho < a$$

where V_b is the potential of the wire. This expression can be rewritten as

$$\vec{E} = \left\{ \frac{K\rho}{2\varepsilon_r\varepsilon_o} - \frac{K(a^2 - b^2)}{4\varepsilon_r\varepsilon_o\, \rho \ln\left(\dfrac{a}{b}\right)} \left[1 - \frac{4\varepsilon_r\varepsilon_o V_b}{K(a^2 - b^2)} \right] \right\} \hat{a}_\rho \quad b < \rho < a \qquad (1.104)$$

Depending on the sign and magnitude of V_b, at either the silo's or wire's surface the electric field magnitude can be greater (relative to no center wire). The location of any maximums or minimums of the field between the wire and silo are determined as before:

$$\frac{d\vec{E}}{d\rho} = \left[\frac{K}{2\varepsilon_r\varepsilon_o} - \frac{V_b}{\rho^2 \ln\left(\frac{a}{b}\right)} + \frac{K(a^2-b^2)}{4\varepsilon_r\varepsilon_o\rho^2 \ln\left(\frac{a}{b}\right)} \right] \hat{a}_\rho = 0$$

$$\Rightarrow \rho^2 = \frac{2\varepsilon_r\varepsilon_o}{K \ln\left(\frac{a}{b}\right)} \left[V_b - \frac{K(a^2-b^2)}{4\varepsilon_r\varepsilon_o} \right]$$

(1.105)

Therefore, a maximum (or minimum) could exist between the wire and silo when the potential of the wire (relative to the silo) is sufficiently large in magnitude:

$$V_b > \frac{K(a^2-b^2)}{4\varepsilon_r\varepsilon_o} \quad \text{if } K > 0, \quad V_b < \frac{K(a^2-b^2)}{4\varepsilon_r\varepsilon_o} \quad \text{if } K < 0$$

(1.106)

The possibility of a maximum (or minimum) did not exist when both the silo and center wire were grounded. The second derivative can be used to determine whether the given position corresponds to a maximum or minimum (since the curvature is a function of the second derivative):

$$\frac{d^2\vec{E}}{d\rho^2} = \left[\frac{2V_b}{\rho^3 \ln\left(\frac{a}{b}\right)} - \frac{2K(a^2-b^2)}{4\varepsilon_r\varepsilon_o\rho^3 \ln\left(\frac{a}{b}\right)} \right] \hat{a}_\rho = \frac{2}{\rho^3 \ln\left(\frac{a}{b}\right)} \left[V_b - \frac{K(a^2-b^2)}{4\varepsilon_r\varepsilon_o} \right] \hat{a}_\rho$$

(1.107)

For $K > 0$, the second derivative is positive, which corresponds to a minimum. For $K < 0$, the second derivative is negative, which corresponds to a maximum. This minimum and maximum occur at the radial position

$$\rho = \sqrt{\frac{2\varepsilon_r\varepsilon_o}{K \ln\left(\frac{a}{b}\right)} \left[V_b - \frac{K(a^2-b^2)}{4\varepsilon_r\varepsilon_o} \right]}$$

(1.108)

Of course, for the maximum or minimum to exist inside the silo, (1.108) should be less than a. In Mathcad 1.9, the maximum field occurs at the silo's surface for the given parameters. Equation (1.108) is also checked. The wire voltage necessary to generate a minimum between the wire and silo is extremely large. For much smaller values of K or volume charge density, the applied voltage is much more reasonable. Also, when a voltage is applied across the silo and wire, the maximum field magnitude can occur at either the silo or wire. Although not shown, many other cases were analyzed. In all of the cases for nonzero V_b's, the maximum field was greater with a center wire.

1.21 All of the Electric Field Boundary Conditions

Discuss, but do not derive, the boundary conditions for electric fields. Clearly explain the effect of single and double layers of surface charge on these boundary conditions. [Haus; Hayt; Melcher]

   All real regions have boundaries. The distribution of the charge or potential along these boundaries will have some effect on the electric field inside the boundaries. This should not be forgotten, especially when computer programs are used to determine the solution for electric fields, magnetic fields, inductance, capacitance, and other electromagnetic-related quantities. Unfortunately, individuals sometimes

$\varepsilon_r := 1 \qquad \varepsilon_o := 8.854 \cdot 10^{-12} \qquad K := 10^{-5} \qquad a := 3 \qquad b := \dfrac{a}{1000} \qquad b = 3 \times 10^{-3}$

$\dfrac{K \cdot (a^2 - b^2)}{4 \cdot \varepsilon_r \cdot \varepsilon_o} = 2.541 \times 10^6 \qquad V_b := 2.55 \cdot 10^6 \qquad \rho_c := \sqrt{\dfrac{2 \cdot \varepsilon_r \cdot \varepsilon_o}{K \cdot \ln\left(\dfrac{a}{b}\right)} \cdot \left[V_b - \dfrac{K \cdot (a^2 - b^2)}{4 \cdot \varepsilon_r \cdot \varepsilon_o}\right]} \qquad \rho_c = 0.047$

$\rho := b, \dfrac{b}{0.95} .. a$

$E_{wire}(\rho, V_b) := \dfrac{K \cdot \rho}{2 \cdot \varepsilon_r \cdot \varepsilon_o} + \dfrac{V_b}{\rho \cdot \ln\left(\dfrac{a}{b}\right)} - \dfrac{K \cdot (a^2 - b^2)}{4 \cdot \varepsilon_r \cdot \varepsilon_o \cdot \rho \cdot \ln\left(\dfrac{a}{b}\right)} \qquad\qquad E_{without}(\rho) := \dfrac{K \cdot \rho}{2 \cdot \varepsilon_r \cdot \varepsilon_o}$

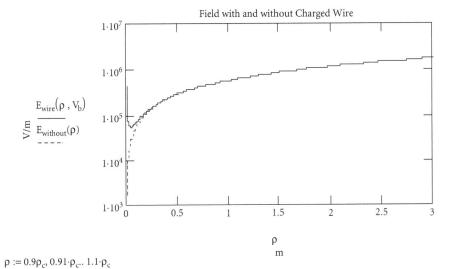

$\rho := 0.9 \rho_c, 0.91 \cdot \rho_c .. 1.1 \cdot \rho_c$

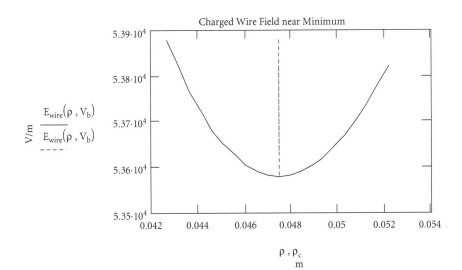

MATHCAD 1.9   Electric field with and without an "electrified" center wire vs. radial distance.

falsely believe that the actual boundaries are nonexistent or at "infinity," resulting in misleading or incorrect solutions to problems. Therefore, it is crucial that boundary conditions be fully understood. The electric field boundary conditions are discussed in this section.

The electric field, $\vec{E}$, is a fundamental quantity, and the boundary conditions can be written in terms of the electric field. In some cases, though, it is easier and more insightful to write some of the boundary conditions in terms of the flux density vector, $\vec{D}$, and the electric potential, Φ (for electrostatic or electroquasistatic conditions). Since the electric field, $\vec{E}$, is a more fundamental quantity, it might seem unnecessary or wasteful to introduce another field vector or scalar quantity. To understand why the introduction of the $\vec{D}$ vector can sometimes be helpful, an elementary understanding of a few concepts in device physics is necessary.

Ideal dielectric materials are usually considered perfectly insulating. Perfectly insulating materials have zero conductivity. The conductivity is a measure of the number of (free) charges that are free to move about in the material. When a voltage is placed across a material with zero conductivity, the conduction current through the material is zero. However, even when a perfectly insulating, charge-neutral dielectric is placed in an electric field, charges referred to as bound charges can be present along the surfaces of the insulator.[14] To understand

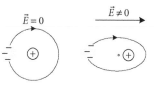

FIGURE 1.38   Polarization of an electron cloud and a nucleus.

this concept, recall that materials are composed of atoms. These atoms have nuclei with electrons (or electron clouds) circulating around the nuclei. When a neutral material is examined on a scale that is large compared to the distance between atoms (i.e., macroscopic scale), the inside of the material appears charge neutral: the total number of positive charges in the nuclei equal the total number of negative charges in the electron clouds. For materials with no permanent polarization, for each atom, the "center of charge" for each electron cloud coincides with the center of the nucleus. As is illustrated in Figure 1.38, when this same material is exposed to an external electric field in the direction shown, each negatively charged electron cloud shifts slightly toward the left while each positively charged nucleus shifts slightly toward the right. Because of this shift, the "center of charge" for the electron cloud and the nucleus for each atom do not coincide. Each atom can then be viewed as a charge dumbbell or dipole. The number of these charge dipoles is a function of the dielectric constant or relative permittivity of the material.

Assuming the insulator was initially charge neutral, the application of an external electric field does not change the neutrality of the material. However, along certain surfaces, a surface charge can exist. These charges exist along surfaces because of the slight shift of the electron clouds and nuclei in the applied field. Along certain surfaces, the surface charge appears negative, corresponding to greater than normal electron clouds. Along other surfaces, the surface charge is positive, corresponding to the uncovering of or excess positive nuclei. Of course, if the dielectric was initially neutral, the total surface charge must sum to zero. These surface charges are referred to as bound charges in ideal dielectrics since they are bound to their "mother" nuclei. Unless they are stripped from the surface (e.g., via rubbing or electrical breakdown of the material), these charges cannot leave the perfect insulator. For conducting materials, the free charges can leave the material when they make contact with another object. The bound charge along several ideal dielectrics exposed to an applied electric field is crudely illustrated in Figure 1.39. The "−" symbols are representing the negative charge resulting from the shift in the electron clouds, and the "+" symbols are representing positive charge resulting from the shift in the nuclei (or exposed positive nuclei). Notice that the insulating materials have been polarized by the external field. The net bound charge along each interface is a function of the dielectric constant of the neighboring materials. It is also a function of the surface area of the interface. The greater the surface area of the interface and the change in dielectric constant across

---

[14]Bound charge can actually exist wherever the permittivity is spatially varying and the electric field is in the direction of this change. Starting with $\nabla \cdot (\varepsilon \vec{E}) = 0$ and allowing ε to be a function of position, it can be readily shown that the bound volume charge is equal to $\rho_{Vb} = -[(\nabla \varepsilon) \cdot \vec{E}] / \varepsilon_r$.

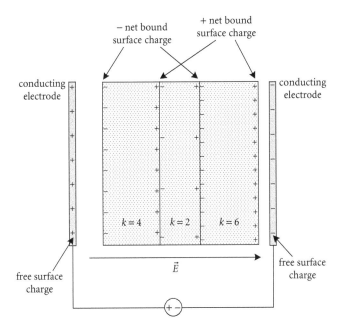

FIGURE 1.39 Crude representation of the bound charge distribution along several dielectric interfaces exposed to an electric field (k = dielectric constant).

the interface, the greater the bound charge.[15] At the interfaces shown, when the dielectric constant steps up in value (in the direction of the applied electric field), the bound surface charge is negative. When the dielectric constant steps down in value, the bound surface charge is positive. If the sign of the applied field changes, the sign of the bound charges change. If the applied field disappears, the bound charges disappear.[16] Although the dielectrics are shown not in direct contact with the electrodes, the dielectrics would still be polarized if they were in direct contact with the electrodes. The bound charge would still be present at these electrode-dielectric interfaces, but the net surface charge along these interfaces would be the sum of the free and bound charges.

What is not shown in Figure 1.39 is the electric field generated by the bound charges. Although these charges are bound, they still generate an electric field. The electric field generated by these bound charges is proportional to a variable referred to as $\vec{P}$, the polarization vector. However, the $\vec{P}$ as will be formally defined, *originates on the negative bound surface charge and terminates on the positive bound surface charge* as shown in Figure 1.40. The electric field from free surface charge originates on positive charge and terminates on negative charge.

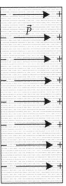

FIGURE 1.40 Polarization vector for a dielectric slab.

---

[15]Bound charge along a surface is also referred to as surface polarization charge.

[16]For some materials, a fraction of the bound charge can remain even after the external field is turned off; that is, the bound charges can be "frozen" in place. These materials are referred to as electrets, and they have a permanent electric field (without a power source) somewhat analogous to magnets. These electrets are used in microphones and exotic audio speakers.

For many materials, the polarization vector is linearly proportional to the total electric field:

$$\vec{P} = \chi_e \varepsilon_o \vec{E} \tag{1.109}$$

where χ_e is the electric (or dielectric) susceptibility and ε_o is the free-space permittivity.[17] It is clear from this relationship that $\vec{P}$ is in the same direction as $\vec{E}$. Again, $\vec{P}$ originates on negative bound charge and terminates on positive bound charge. For many materials, it is much more convenient to lump together the applied external field and the field generated by the bound polarized charges. Thus, the flux density vector or displacement flux density vector, $\vec{D}$, is born:

$$\vec{D} = \varepsilon_o \vec{E} + \vec{P} = \varepsilon_o \vec{E} + \chi_e \varepsilon_o \vec{E} \tag{1.110}$$

The vector $\vec{D}$ has units of C/m². Assuming that the electric susceptibility is positive, it might appear that $\vec{P}$ as defined always increases the flux density. For a given $\vec{E}$, this statement is true. However, when a dielectric object is placed in an applied electric field, it will affect the electric field distribution. The polarization vector is a function of the total electric field in the dielectric from both the applied field and bound surface-charge generated field. Depending on the sign, magnitude, and distribution of the bound charge, the resultant electric field can be intensified or rarefied (but not reversed in direction). For example, referring to Figure 1.39, the electric field is directed from the left electrode to the right electrode in all regions. Compared to the electric field between the two plates with no dielectrics present, the electric field is intensified in both free-space regions. When passing from the left-most free-space region to the right-most free-space region, the electric field steps down at the $k = 1\text{-}4$ interface, steps up at the $k = 4\text{-}2$ interface, steps down at the $k = 2\text{-}6$ interface, and steps up at the $k = 6\text{-}1$ interface.

For linear dielectrics, it is common not to work directly with $\vec{P}$. Instead, a dimensionless quantity, the relative permittivity, ε_r, is used:

$$\vec{D} = \varepsilon_o \vec{E} + \chi_e \varepsilon_o \vec{E} = (1 + \chi_e) \varepsilon_o \vec{E} = k \varepsilon_o \vec{E} = \varepsilon_r \varepsilon_o \vec{E} \quad \text{where } k = \varepsilon_r = 1 + \chi_e \tag{1.111}$$

where the permittivity of the material is defined as $\varepsilon = k\varepsilon_o = \varepsilon_r \varepsilon_o$. The dielectric constant, k, which is equal to the relative permittivity, ε_r, is frequently used to describe the dielectric property of a material. The dielectric constant of free space is one. Most materials have dielectric constants ranging from one to ten. The $\vec{D}$ vector includes both the applied external and internal polarization fields.

Now, several boundary conditions can be introduced. Unless specifically labeled, the bound charge is not shown in the figures. Free surface charge is given the label ρ_S. The first boundary condition, which is easily derived using Gauss' law, relates the normal components of $\vec{D}$ at the boundary between two interfaces to the free surface charge along the interface:

$$D_{n2} - D_{n1} = \rho_S \tag{1.112}$$

This boundary condition is valid for static and time-varying electric fields. When there is no surface charge, the normal component of the flux density vector is continuous at the interface. Even when the surface charge is zero, the normal component of the electric field, $\vec{E}$, is not continuous unless the dielectric constants on both sides are equal:

$$D_{n2} = k_2 \varepsilon_o E_{n2} = D_{n1} = k_1 \varepsilon_o E_{n1} \quad \text{if } \rho_S = 0 \quad \Rightarrow \frac{E_{n2}}{E_{n1}} = \frac{k_1}{k_2} \tag{1.113}$$

---

[17]Under intense electric fields, materials will become nonlinear. In the field of nonlinear optics, this property is important.

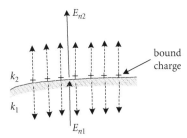

FIGURE 1.41   Net electric field on both sides of a dielectric interface. Also shown is the electric field from the bound, positive surface charge.

This relationship is quite reasonable since any net bound charge along the dielectric interface generates an electric field, which is the source of the discontinuity in the electric field. In Figure 1.41, the net bound surface charge is assumed positive; hence, the field is directed away from the charge. The electric field from this bound charge is shown as dotted-line arrows in the figure. In the k_2 region, the bound-charge field is in the same direction as the external electric field, while in the k_1 region, the bound-charge field is in the opposite direction. The vector $\vec{D}$ includes the effects of this additional bound-charge field. When $k_1 = k_2$, this additional bound-charge generated field normal to the interface would be zero.

   Surface charge, charge existing only along an interface, is common. This surface charge density, ρ_s, in Equation (1.112) is free surface charge. The effects of any bound surface charge have already been incorporated into this expression via the $\vec{D}$ vector. Any free surface charge will cause a discontinuity or jump in the $\vec{D}$ field since the free surface charge is a source of an electric field. The discontinuity in $\vec{D}$ increases with the size of this free surface charge. In Figure 1.42, this free charge is shown as positive. The dielectric constants are set equal so that only free charge is present along the interface. Before leaving this boundary condition, an equivalent form will be given:

$$\hat{a}_n \cdot (\vec{D}_2 - \vec{D}_1) = \rho_s \qquad (1.114)$$

where $\hat{a}_n$ is the unit vector normal or perpendicular to the interface directed from region 1 to region 2. By taking the dot product of this normal vector with each of the $\vec{D}$ vectors, the normal components are obtained.

   The second boundary condition involves the tangential components of the electric field vector, $\vec{E}$, on both sides of the interface:

$$E_{t1} = E_{t2} \qquad (1.115)$$

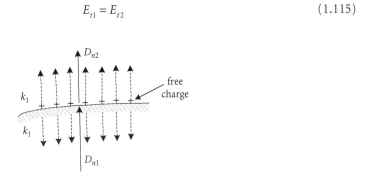

FIGURE 1.42   Flux density on both sides of an interface containing free charge. Also shown is the electric field from the free surface charge.

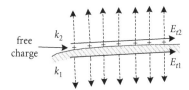

FIGURE 1.43 Electric fields tangential to an interface. Also shown is the electric field generated by the free (or bound) surface charge.

This boundary condition is also valid for static or time-varying electric fields and when surface charge is present. As is shown in Figure 1.43, this boundary condition is not a function of the free surface charge since the free surface charge tangential to the interface generates an electric field normal to the interface. (This assumes that the interface is relatively flat over the region shown, and the variation in the surface charge is small over this region.) This boundary condition is not a function of the dielectric constants because the bound charges also generate a field normal to the interface. An equivalent method of writing this boundary condition involves the cross product of the unit normal vector directed from region 1 to region 2:

$$\hat{a}_n \times (\vec{E}_2 - \vec{E}_1) = 0 \qquad (1.116)$$

Before the boundary conditions associated with a double layer of surface charge are discussed, the boundary conditions in terms of the potential will be provided. The electric potential, Φ, is the negative gradient of the electric field, $\vec{E} = -\nabla\Phi$. This relationship should only be used for electrostatic or nearly electrostatic conditions. Therefore, the following boundary conditions involving the potential should only be used for electroquasistatic conditions. The first condition is very easy to apply and understand:

$$\Phi_2 = \Phi_1 \qquad (1.117)$$

The potential is continuous across an interface even when (a single layer of) surface charge is present. The voltage difference between locations a and b is defined as the negative line integral of the electric field:

$$V = -\int_a^b \vec{E} \cdot d\vec{L}$$

Since integration is a smoothing operation, any discontinuity in the electric field caused by surface or bound charges, is smoothed over. If the electric field is given by $Au(x)\hat{a}_x$, where $u(x)$ is the discontinuous unit step function, which turns on at $x = 0$, then the potential from -1 to 3 in the x direction is $-3A$:

$$V = -\int_{-1}^{3} Au(x)\,\hat{a}_x \cdot dx\hat{a}_x = -\int_{-1}^{3} Au(x)\,dx = -\int_0^3 A\,dx = -3A$$

There is no sudden jump in the potential even if the upper limit of integration is just slightly past $x = 0$. Because the potential is a scalar quantity, the adjectives tangential or normal should not be used with potential. Equation (1.117) is equivalent to $E_{t2} = E_{t1}$. The second boundary condition involving the potential indicates that the derivative of the potential in the direction normal to the interface is discontinuous:

$$k_2 \frac{d\Phi_2}{dn} - k_1 \frac{d\Phi_1}{dn} = \frac{-\rho_s}{\varepsilon_o} \qquad (1.118)$$

This expression, unless the derivatives are readily available, is not as convenient to apply. It is equivalent to $D_{n2} - D_{n1} = \rho_S$. It can also be written as

$$\hat{a}_n \cdot (k_2 \nabla \Phi_2 - k_1 \nabla \Phi_1) = \frac{-\rho_S}{\varepsilon_o} \qquad (1.119)$$

where $\hat{a}_n$ is as previously defined. Rather than working with Equations (1.118) or (1.119), sometimes Equation (1.117) is used along with $D_{n2} - D_{n1} = \rho_S$.

A dipole layer consists of two layers of surface charge of equal magnitude but opposite sign as shown in Figure 1.44. During the tribocharging of a flowing liquid inside a pipe, for example, a dipole layer can exist along the liquid and pipe interface. If the distance between the surface charges is d, the potential difference across the positively charged surface layer and the negatively charged surface layer is given by

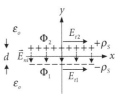

FIGURE 1.44   Double or dipole layer of surface charge.

$$\Delta \Phi = \Phi_2 - \Phi_1 = \frac{\rho_S d}{\varepsilon_o} \qquad (1.120)$$

This expression is also valid if the surface charge is varying along the interface (e.g., $\rho_S(x)$). The potential is continuous, even across a single layer of surface charge. Thus, if the distance between the two layers, d, is nonzero, the potential across each of the individual surface layers is also continuous. However, frequently the surface charge density, ρ_S, is assumed very large while the distance between the layers, d, is assumed very small. In the limit where $d \to 0$ and $\rho_S \to \infty$, the surface charge distribution approaches a doublet. In this case, there is a jump in the potential across the double layer of surface charge since the electric field between the layers of charge approaches infinity. (The integration of an impulse function, representing the infinite electric field, over the range that the impulse is nonzero is a unit step.) Not surprisingly, for nonzero values of d, the internal electric field directed from the positive to the negative surface charge is

$$\vec{E}_{ni} = \frac{\rho_S}{\varepsilon_o} \hat{a}_n \qquad (1.121)$$

Equation (1.121) is the standard expression for the field between two infinite sheets of surface charge, each of opposite sign. If ρ_S is infinite, then the electric field is infinite "between" the two layers. There is a jump in the normal component of the electric field, even with free-space conditions on both sides of the dipole layer. If this electric field is integrated across the layers, (1.120) is obtained. With a single layer of surface charge, the tangential components of the electric field are continuous. With a double layer of surface charge, the tangential components are not necessarily continuous. If the surface charge is varying along the x direction, for example, then the jump in the tangential components of the electric field near the entire double layer is

$$E_{x2} - E_{x1} = -\frac{d}{\varepsilon_o} \frac{d\rho_S(x)}{dx} \qquad (1.122)$$

According to this expression, if the positive surface charge is increasing in the x direction, then $E_{x2} < E_{x1}$.

Although boundary conditions are used throughout this book, a few classical problems and solutions will be given. In these examples, either the $\vec{E}$ or $\vec{D}$ field is provided on one side of a boundary or interface

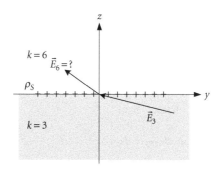

FIGURE 1.45   Flat dielectric interface containing free surface charge. The electric field is given below the interface.

and both $\vec{E}$ and $\vec{D}$ are desired immediately on the other side of the interface. Both sides of the interface are considered perfectly insulating so that their conductivities are zero. For the first example, the dielectric interface will be along the $z = 0$ plane (i.e., xy plane). For $z < 0$, the dielectric constant k (or relative permittivity) is equal to 3 while for $z > 0$ the dielectric constant is equal to 6. A surface charge equal to ρ_S is present along this interface at $z = 0$ as shown in Figure 1.45. The electric field is given immediately below the $z = 0$ plane as

$$\vec{E}_3 = -3\hat{a}_x - 4\hat{a}_y + \hat{a}_z \quad @\, z = 0^- \text{ where } k = 3$$

In Figure 1.45, only the components in the yz plane are shown. Also, even though the surface charge is denoted by "+" symbols, this does not imply that the surface charge is positive. It can be of either sign. Since the electric field is given, the electric flux density in this same region can be written down immediately using the dielectric constant for that region:

$$\vec{D}_3 = 3\varepsilon_o\vec{E}_3 = -9\varepsilon_o\hat{a}_x - 12\varepsilon_o\hat{a}_y + 3\varepsilon_o\hat{a}_z \quad @\, z = 0^- \text{ where } k = 3$$

If $\vec{D}$ were initially given, then the electric field for that region would be $\vec{E} = \vec{D}/(k\varepsilon_o)$. To determine the electric field and electric flux density immediately above the interface, the normal and tangential directions for the interface must be identified. These directions are easily determined since the interface is flat (and along the $z = 0$ plane). The x and y components are both tangential to the interface. Therefore, since the tangential components of the $\vec{E}$ are continuous along the interface (that is, the same immediately below or above the interface), the tangential $\vec{E}$ components directly above the plane are immediately determined by inspection:

$$\vec{E}_6 = -3\hat{a}_x - 4\hat{a}_y + E_{z6}\hat{a}_z \quad @\, z = 0^+ \text{ where } k = 6$$

Again, the tangential components of $\vec{E}$ not $\vec{D}$ are continuous across a dielectric interface. Two of the $\vec{D}$ field terms can be obtained from this result:

$$\vec{D}_6 = 6\varepsilon_o\vec{E}_6 = -18\varepsilon_o\hat{a}_x - 24\varepsilon_o\hat{a}_y + D_{z6}\hat{a}_z \quad @\, z = 0^+ \text{ where } k = 6$$

The normal component is in the z direction. The expression that relates the normal components of $\vec{D}$ to the surface charge is

$$\hat{a}_n \cdot (\vec{D}_2 - \vec{D}_1) = \rho_S$$

where $\hat{a}_n$ is the unit vector normal or perpendicular to the interface directed from region 1 to region 2. Either side of the interface can be designated as the $\vec{D}_2$ or $\vec{D}_1$ vector in this expression as long as the normal vector is from region 1 to region 2. If $\vec{D}_2$ corresponds to the $z < 0$ region, then $\hat{a}_n$ is equal to $-\hat{a}_z$:

$$-\hat{a}_z \cdot [(-9\varepsilon_o\hat{a}_x - 12\varepsilon_o\hat{a}_y + 3\varepsilon_o\hat{a}_z) - (-18\varepsilon_o\hat{a}_x - 24\varepsilon_o\hat{a}_y + D_{z6}\hat{a}_z)] = \rho_S$$

$$-3\varepsilon_o + D_{z6} = \rho_S \quad \Rightarrow D_{z6} = \rho_S + 3\varepsilon_o$$

Therefore,

$$\vec{D}_6 = -18\varepsilon_o\hat{a}_x - 24\varepsilon_o\hat{a}_y + (\rho_S + 3\varepsilon_o)\hat{a}_z \quad @z = 0^+ \text{ where } k = 6$$

$$\vec{E}_6 = \frac{\vec{D}_6}{6\varepsilon_o} = -3\hat{a}_x - 4\hat{a}_y + \frac{(\rho_S + 3\varepsilon_o)}{6\varepsilon_o}\hat{a}_z \quad @z = 0^+ \text{ where } k = 6$$

The next example is in the cylindrical coordinate system. As long as the interface or boundary is along a surface "easily generated" in the coordinate system, then the tangential and normal components are readily determined. In the cylindrical coordinate system, cylindrical surfaces centered on the z axis are easily generated. Imagine that a very long solid cylindrical rod with a relative permittivity of 4 is surrounded by free space ($\varepsilon_r = 1$) as shown in Figure 1.46. A layer of surface charge equal to ρ_S is along the surface of the rod near the location of interest. The flux density is given as

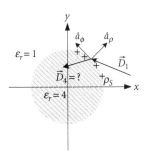

FIGURE 1.46 Cylindrical dielectric interface containing free surface charge. The flux density vector is given outside the rod.

$$\vec{D}_1 = -2\hat{a}_\rho + 4\hat{a}_\phi + 3\hat{a}_z$$

near the location of interest along the outer surface of the rod in free space. The components of the electric field and flux density are desired along the inner surface of the rod. The electric field corresponding to $\vec{D}_1$ is

$$\vec{E}_1 = \frac{\vec{D}_1}{\varepsilon_o} = \frac{-2}{\varepsilon_o}\hat{a}_\rho + \frac{4}{\varepsilon_o}\hat{a}_\phi + \frac{3}{\varepsilon_o}\hat{a}_z$$

Both the $\hat{a}_\phi$ and $\hat{a}_z$ components are tangential to the cylindrical surface. Therefore, these electric field terms must be continuous no matter what the value of the single layer of surface charge:

$$\vec{E}_4 = E_{\rho 4}\hat{a}_\rho + \frac{4}{\varepsilon_o}\hat{a}_\phi + \frac{3}{\varepsilon_o}\hat{a}_z$$

The flux density in the rod along the inner surface is then

$$\vec{D}_4 = 4\varepsilon_o\vec{E}_4 = 4\varepsilon_o E_{\rho 4}\hat{a}_\rho + 16\hat{a}_\phi + 12\hat{a}_z$$

The surface charge density affects the normal component of the $\vec{D}$ field:

$$\hat{a}_n \cdot (\vec{D}_4 - \vec{D}_1) = \rho_S$$

where $\hat{a}_n = -\hat{a}_\rho$ is the unit vector normal to the interface directed from region 1 to region 4. This unit vector is directed toward the origin. Substituting the $\vec{D}$ fields into this expression and taking the dot product with $-\hat{a}_\rho$,

$$-4\varepsilon_o E_{\rho 4} - 2 = \rho_S \quad \Rightarrow E_{\rho 4} = \frac{-\rho_S - 2}{4\varepsilon_o}$$

The desired fields are then

$$\vec{D}_4 = -(\rho_S + 2)\,\hat{a}_\rho + 16\hat{a}_\phi + 12\hat{a}_z, \quad \vec{E}_4 = -\frac{(\rho_S + 2)}{4\varepsilon_o}\,\hat{a}_\rho + \frac{4}{\varepsilon_o}\hat{a}_\phi + \frac{3}{\varepsilon_o}\hat{a}_z$$

The last example is the most complex. The electric field is given below a curved interface by the expression[18]

$$\vec{D}_4 = 3\hat{a}_x + 2\hat{a}_y - 4\hat{a}_z \quad \text{where } k = 4$$

Therefore, the electric field in this same region along this interface is

$$\vec{E}_4 = \frac{\vec{D}_4}{4\varepsilon_o} = \frac{3}{4\varepsilon_o}\hat{a}_x + \frac{1}{2\varepsilon_o}\hat{a}_y - \frac{1}{\varepsilon_o}\hat{a}_z \quad \text{where } k = 4$$

The dielectric interface is not along a flat plane but follows the curve given by $y = x^3$ as shown in Figure 1.47. This interface is infinite in both the $\pm z$ directions, and there is no surface charge along the interface. The $\hat{a}_n$ vector clearly varies with x. To determine this normal vector, the fabulous gradient operator can be used. The function f is first defined using the expression describing the interface:

$$f = y - x^3$$

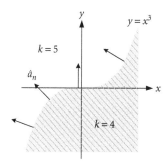

FIGURE 1.47 Curved dielectric interface. The direction of the unit normal vector along the interface changes with x.

---

[18]Although the electric field is assumed in the same direction all along this interface, it is likely that the field will be distorted or not uniform near the interface.

The gradient of f is normal to this interface. Because $\hat{a}_n$ is a unit vector, it has a magnitude of one. The gradient of f, ∇f, is normalized by dividing it by the magnitude of ∇f:

$$\hat{a}_n = \frac{\nabla f}{|\nabla f|} = \frac{-3x^2\hat{a}_x + \hat{a}_y}{\sqrt{(-3x^2)^2 + (1)^2}} = \frac{-3x^2\hat{a}_x + \hat{a}_y}{\sqrt{9x^4 + 1}}$$

By examining this unit vector at various x values, it is clear that the vector is in the direction shown. (If the other direction is desired, the $\hat{a}_n$ expression can be multiplied by -1.) The $\vec{D}$ field in the region of dielectric constant 5 is

$$\vec{D}_5 = D_{x5}\hat{a}_x + D_{y5}\hat{a}_y + D_{z5}\hat{a}_z$$

and the corresponding electric field is

$$\vec{E}_5 = \frac{\vec{D}_5}{5\varepsilon_o} = \frac{D_{x5}}{5\varepsilon_o}\hat{a}_x + \frac{D_{y5}}{5\varepsilon_o}\hat{a}_y + \frac{D_{z5}}{5\varepsilon_o}\hat{a}_z$$

Several equations are required to solve for these flux density components. One equation is obtained from the normal component continuity expression, with zero surface charge:

$$\hat{a}_n \cdot (\vec{D}_5 - \vec{D}_4) = \frac{-3x^2\hat{a}_x + \hat{a}_y}{\sqrt{9x^4 + 1}} \cdot [(D_{x5}\hat{a}_x + D_{y5}\hat{a}_y + D_{z5}\hat{a}_z) - (3\hat{a}_x + 2\hat{a}_y - 4\hat{a}_z)] = 0$$

$$\Rightarrow \frac{-3x^2}{\sqrt{9x^4 + 1}}(D_{x5} - 3) + \frac{1}{\sqrt{9x^4 + 1}}(D_{y5} - 2) = 0 \qquad (1.123)$$

Two more (independent) equations are required to determine D_{x5}, D_{y5}, and D_{z5}. Both of these equations are obtained from the tangential component boundary condition:

$$\hat{a}_n \times (\vec{E}_5 - \vec{E}_4) = 0$$

$$\frac{-3x^2\hat{a}_x + \hat{a}_y}{\sqrt{9x^4 + 1}} \times \left[\left(\frac{D_{x5}}{5\varepsilon_o}\hat{a}_x + \frac{D_{y5}}{5\varepsilon_o}\hat{a}_y + \frac{D_{z5}}{5\varepsilon_o}\hat{a}_z \right) - \left(\frac{3}{4\varepsilon_o}\hat{a}_x + \frac{1}{2\varepsilon_o}\hat{a}_y - \frac{1}{\varepsilon_o}\hat{a}_z \right) \right] = 0$$

$$\frac{-3x^2\hat{a}_x + \hat{a}_y}{\sqrt{9x^4 + 1}} \times \left[\left(\frac{D_{x5}}{5\varepsilon_o} - \frac{3}{4\varepsilon_o} \right)\hat{a}_x + \left(\frac{D_{y5}}{5\varepsilon_o} - \frac{1}{2\varepsilon_o} \right)\hat{a}_y + \left(\frac{D_{z5}}{5\varepsilon_o} + \frac{1}{\varepsilon_o} \right)\hat{a}_z \right] = 0$$

Taking the cross products,

$$\frac{-3x^2}{\sqrt{9x^4 + 1}}\left(\frac{D_{y5}}{5\varepsilon_o} - \frac{1}{2\varepsilon_o} \right)\hat{a}_z + \frac{3x^2}{\sqrt{9x^4 + 1}}\left(\frac{D_{z5}}{5\varepsilon_o} + \frac{1}{\varepsilon_o} \right)\hat{a}_y - \frac{1}{\sqrt{9x^4 + 1}}\left(\frac{D_{x5}}{5\varepsilon_o} - \frac{3}{4\varepsilon_o} \right)\hat{a}_z$$

$$+ \frac{1}{\sqrt{9x^4 + 1}}\left(\frac{D_{z5}}{5\varepsilon_o} + \frac{1}{\varepsilon_o} \right)\hat{a}_x = 0$$

All components of the resulting vector must be zero:

$$\left[\frac{-3x^2}{\sqrt{9x^4+1}}\left(\frac{D_{y5}}{5\varepsilon_o}-\frac{1}{2\varepsilon_o}\right)-\frac{1}{\sqrt{9x^4+1}}\left(\frac{D_{x5}}{5\varepsilon_o}-\frac{3}{4\varepsilon_o}\right)\right]\hat{a}_z=0 \qquad (1.124)$$

$$\frac{3x^2}{\sqrt{9x^4+1}}\left(\frac{D_{z5}}{5\varepsilon_o}+\frac{1}{\varepsilon_o}\right)\hat{a}_y=0$$

$$\frac{1}{\sqrt{9x^4+1}}\left(\frac{D_{z5}}{5\varepsilon_o}+\frac{1}{\varepsilon_o}\right)\hat{a}_x=0$$

For the last two expressions to be zero for all x, $D_{z5}=-5$. Therefore, $E_z=-5/5\varepsilon_o=-1/\varepsilon_o$. This result is reasonable since it is equal to the z component for $\vec{E}_4$, which is entirely tangential to the surface. To determine D_{x5} and D_{y5}, Equations (1.123) and (1.124) must be solved simultaneously. Not surprisingly, D_{x5} and D_{y5} are both a function of x since the direction of the normal to the interface changes with x. Using a symbolic package, these variables are readily solved. With $D_{z5}=-5$,

$$\vec{D}_5=\frac{3(36x^4+2x^2+5)}{4(9x^4+1)}\hat{a}_x+\frac{(90x^4+9x^2+8)}{4(9x^4+1)}\hat{a}_y-5\hat{a}_z$$

$$\vec{E}_5=\frac{\vec{D}_5}{5\varepsilon_o}=\frac{3(36x^4+2x^2+5)}{20\varepsilon_o(9x^4+1)}\hat{a}_x+\frac{(90x^4+9x^2+8)}{20\varepsilon_o(9x^4+1)}\hat{a}_y-\frac{1}{\varepsilon_o}\hat{a}_z$$

As a simple check on $\vec{D}_5$,

$$\lim_{x\to\infty}\vec{D}_5=3\hat{a}_x+\frac{5}{2}\hat{a}_y-5\hat{a}_z$$

For very large (positive or negative) values of x, the normal to the interface is nearly entirely in the $-\hat{a}_x$ direction. As expected, the x components of $\vec{D}_4$ and $\vec{D}_5$ are equal at these large x values. As another check on $\vec{D}_5$

$$\vec{D}_5\Big|_{x=0}=\frac{15}{4}\hat{a}_x+2\hat{a}_y-5\hat{a}_z$$

At $x=0$, the normal to the interface is in the $\hat{a}_y$ direction. As expected, at this x location, the y components of $\vec{D}_4$ and $\vec{D}_5$ are equal. As a check on the tangential electric field boundary condition,

$$\lim_{x\to\infty}\vec{E}_5=\frac{3}{5\varepsilon_o}\hat{a}_x+\frac{1}{2\varepsilon_o}\hat{a}_y-\frac{1}{\varepsilon_o}\hat{a}_z$$

For these large values of x, both the y and z components are tangential to the interface. These components are indeed equal to the y and z components of $\vec{E}_4$.

1.22 Powder Bed

Using a simple one-dimensional model for linearly varying charged powder in a grounded container, determine the expressions for the potential and electric field both in and above the powder bed. [Jones, '87; Jones, '89]

Many insulating powders and liquids when transported and mixed become highly charged due to tribocharging. The electric field generated by this charge, if sufficiently high, can be the ignition source for an explosion or fire. When liquids are pumped into a storage vessel, the volume charge density can sometimes be assumed to be uniform because of the turbulent mixing that occurs. When powders are pumped into a vessel, though, stratification of the volume charge can occur. The actual process is dynamic, with charge and material increasing via the powder inlet and with charge dissipating via the grounded walls of the vessel.

In this analysis, the fields will be determined for one instant of time. The volume charge distribution of the powder of relative permittivity ε_r is assumed to vary according to the expression

$$\rho_V = Ky + F \qquad (1.125)$$

inside the powder. Furthermore, a surface charge density ρ_S is assumed on the top of the powder's surface as shown in Figure 1.48. Being a one-dimensional model, the fringing of the fields near the sides of the vessel are neglected. The top and bottom of the vessel are assumed grounded or at 0 V. Air, with no volume charge, is assumed above the powder's surface. For situations involving multiple regions, for each region, Poisson's equation is solved to determine the general solution for the potentials. Then, boundary conditions are used to determine the constants of integration for each of the regions. If there are two distinct regions, there will be four constants of integration, and four independent boundary conditions are required to solve these constants of integration. If there are three regions, six boundary conditions are required. When there is a common interface between regions, boundary conditions are used to couple the solutions (if there is any coupling).

Poisson's equation is used to determine the general solution in the charged powder bed, while Laplace's equation is used to determine the general solution in the air, zero volume-charge region:

$$\nabla^2\Phi_p = \frac{d^2\Phi_p}{dy^2} = -\frac{\rho_V}{\varepsilon_r\varepsilon_o} = -\frac{Ky+F}{\varepsilon_r\varepsilon_o} \quad 0 \le y \le a, \quad \nabla^2\Phi_a = \frac{d^2\Phi_a}{dy^2} = 0 \quad a \le y \le a+b$$

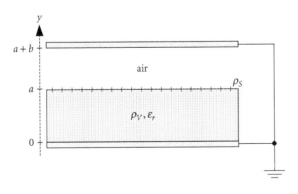

FIGURE 1.48 One-dimensional static model for charged powder inside a grounded storage vessel.

Both of these two expressions can be easily integrated twice (or the results in Table 1.4 used) to obtain

$$\Phi_p = -\frac{Ky^3}{6\varepsilon_r\varepsilon_o} - \frac{Fy^2}{2\varepsilon_r\varepsilon_o} + C_1 y + C_2, \quad \vec{E}_p = -\nabla\Phi_p = \left(\frac{Ky^2}{2\varepsilon_r\varepsilon_o} + \frac{Fy}{\varepsilon_r\varepsilon_o} - C_1\right)\hat{a}_y$$

$$\Phi_a = C_3 y + C_4, \quad \vec{E}_a = -\nabla\Phi_a = -C_3\hat{a}_y$$

There are four constants of integration, as expected for two second-order differential equations. Three of the boundary conditions directly involve the potential. The potential is 0 V at $y = 0$ and $y = a + b$:

$$\Phi_p(0) = -\frac{K(0)^3}{6\varepsilon_r\varepsilon_o} - \frac{F(0)^2}{2\varepsilon_r\varepsilon_o} + C_1(0) + C_2 = C_2 = 0$$

$$\Phi_a(a+b) = C_3(a+b) + C_4 = 0 \quad \Rightarrow C_4 = -C_3(a+b)$$

Also, the potential is continuous at the powder-air interface, even when a single layer of surface charge is present:

$$\Phi_p(a) = -\frac{Ka^3}{6\varepsilon_r\varepsilon_o} - \frac{Fa^2}{2\varepsilon_r\varepsilon_o} + C_1 a + C_2 = \Phi_a(a) = C_3 a + C_4$$

$$\Rightarrow C_1 = -C_3\frac{b}{a} + \frac{Ka^2}{6\varepsilon_r\varepsilon_o} + \frac{Fa}{2\varepsilon_r\varepsilon_o}$$

Since Poisson's equation only involves volume charge, the effect of the surface charge on the potential must be included via a boundary condition. The surface charge density affects the normal component of the $\vec{D}$ fields:

$$\hat{a}_n \cdot (\vec{D}_2 - \vec{D}_1) = \rho_S$$

where $\hat{a}_n$ is the unit vector normal to the interface directed from region 1 to region 2. If region 2 is selected as the air region above the powder, then $\hat{a}_n = \hat{a}_y$:

$$\hat{a}_y \cdot (\varepsilon_o\vec{E}_a - \varepsilon_r\varepsilon_o\vec{E}_p) = \hat{a}_y \cdot (-\varepsilon_o\nabla\Phi_a + \varepsilon_r\varepsilon_o\nabla\Phi_p) = \hat{a}_y \cdot \left(-\varepsilon_o\frac{d\Phi_a}{dy}\hat{a}_y + \varepsilon_r\varepsilon_o\frac{d\Phi_p}{dy}\hat{a}_y\right)$$

$$= -\varepsilon_o\frac{d\Phi_a}{dy} + \varepsilon_r\varepsilon_o\frac{d\Phi_p}{dy} = \rho_S \quad @ \; y = a$$

As a boundary condition, this relationship must be evaluated at the boundary $y = a$. Substituting the expressions for the potential,

$$\frac{d}{dy}(C_3 y + C_4) - \varepsilon_r\frac{d}{dy}\left(-\frac{Ky^3}{6\varepsilon_r\varepsilon_o} - \frac{Fy^2}{2\varepsilon_r\varepsilon_o} + C_1 y + C_2\right) = -\frac{\rho_S}{\varepsilon_o} \quad @ \; y = a$$

$$C_3 - \varepsilon_r\left(-\frac{Ky^2}{2\varepsilon_r\varepsilon_o} - \frac{Fy}{\varepsilon_r\varepsilon_o} - C_3\frac{b}{a} + \frac{Ka^2}{6\varepsilon_r\varepsilon_o} + \frac{Fa}{2\varepsilon_r\varepsilon_o}\right) = -\frac{\rho_S}{\varepsilon_o} \quad @ \; y = a$$

$$C_3 = -\frac{\rho_S}{\varepsilon_o}\left(\frac{a}{a+\varepsilon_r b}\right) - \frac{Ka^2}{3\varepsilon_o}\left(\frac{a}{a+\varepsilon_r b}\right) - \frac{Fa}{2\varepsilon_o}\left(\frac{a}{a+\varepsilon_r b}\right)$$

Since C_1 and C_4 are a function of C_3, all of the constants are now known.

$$a := 1 \quad b := 3 \quad \varepsilon_r := 4 \quad \varepsilon_o := 8.854 \cdot 10^{-12} \quad K := 10^{-7} \quad F := 10^{-9} \quad \rho_S := 10^{-7}$$

$$C_3 := \frac{-\rho_S}{\varepsilon_o} \cdot \left(\frac{a}{a + \varepsilon_r \cdot b} \right) - \frac{K \cdot a^2}{3 \cdot \varepsilon_o} \cdot \left(\frac{a}{a + \varepsilon_r \cdot b} \right) - \frac{F \cdot a}{2 \cdot \varepsilon_o} \cdot \frac{a}{a + \varepsilon_r \cdot b}$$

$$C_4 := -C_3 \cdot (a + b) \qquad C_1 := -C_3 \cdot \frac{b}{a} + \frac{K \cdot a^2}{6 \cdot \varepsilon_r \cdot \varepsilon_o} + \frac{F \cdot a}{2 \cdot \varepsilon_r \cdot \varepsilon_o}$$

$$y := 0, \frac{a + b}{1000} \,.. \, a + b$$

$$\Phi_p(y) := \frac{-K \cdot y^3}{6 \cdot \varepsilon_r \cdot \varepsilon_o} - \frac{F \cdot y^2}{2 \cdot \varepsilon_r \cdot \varepsilon_o} + C_1 \cdot y \qquad E_p(y) := \frac{K \cdot y^2}{2 \cdot \varepsilon_r \cdot \varepsilon_o} + \frac{F \cdot y}{\varepsilon_r \cdot \varepsilon_o} - C_1$$

$$\Phi_a(y) := C_3 \cdot y + C_4 \qquad E_a(y) := -C_3$$

$$\Phi(y) := \text{if}\big(y < a, \Phi_p(y), \Phi_a(y)\big) \qquad E(y) := \text{if}\big(y < a, E_p(y), E_a(y)\big)$$

Potential between Grounded Electrodes

Field between Grounded Electrodes

MATHCAD 1.10   Potential and electric field vs. position inside a grounded vessel containing charged powder.

As shown in Mathcad 1.10, the potential clearly drops to zero at both electrodes and is continuous at the interface ($y = 1$). The electric field is also discontinuous at the interface due to the change in the dielectric constant and the surface charge. Although not shown, the $\vec{D}$ field is also discontinuous at this interface because of the nonzero surface charge.

1.23   The Field from Corona

A corona generated current of I is measured between an electrified thin wire of radius a and a surrounding grounded, concentric cylindrical conductor of radius R. Assuming the mobility of the charged particles is b and the corona onset electric field at the wire's surface is E_c, determine the relationship for the electric field between the wire and outer grounded conductor. Under low-current and high-current conditions,

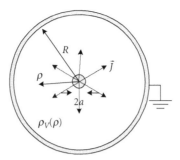

FIGURE 1.49 Generation of volume charge using a high-voltage wire surrounded by a cylindrical grounded conductor.

determine the voltage-current relationship. [Moore, '73; Cooperman, '60; Cooperman, '81; Feng; Thomson; Mizuno]

In electrostatic devices such as "smoke eaters," electrophotographic printers, charge neutralizers, and industrial precipitators, charged particles are intentionally generated via corona. With "smoke eaters," for example, the smoke particles are charged by passing the polluted air through an ionized or a charged region generated by corona discharge. The air is then cleaned by encouraging the charged smoke particles to pass and adhere to a grounded "collecting" electrode. (The scum on the grounded object is periodically cleaned.)

In practice, many different shapes are used for the high-voltage electrode generating the corona and the return or ground electrode. Unfortunately, explicit relationships for the electric field and potential between the electrodes are only available for certain highly symmetrical shapes. One of these shapes, the one discussed in this section, is a long wire surrounded by a concentric outer conductor as shown in Figure 1.49.

The motion of the charges between the conductors due to (1) the electric field generated by the charges themselves (sometimes referred to as the self or space charge field), (2) the electric field generated by the voltage difference between the electrodes (sometimes referred to as the external or applied field), or (3) other processes (such as wind) result in a current between the electrodes. For some special low-current and high-current situations, relatively simple voltage-current relationships exists for these electrostatic devices. Since most electrical engineers are accustomed to working with the *v-i* characteristics of electrical and electronic devices, energy will be devoted to obtaining a few *v-i* relationships.

If the charge density was given, then Poisson's equation could be used to determine the electric field between the inner wire and outer concentric conductor in the cylindrical coordinate system:

$$\frac{1}{\rho}\frac{\partial}{\partial\rho}\left(\rho\frac{\partial\Phi}{\partial\rho}\right)+\frac{1}{\rho^2}\frac{\partial^2\Phi}{\partial\phi^2}+\frac{\partial^2\Phi}{\partial z^2}=-\frac{\rho_V}{\varepsilon_o}$$

where ρ_V is the volume charge density between the conductors in free space (which, unfortunately, is described by the same basic variable ρ for the radial distance). The length of the conductors is assumed large in this analysis compared to the radius of the outer conductor so that end effects can be neglected. Furthermore, assuming the charge and electric field only vary in the radial direction, this partial differential equation reduces[19] to an ordinary differential equation:

$$\frac{1}{\rho}\frac{d}{d\rho}\left(\rho\frac{d\Phi}{d\rho}\right)=-\frac{\rho_V}{\varepsilon_o}$$

[19]Thank goodness!

The volume charge density in the radial direction can be related to the steady-state current, I, between the inner and outer conductors. The current is assumed entirely due to conduction. The volume current density, $\vec{J}$, is related to the electric field and charged particles (e.g., ions) flowing between the two conductors via the expression

$$J_\rho(\rho) = b|\rho_V(\rho)|E_\rho(\rho) \tag{1.126}$$

where b is the mobility of the charged particles, assumed to be constant. Notice that both the steady-state volume charge density and electric field can vary with the radial distance from the center of the wire. The volume current density is simply related to the steady-state current, I, by multiplying $J_\rho(\rho)$ by the area that the current density is passing through:

$$I = 2\pi\rho L J_\rho(\rho) = 2\pi\rho Lb|\rho_V(\rho)|E_\rho(\rho) \tag{1.127}$$

where L is the length of the conductors. Solving for the volume charge density, assuming that the volume charge density is positive (i.e., positive unipolar charge), and then substituting into Poisson's equation,

$$\frac{1}{\rho}\frac{d}{d\rho}\left(\rho\frac{d\Phi}{d\rho}\right) = -\frac{1}{\varepsilon_o}\left(\frac{I}{2\pi\rho LbE_\rho(\rho)}\right) \tag{1.128}$$

To eliminate the potential function, substitute $E_\rho(\rho) = -d\Phi/d\rho$ into Equation (1.128):

$$\frac{1}{\rho}\frac{d}{d\rho}[-\rho E_\rho(\rho)] = -\frac{1}{\varepsilon_o}\left(\frac{I}{2\pi\rho LbE_\rho(\rho)}\right)$$

Simplifying,

$$\frac{d}{d\rho}[\rho E_\rho(\rho)] = \frac{I}{2\pi\varepsilon_o LbE_\rho(\rho)}$$

One way of determining the solution of this differential equation is to let $m = \rho E_\rho(\rho)$:

$$\frac{dm}{d\rho} = \frac{I\rho}{2\pi\varepsilon_o Lbm}$$

Then, integrating in the standard way

$$\int_{m(a)}^{m(\rho)} mdm = \int_a^\rho \frac{I\rho d\rho}{2\pi\varepsilon_o Lb} \Rightarrow \frac{1}{2}[m^2(\rho) - m^2(a)] = \frac{I}{4\pi\varepsilon_o Lb}(\rho^2 - a^2)$$

and solving for m,

$$m(\rho) = \sqrt{\frac{I}{2\pi\varepsilon_o Lb}(\rho^2 - a^2) + m^2(a)}$$

In terms of the electric field,

$$E_\rho(\rho) = \sqrt{\frac{I}{2\pi\varepsilon_o Lb}\left[1-\left(\frac{a}{\rho}\right)^2\right] + \frac{m^2(a)}{\rho^2}}$$

Since $m(a) = aE_\rho(a) = aE_c$,

$$E_\rho(\rho) = \sqrt{\frac{I}{2\pi\varepsilon_o Lb} + \left(\frac{a}{\rho}\right)^2\left(E_c^2 - \frac{I}{2\pi\varepsilon_o Lb}\right)} \qquad (1.129)$$

The fields generated by the electrodes (i.e., the inner wire and outer conductor) and space charge are included in this result. Not too close to the wire and for larger currents, the field is approximately uniform and independent of position:

$$E_\rho(\rho) = \sqrt{\frac{I}{2\pi\varepsilon_o Lb} - \left(\frac{a}{\rho}\right)^2\frac{I}{2\pi\varepsilon_o Lb} + \left(\frac{a}{\rho}\right)^2 E_c^2}$$

$$\approx \sqrt{\frac{I}{2\pi\varepsilon_o Lb}} \quad \text{if } \frac{I}{2\pi\varepsilon_o Lb} \gg \left(\frac{a}{\rho}\right)^2\frac{I}{2\pi\varepsilon_o Lb} \quad \text{and} \quad \frac{I}{2\pi\varepsilon_o Lb} \gg \left(\frac{a}{\rho}\right)^2 E_c^2 \qquad (1.130)$$

$$= \sqrt{\frac{I}{2\pi\varepsilon_o Lb}} \quad \text{if } \rho \gg a \text{ and } I \gg \left(\frac{a}{\rho}\right)^2 E_c^2 2\pi\varepsilon_o Lb$$

The electric field is plotted vs. ρ in Mathcad 1.11. The radius of the inner wire is 54.5 mils (≈ 1.38 mm), and the radius of the concentric outer conductor is 4.5 in (≈ 114 mm). The length of the conductors is 12 ft. Equation (1.74) was used, with a change in notation, for the corona onset voltage for two concentric cylinders:

$$V_c = 3.08a\left(1 + \frac{0.034}{\sqrt{a}}\right)\ln\left(\frac{R}{a}\right) \times 10^6 \text{ V}$$

As seen, for larger currents far from the wire, the electric field is indeed approximately constant. As expected, the electric field is greatest near the high-curvature thin wire. The variable E_c is defined later in terms of the corona onset voltage.

To obtain the relationship between the voltage of the center wire, relative to the outer conductor, and the current, it is first necessary to integrate Equation (1.129):

$$\frac{d\Phi}{d\rho} = -\sqrt{\frac{I}{2\pi\varepsilon_o Lb} + \left(\frac{a}{\rho}\right)^2\left(E_c^2 - \frac{I}{2\pi\varepsilon_o Lb}\right)}$$

$$\int_{V_o}^{0} d\Phi = -\int_a^R \sqrt{\frac{I}{2\pi\varepsilon_o Lb} + \left(\frac{a}{\rho}\right)^2\left(E_c^2 - \frac{I}{2\pi\varepsilon_o Lb}\right)}\, d\rho$$

$$V_o = \int_a^R \sqrt{\frac{I}{2\pi\varepsilon_o Lb} + \left(\frac{a}{\rho}\right)^2\left(E_c^2 - \frac{I}{2\pi\varepsilon_o Lb}\right)}\, d\rho \qquad (1.131)$$

$\varepsilon_o := 8.854 \cdot 10^{-12}$ $b := 2.1 \cdot 10^{-4}$ $a := 54.5 \cdot 2.54 \cdot 10^{-5}$ $R := 4.5 \cdot 2.54 \cdot 10^{-2}$ $L := 12 \cdot 12 \cdot 2.54 \cdot 10^{-2}$

$V_c := 3.08 \cdot a \cdot \left(1 + \dfrac{0.034}{\sqrt{a}}\right) \cdot \ln\left(\dfrac{R}{a}\right) \cdot 10^6$ $V_c = 3.601 \times 10^4$ $a = 1.384 \times 10^{-3}$ $R = 0.114$

$\rho := a, 1.1 \cdot a .. R$

$$E_\rho(\rho, I) := \sqrt{\frac{I}{2 \cdot \pi \cdot \varepsilon_o \cdot L \cdot b} + \left(\frac{a}{\rho}\right)^2 \cdot \left[\left(\frac{V_c}{a \cdot \ln\left(\frac{R}{a}\right)}\right)^2 - \frac{I}{2 \cdot \pi \cdot \varepsilon_o \cdot L \cdot b}\right]}$$

$E_\rho(\rho, 10^{-2})$

$E_\rho(\rho, 10^{-5})$ (V/m)

$E_\rho(\rho, 10^{-5}) - E_\rho(\rho, 10^{-8})$

MATHCAD 1.11   Electric field vs. radial distance for the situation in Figure 1.49.

where R is the radius of the outer conductor and V_o is the voltage of the wire relative to the outer conductor. One voltage vs. current curve is provided in Mathcad 1.12. At low current levels, a relatively simple relationship can be obtained for the voltage between the wire and outer conductor:

$$V_o \approx \int_a^R \sqrt{\frac{I}{2\pi\varepsilon_o Lb} + \left(\frac{a}{\rho}\right)^2 E_c^2}\, d\rho = \int_a^R \frac{a}{\rho} E_c \sqrt{\frac{I}{2\pi\varepsilon_o Lb}\left(\frac{\rho}{a}\right)^2 \frac{1}{E_c^2} + 1}\, d\rho \quad \text{if } E_c^2 \gg \frac{I}{2\pi\varepsilon_o Lb}$$

$$\approx \int_a^R \frac{a}{\rho} E_c \left[1 + \frac{1}{2}\frac{I}{2\pi\varepsilon_o Lb}\left(\frac{\rho}{a}\right)^2 \frac{1}{E_c^2}\right] d\rho \quad \text{if } \frac{I}{2\pi\varepsilon_o Lb}\left(\frac{\rho}{a}\right)^2 \frac{1}{E_c^2} \ll 1$$

The binomial expansion, $(1+x)^{1/2} \approx 1 + x/2$ if $|x| \ll 1$, was used for this last approximation. Integrating,

$$V_o = \int_a^R \left(\frac{aE_c}{\rho} + \frac{I\rho}{4\pi\varepsilon_o LbE_c a}\right) d\rho \quad \text{if } \left(\frac{a}{\rho}\right)^2 E_c^2 \gg \frac{I}{2\pi\varepsilon_o Lb},\ R > a$$

$$= \left(aE_c \ln\rho + \frac{I\rho^2}{8\pi\varepsilon_o LbE_c a}\right)\Bigg|_a^R \tag{1.132}$$

$$= aE_c \ln\left(\frac{R}{a}\right) + \frac{I(R^2 - a^2)}{8\pi\varepsilon_o LbE_c a} \quad \text{if } \left(\frac{a}{R}\right)^2 E_c^2 \gg \frac{I}{2\pi\varepsilon_o Lb}$$

$\varepsilon_o := 8.854 \cdot 10^{-12}$ $b := 2.1 \cdot 10^{-4}$ $a := 54.5 \cdot 2.54 \cdot 10^{-5}$ $R := 4.5 \cdot 2.54 \cdot 10^{-2}$ $L := 12 \cdot 12 \cdot 2.54 \cdot 10^{-2}$

$V_c := 3.08 \cdot a \cdot \left(1 + \dfrac{0.034}{\sqrt{a}}\right) \cdot \ln\left(\dfrac{R}{a}\right) \cdot 10^6$ $V_c = 3.601 \times 10^4$

$I := 1 \cdot 10^{-7}, \; 2 \cdot 10^{-7} .. \; 1 \cdot 10^{-4}$

$$V_o(I) := \int_a^R \sqrt{\frac{I}{2 \cdot \pi \cdot \varepsilon_o \cdot L \cdot b} + \left(\frac{a}{\rho}\right)^2 \cdot \left[\left(\frac{V_c}{a \cdot \ln\left(\frac{R}{a}\right)}\right)^2 - \frac{I}{2 \cdot \pi \cdot \varepsilon_o \cdot L \cdot b}\right]} \; d\rho$$

MATHCAD 1.12   Voltage-current curve for the situation in Figure 1.49.

The critical voltage for the onset of corona near the surface of the inner wire is related to the electric field near the inner wire. It is obtained from Equation (1.87) with a change in notation:

$$E_c = \frac{V_c}{a \ln\left(\dfrac{R}{a}\right)} \tag{1.133}$$

Equation (1.133) is the standard electric field near a wire's surface assuming zero volume charge and a concentric, grounded outer conductor. Before the corona begins, there is no charge between the wire and outer conductor. Substituting into Equation (1.132) and rearranging,

$$I = \frac{8\pi\varepsilon_o Lb}{(R^2 - a^2)\ln\left(\dfrac{R}{a}\right)} V_c (V_o - V_c) \quad \text{if} \; \left(\frac{a}{R}\right)^2 \left[\frac{V_c}{a\ln\left(\dfrac{R}{a}\right)}\right]^2 \gg \frac{I}{2\pi\varepsilon_o Lb} \tag{1.134}$$

Adding the additional reasonable assumption that $R \gg a$,

$$I \approx \frac{8\pi\varepsilon_o Lb}{R^2 \ln\left(\dfrac{R}{a}\right)} V_c (V_o - V_c) \quad \text{if} \quad \left[\frac{V_c}{R\ln\left(\dfrac{R}{a}\right)}\right]^2 \gg \frac{I}{2\pi\varepsilon_o Lb}, R \gg a \tag{1.135}$$

where V_o is the applied voltage (relative to the outer conductor) to the thin inner wire. The commonly used low-current approximation (from Townsend and Warburg) is

$$I \approx \frac{8\pi\varepsilon_o Lb}{R^2 \ln\left(\dfrac{R}{a}\right)} V_o (V_o - V_c) \tag{1.136}$$

The voltage at the inner wire is very close to the onset voltage, V_c, for small currents. As seen in Mathcad 1.13, approximations (1.132) and (1.136) are excellent for small currents. For large values of

$\varepsilon_o := 8.854 \cdot 10^{-12} \qquad b := 2.1 \cdot 10^{-4} \qquad a := 54.5 \cdot 2.54 \cdot 10^{-5} \qquad R := 4.5 \cdot 2.54 \cdot 10^{-2} \qquad L := 12 \cdot 12 \cdot 2.54 \cdot 10^{-2}$

$V_c := 3.08 \cdot a \cdot \left(1 + \dfrac{0.034}{\sqrt{a}}\right) \cdot \ln\left(\dfrac{R}{a}\right) \cdot 10^6 \qquad V_c = 3.601 \times 10^4 \qquad TOL := 10^{-12} \qquad V_{o2} := 3 \cdot 10^6$

$I := 1 \cdot 10^{-7}, 2 \cdot 10^{-7} .. 1 \cdot 10^{-4} \qquad\qquad \dfrac{10^{-5}}{2 \cdot \pi \cdot \varepsilon_o \cdot L \cdot b} = 2.34 \times 10^8 \qquad \left(\dfrac{V_c}{R \cdot \ln\left(\dfrac{R}{a}\right)}\right)^2 = 5.097 \times 10^9$

$V_o(I) := \displaystyle\int_a^R \sqrt{\frac{I}{2 \cdot \pi \cdot \varepsilon_o \cdot L \cdot b} + \left(\frac{a}{\rho}\right)^2 \cdot \left[\left(\frac{V_c}{a \cdot \ln\left(\dfrac{R}{a}\right)}\right)^2 - \frac{I}{2 \cdot \pi \cdot \varepsilon_o \cdot L \cdot b}\right]} \, d\rho$

$V_{o1}(I) := V_c + \dfrac{I \cdot (R^2 - a^2)}{8 \cdot \pi \cdot \varepsilon_o \cdot L \cdot b \cdot \dfrac{V_c}{\ln\left(\dfrac{R}{a}\right)}} \qquad\qquad V_{o2}(I) := \text{root}\left[\dfrac{8 \cdot \pi \cdot \varepsilon_o \cdot b}{R^2 \cdot \ln\left(\dfrac{R}{a}\right)} \cdot V_{o2} \cdot (V_{o2} - V_c) - \dfrac{I}{L}, V_{o2}\right]$

Percent Error with V-I Approximations

$\left|\dfrac{V_o(I) - V_{o1}(I)}{V_o(I)}\right| \cdot 100$

$\left|\dfrac{V_o(I) - V_{o2}(I)}{V_o(I)}\right| \cdot 100$

MATHCAD 1.13 Percent error between the exact voltage-current expression and the approximations given in Equations (1.132) and (1.136).

current and $R \gg a$, an approximate relationship between the voltage and current is quickly obtained from (1.131):

$$V_o \approx \sqrt{\frac{I}{2\pi\varepsilon_o Lb}} \int_a^R \sqrt{1-\left(\frac{a}{\rho}\right)^2}\, d\rho \quad \text{if } \frac{I}{2\pi\varepsilon_o Lb} \gg E_c^2,\, R \gg a$$

$$V_o \approx R\sqrt{\frac{I}{2\pi\varepsilon_o Lb}} \quad \text{if } \frac{I}{2\pi\varepsilon_o Lb} \gg \left[\frac{V_c}{a\ln\left(\frac{R}{a}\right)}\right]^2,\, R \gg a \tag{1.137}$$

$$I = \frac{2\pi\varepsilon_o Lb}{R^2}V_o^2 \quad \text{if } \frac{I}{2\pi\varepsilon_o Lb} \gg \left[\frac{V_c}{a\ln\left(\frac{R}{a}\right)}\right]^2,\, R \gg a \tag{1.138}$$

To understand the source of the approximation for the integral of $\sqrt{1-(a/\rho)^2}$, realize that this square root function is equal to zero at $\rho = a$ and increases to one for $\rho = R \gg a$. Thus, ignoring the nonzero rise time in the square root function, the area is about $1 \times (R-a) \approx R$.

The expressions for the electric field between the wire and outer conductor can be improved by allowing for stationary or suspended volume charge between the conductors. Also, although this analysis seems to indicate that the v-i relationship is not a function of the sign of the applied voltage, actually there are slight differences as with the corona onset voltage.

For other electrode configurations, such as the N equally spaced parallel wires centered between two flat, plane grounded electrodes shown in Figure 1.50, the mathematics is more challenging. Approximations are available for this configuration based on the concentric cylindrical analysis. For small current levels, which are common for precipitators,

$$I = \frac{4\pi\varepsilon_o Lb}{s^2\ln\left(\frac{d}{a}\right)}V_o(V_o - V_c), \quad J_{avg} = \frac{\pi\varepsilon_o b}{cs^2\ln\left(\frac{d}{a}\right)}V_o(V_o - V_c) \tag{1.139}$$

where J_{avg} is the average current density at the plates and

$$V_c = aE_c\ln\left(\frac{d}{a}\right) \tag{1.140}$$

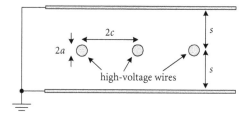

FIGURE 1.50   "Duct" precipitator.

The effective plate area per wire is $A_{eff} = 2(2c)L$ since there are two plates. The variable d is a characteristic length similar to R (or an equivalent cylinder radius) given by

$$d = \begin{cases} \dfrac{4s}{\pi} & \text{if } \dfrac{s}{c} \leq 0.6 \\[2mm] \approx c\left[1.98 - 4.01\dfrac{s}{c} + 3.35\left(\dfrac{s}{c}\right)^2\right] & \text{if } 0.6 < \dfrac{s}{c} < 2.0 \\[2mm] \dfrac{c}{\pi}e^{\frac{\pi s}{2c}} & \text{if } \dfrac{s}{c} \geq 2.0 \end{cases} \tag{1.141}$$

or, in terms of the product (*not* sum) function,

$$d = \frac{4s}{\pi} \prod_{m=1}^{\infty} \frac{\cosh\left(\dfrac{m\pi c}{s}\right)+1}{\cosh\left(\dfrac{m\pi c}{s}\right)-1} \tag{1.142}$$

$\varepsilon_0 := 8.854 \cdot 10^{-12}$ $b := 2.1 \cdot 10^{-4}$ $a := 54.5 \cdot 2.54 \cdot 10^{-5}$ $s := 4.5 \cdot 2.54 \cdot 10^{-2}$ $L := 12 \cdot 2.54 \cdot 10^{-2}$

$E_c := 3.0 \cdot \left(1 + \dfrac{0.030}{\sqrt{a}}\right) \cdot 10^6$ $E_c = 5.419 \times 10^6$ $TOL := 10^{-12}$ $V_0 := 5 \cdot 10^5$

$c := \dfrac{s}{10}, \dfrac{s}{9} .. 4 \cdot s$

$d(c) := \text{if}\left[\dfrac{s}{c} \leq 0.6, \dfrac{4 \cdot s}{\pi}, \text{if}\left[\dfrac{s}{c} \geq 2, \dfrac{c}{\pi} \cdot e^{\frac{\pi \cdot s}{2 \cdot c}}, c \cdot \left[1.98 - 4.01 \cdot \dfrac{s}{c} + 3.35 \cdot \left(\dfrac{s}{c}\right)^2\right]\right]\right]$

$I(V_0, c) := \dfrac{4 \cdot \pi \cdot \varepsilon_0 \cdot b \cdot L}{s^2 \cdot \ln\left(\frac{d(c)}{a}\right)} \cdot V_0 \cdot \left(V_0 - a \cdot E_c \cdot \ln\left(\dfrac{d(c)}{a}\right)\right)$

MATHCAD 1.14 Current vs. wire-to-wire spacing for three applied voltages.

For large current densities where the field due to the space charges between the electrodes is large compared to the applied field from the electrodes, the average current density is

$$J_{avg} = \frac{9\varepsilon_o b}{8s^3}(V_o - V_c)^2 \tag{1.143}$$

For this "duct" precipitator and large values of the wire-to-wire spacing, $2c$, the current is essentially independent of the wire spacing:

$$I = \frac{4\pi\varepsilon_o Lb}{s^2 \ln\left(\dfrac{4s}{\pi a}\right)}V_o\left[V_o - aE_c \ln\left(\frac{4s}{\pi a}\right)\right] \quad \text{if } c > 1.7s \tag{1.144}$$

For small values of the wire-to-wire spacing, the current is small since d is large. As seen in Mathcad 1.14, there is also a maximum in the current vs. spacing relationship.

In the design of electrostatic precipitators, there are other parameters that are considered. According to the literature, a negative voltage for the inner wire(s) provides a greater margin of operating voltage while a positive voltage generates a lower ozone level. The collection efficiency is a function of the air velocity, area of the collecting plates, and geometry of the electrodes.

2

Principles and Applications

The buildup of charge on a body changes its potential and the electric field from it. If the charge continues to build up, breakdown of the medium near the body eventually occurs. This breakdown is a source of conducted and radiated emissions. This electrostatic discharge can also be the ignition source for a hazardous explosion.

2.1 What Is ESD?

Briefly, what is ESD, and why is it important?

An electrostatic discharge (ESD) is the loss of charge that can occur when a sufficiently high quantity of charge is accumulated on an object. For an ESD event to occur, charge must be generated, accumulated, and discharged. A net charge generates an electric field. This electric field, when sufficiently intense, can cause a discharge or breakdown of the medium surrounding the charged object.

ESD can be viewed as a bantam lightning discharge. When the electric field becomes sufficiently intense between two objects, whether conducting or insulating, the insulation between the objects breaks down. To control ESD, the charge accumulation must be controlled.

In electronic devices, as the distance between conductors or devices decreases, the sensitivity of the device to ESD increases. All devices have a maximum voltage threshold. When this threshold is exceeded, either immediate or latent damage occurs. Electrostatic discharge can also be a source of danger to personnel: the energy in an electrostatic discharge can ignite vapors, gases, and liquids.

2.2 Methods of Charging

Describe the major methods of charging an object and provide examples of each. [Moore, '73]

Since charge is responsible for ESD, it is essential to know how charge can build up on an object. Although rarely is a single unit of charge of importance, the smallest fundamental unit of charge has a magnitude of 0.16×10^{-18} C. An object can accumulate or be charged by positive or negative integer multiples of this smallest unit of charge. The accumulation of some net charge on an object is frequently referred to as static electricity, even if the charge is in motion.

There are a number of methods of charging an object. Probably the most familiar way of charging is through a battery (or other voltage source). By connecting a battery between two objects, charge is transferred between them. The battery supplies the energy required to transfer this charge. It does not generate charge. The quantity of charge transferred is a function of the capacitance, C, between and voltage, V, across the two objects. Referring to Figure 2.1, assuming that both objects were initially charge neutral, the charge transferred between the objects is

$$Q = CV = C(10)$$

When a bird sits on a high-voltage ac line, the bird is charged to the voltage of the line. Since the line voltage is varying in a sinusoidal manner, the charge on the bird is also varying in a sinusoidal manner.

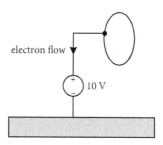

FIGURE 2.1   Charging a neutral object via a battery.

In many electrostatic painting operations, the paint is charged by direct contact with a high-voltage electrode.

If two objects, one of which is charged, are brought into electrical contact, charge is transferred between the objects until the potential difference across them is zero.[1] The charge transferred is a function of the geometry of the two objects. As a general guideline, the larger of the two objects will carry the greater charge. Although a conducting path is shown in Figure 2.2 connecting the two objects, direct electrical contact between the two objects through physical contact or via a conducting wire is not required for charge transfer. All real insulators, including air, have a finite resistance. Eventually, charge will transfer even between two "isolated" objects. The time required for this charge transfer to reach a steady-state value is a function of the time constant

$$\tau = \frac{\varepsilon_r \varepsilon_o}{\sigma} \tag{2.1}$$

where $\varepsilon_o = 8.854 \times 10^{-12}$ F/m, ε_r is the relative permittivity, and σ is the conductivity of the medium, assumed homogeneous or the same everywhere, between the objects.

The oldest methods of charging an object are probably through contact and tribo (frictional) charging. Usually, the term tribocharging is used to describe both contact and frictional charging since in practice the effects are difficult to separate. When two initially charge-neutral objects are touched and then separated, one of the objects becomes positively charged and the other becomes negatively charged as illustrated in Figure 2.3. This charge transfer mechanism is often demonstrated in elementary schools by rubbing a plastic rod against a wool cloth. The plastic rod and cloth both become charged but of opposite signs. There are several explanations given for why this transfer occurs, such as the actual transfer of material and ions between the two objects and the transfer of electrons, via atomic attractions, between

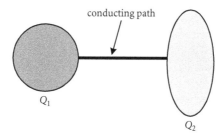

FIGURE 2.2   Charge transfer between two objects via conduction.

---

[1] Actually, when two different metallic materials are brought into contact, a voltage difference can still exist between them referred to as the contact potential. This potential is due to the different work functions for the two metals. The work function is the energy required to remove a free electron from the metal.

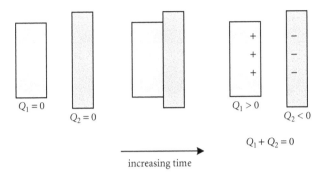

FIGURE 2.3   Charging neutral objects via tribocharging.

the two objects. With tribocharging, the degree of charge transfer *generally* increases with contact area, contact roughness, and contact velocity. It is not necessary that the objects be different materials. A table referred to as the triboelectric series attempts to indicate the qualitative strength and sign of the charge transfer between various materials due to contact charging and/or tribocharging. Unfortunately, the results given in the series are not always reproducible because of the sensitivity of the charge transfer to contaminates and other factors. Most materials, including metals and dirt particles in the air, can be tribocharged. In electrostatic powder coating operations, powder is blown through plastic tubes. This powder becomes charged due to tribocharging with the tube surface. During grinding operations in food manufacturing, the resultant particles can be tribocharged. When insulating fuel oil is pumped through pipes, the oil can be tribocharged. Even coaxial cables can be tribocharged when flexed. With tribocharging, no external electric field or external voltage source is required.

Probably the most "devious" method of charging is by induction. A conducting object when placed in an electric field will become polarized (the charges on the object will separate). If the object is initially isolated and charge neutral, it will remain charge neutral in the polarizing electric field. However, if it makes direct contact with certain metallic bodies, it can become charged. The metallic body could be another charged object, a ground plane, or even a person. The process of charging by induction is illustrated in Figure 2.4. To charge by induction, an external electric field should be present, and *the object* should have a small time constant relative to the period of the electric field and the time of contact with the other metallic body. Physical contact between the source of the electric field and the object is not required with induction charging. Induction charging can also occur when two isolated charge-neutral metallic objects in a polarizing electric field come in contact. After contact, the two objects appear

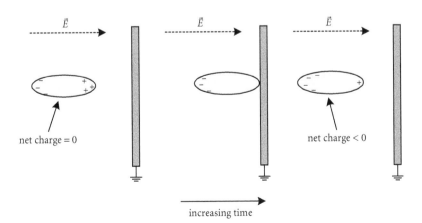

FIGURE 2.4   Charging an object via induction and contact with a grounded metal plate.

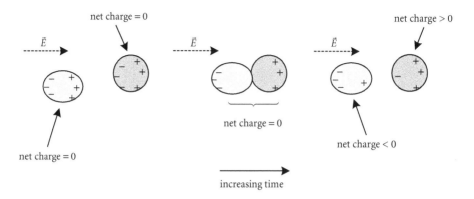

FIGURE 2.5   Charging an object via induction and contact with a floating metallic object.

as one to the electric field, and this one "larger" object becomes polarized. If the objects are then separated, they both become charged as is illustrated in Figure 2.5. This mechanism is used to charge paint in some electrostatic paint operations by exposing a liquid jet to an electric field and then separating a droplet from the jet. Another common scenario involving induction charging is an individual, with insulating shoes, walking through an electric field (generated, for example, by a negatively charged moving belt). The charge distribution on the individual becomes polarized when in the field (e.g., positive charges on one hand near the negatively charged belt while negative charges on the other hand far from the belt). If the individual then touches a grounded object, one sign of charge will dissipate to the grounded object leaving the individual with a net charge.

Another method of charging is corona charging. When a sufficiently high voltage is applied across two electrodes, eventually the medium between the electrodes breaks down. When one of the electrodes is very sharp (i.e., has a small radius of curvature), the nonuniform electric field between the electrodes is the greatest near the sharp electrode. For a sufficiently high voltage across the electrodes (but not too great as to cause a complete breakdown across the entire gap between the electrodes), the medium near the sharp electrode breaks down. This local breakdown generates a glow near the sharp electrode and is referred to as corona. The air is ionized near the sharp electrode resulting in both positively and negatively charged ions. Farther from the sharp electrode, the field is less nonuniform, and this region contains mainly charges of the same polarity as the sharp electrode. These charges can be used to charge powders, surfaces, or fetid gas molecules. Corona charging is used to neutralize charged surfaces that may be an ESD hazard. In electrostatic "smoke eaters," smoke particles come in contact with charged particles that are generated by corona. The smoke particles, which are then charged, are attracted to and make contact with a grounded plate (that is periodically cleaned). Antennas on cold, dry, and windy days can be charged by ions in the surrounding air. Figure 2.6 shows a positive dc voltage applied to a sharp electrode. It is generating positive ions that are deposited on the surface of an insulator. Negative dc voltages can be used to generate negative ions, and ac voltages can be used to generate both polarities of charge.

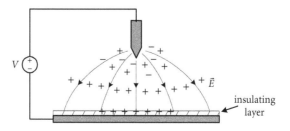

FIGURE 2.6   Charging an insulating layer via corona.

Furthermore, even if the sharp electrode is grounded, the electric field from other sources can be distorted by this sharp electrode resulting in corona. External air fans can be used to direct the charged ions to distant or hard-to-reach locations. In the figure, the charge, as it builds up on the insulating surface, will generate its own electric field. This electric field can break down the medium around it (including the insulating layer) resulting in back ionization. This field and the ions generated by the back ionization will limit the charge buildup on the insulator.

Although the common or traditional ways of charging have been discussed in this section, there are other methods. For example, a material can be charged through freezing, ion and electron beams, thermionic emission (e.g., via hot spots), photoelectric effect, radioactive decay, field emission, and mechanical fracture.

2.3 Triboelectric Series

Provide a list of materials commonly seen in triboelectric series. Can this list be used with a high degree of confidence? [Cross; Harper, '67; Krevelen; Jonassen; Moore, '73; Sclater; Pratt]

When two materials are placed in contact or "rubbed" together, a common question in physics courses is "which material is positive and which is negative after the materials are separated?" The triboelectric series is occasionally referenced to obtain this "answer." This series (in theory) enables the polarity or sign of the charge on the two objects to be determined after their separation.

Some researchers indicate reasonable results when using various triboelectric series tables, while others prefer not to use them even as a guide. Because the processes that occur when two objects are rubbed together are complicated, some workers only use these tables with contact charging as opposed to frictional charging. The materials on the top end of Table 2.1 will lose or donate electrons to acceptor materials lower on the list. For example, if initially neutral (zero net charge) rabbit fur and hard rubber are contacted and then pulled apart, Series 1 indicates that the rabbit fur will be positively charged while the hard rubber will be negatively charged. The greater the separation between two materials on the series, the greater the magnitude of the transferred charge. The triboelectric series from several different sources are given in Table 2.1. The series are listed in no particular order.

The triboelectric series should not be used with a high degree of confidence since the order of the materials in the series can differ. The position of a material can easily change based on factors such as the purity of the material (i.e., type and quantity of contaminants), surface roughness, humidity, and whether the materials are touched or rubbed together.

As previously stated, charge transfer can also occur between two like materials when they are rubbed together. For example, when a plastic sheet is unrolled, the sheet is charged. This frictional charging is a complex process that depends on the energy used in the rubbing, but it is difficult to predict the quantity of charge transferred during rubbing. Often in practice, charge transfer due to contact charging and tribocharging cannot be distinguished from each other.

Contact and tribocharging are most apparent in nonconductors. With conductors, the charge can quickly dissipate or decay to other objects (e.g., a ground) or along the surface of the same conducting object. When one of the materials involved in the tribocharging is an insulator or a poor conductor, the charge takes much longer to dissipate or decay. This is why many problems related to electrostatic discharge occur with insulators or poor conductors. When two neutral conductive materials are placed in contact, there will be a difference in potential between them, referred to as the contact potential. This voltage is usually between a few tenths of a volt and two volts for conductive materials. Generally, little charge is transferred between two conductors after they are separated unless the separation is very fast. (The time constant for charge in conductors is very small. The charge will want to return to its natural, lower energy state.) Interestingly, metal powders when strongly blown on a metal surface can be charged. Charge can easily transfer between conductors and nonconductors. For example, both a plastic sheet and metallic roller can be charged when the sheet passes over the roller.

It is sometimes helpful, especially in consulting work related to electrostatic hazards, to know the typical range of charge-to-mass ratio after tribocharging has occurred. For medium-resistivity organic

TABLE 2.1   Various Triboelectric Series

The left margin of the table is labeled, from top to bottom: *donor of electrons / acquires positive charge* (+), with a downward arrow to *acceptor of electrons / acquires negative charge* (−).

Series 1	Series 2	Series 3	Series 4	Series 5	Series 6	Series 7
Human Hands						
Asbestos						
Rabbit's Fur			Lucite 2041			
Acetate			Dapon			Nylon 6.6
Glass			Lexan 105	Plexiglass		Cellulose
Mica			Formvar	Bakelite	Rabbit's Fur	Cellulose
Human Hair		Polyox	Estane	Cellulose	Lucite	Acetate
Nylon	Wool	Polyethylene	DuPont 49000	Acetate	Bakelite	Polymethyl
Wool	Nylon	Amine	Durex	Glass	Cellulose	Methacrylate
Fur	Viscose	Gelatin	Ethocel 10	Quartz	Acetate	Polyacetal
Lead	Cotton	Vinac	Polystyrene 8X	Nylon	Glass	Polyethylene
Silk	Silk	Lucite 44	Epolene C	Wool	Quartz	Terephthalate
Aluminum	Acetate	Lucite 42	Polysulphone	Silk	Mica	Polyacrylonitrile
Paper	Rayon	Acryloid A101	P-3500	Cotton	Wool	Polyvinyl
Cotton	Lucite,	Zelec DX	Hypalon 30	Paper	Cat's Fur	Chloride
Steel	Perspex	Polyacrylamide	Cyclolac H-1000	Amber	Silk	Polybisphenol
Wood	Polyvinyl	Cellulose	Uncoated Iron	Resins	Cotton	Carbonate
Amber	Alcohol	Acetate/	Cellulose Acetate	Metals	Wood	Polychloroether
Hard	Dacron	Butyrate	Butyrate	Rubber	Amber	Penton
Rubber	Orlon	Acysol	Epon	Acetate	Resins	Polyvinylidine
Mylar	PVC	Carbopol	828/V125	Rayon	Metals	Chloride
Nickel, Copper	Dynel	Polyethylene	Polysulphone	Dacron	Polystyrene	Poly2.6-
Silver	Velon	terephthalate	P-1700	Orlon	Polyethylene	Dimethyl
UV Resist	Polyethylene	Polyvinyl	Cellulose Nitrate	Polystyrene	Teflon	Polyphenylene
Brass, Stainless	Teflon	Butyral	Kynar	Teflon	Cellulose	Oxide
Steel		Polyethylene		Cellulose	Nitrate	Polystyrene
Gold				Nitrate		Polyethylene
Polyester				PVC		Polypropylene
Celluloid						Polytetrafluoro-
Styrene						ethylene
Acrylic						

powders around $\rho \approx 10^{12}$ Ω-m, experimental data is available. The charge-to-mass ratios for several operations on these powders are provided in Table 2.2. For high-resistivity powders, the charge-to-mass ratio is a strong function of the mechanical energy used in changing the form of the powder. The energy required to sift particles is generally far less than grinding the particles. Grinding particles increases the surface area of the material, by generating more particles, which is an important factor in tribocharging.[2]

TABLE 2.2   Charge-to-Mass Ratios for Specific
Mechanical Operations [Gibson]

Operation	C/kg
Sieving	10^{-11} to 10^{-9}
Pouring	10^{-9} to 10^{-7}
Scroll feed transfer	10^{-8} to 10^{-6}
Grinding	10^{-7} to 10^{-6}
Micronizing	10^{-7} to 10^{-4}
Pneumatic transfer	10^{-6} to 10^{-4}

---

[2] For example, the surface area of one sphere of radius a is $4\pi a^2$. If this sphere is split in half, the total surface area of the two hemispheres is $4\pi a^2 + 2\pi a^2$.

2.4　Microphony

What is microphony, and how can it be reduced? [Perls; Ratz; Donovan; Fowler; Wood; Keithley]

With such dramatic improvements in the quality of amplifiers, transducers, and other electronic products, electrical interference is an important factor in limiting the signal-to-noise ratio of a system. When cables are involved, the susceptibility of the cable to external sources is important, but it can be reduced in many cases to an acceptable level by intelligently selecting the cable type. Unfortunately, the cable itself can be an inherent source of noise or interference. When a cable is flexed, compressed, twisted, or mechanically stressed in other ways, electrical noise can be produced. Even hydrodynamic forces on submarine cables can generate annoying noise. The noise that is generated by this charging can be critical in low-voltage signal applications since it can decrease the signal-to-noise ratio and affect the dynamic range of a system. This mechanically induced noise is sometimes referred to as microphony or handling noise.

There are three major sources of mechanically induced noise in a cable:

1. piezoelectric effects from the dielectric located between the conductors in the cable
2. changes in the cable capacitance
3. tribocharging between the dielectric and conductors.

The literature indicates that piezoelectric effects are important after the tribocharging and capacitance change effects have been reduced to a low level. Careful control of the extrusion of the dielectric appears to reduce the noise due to piezoelectricity. Capacitance variations can occur when the position of the cable components move relative to each other. For example, if under mechanical stress the distance between the outer and inner conductors decreases, the capacitance in the local region of stress increases. If the contact area between the insulation and conductors and the stress are well controlled, this effect can usually be made negligible. Tribocharging between the various elements in a cable is considered the major source of self-generated noise during mechanical stressing. This self-generated noise is the main topic of this discussion. Recall that tribocharging can occur between insulators and conductors, and that both rubbing and contact will generate this charge. Many dielectric materials used in cables are in the most negative end of the triboelectric series. Commonly used conductors in a cable tend to tribocharge with the dielectric.

To understand how tribocharging with dimensional changes in a cable can generate noise across a load, a very simplified model of the situation will be analyzed. Due to tribocharging between the dielectric and outer conductor of a coaxial cable *and* the small physical separation of the dielectric from the outer conductor, as shown in Figure 2.7, an electric field is generated in the air-gap region. (The analysis is identical if the separation occurs between the dielectric and inner conductor.) In the region of the tribocharging, the dielectric surface is shown negatively charged and inner surface of the outer conductor is shown positively charged.[3] In reality, these surface charge distributions are probably not uniform, but

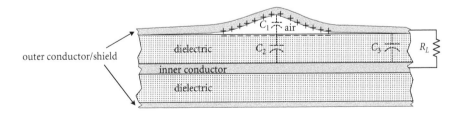

FIGURE 2.7　Small physical separation of the outer conductor from the dielectric of a cable. The charges shown are due to tribocharging.

---

[3]This implies that the dielectric material is lower or more negative on the triboelectric series than the conductor.

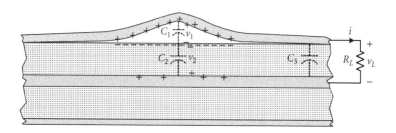

FIGURE 2.8 Charge decay through R_L.

to obtain a simple yet insightful model, the distributions will be assumed constant. Even though the dielectric surface in this air-gap region is probably not an equipotential surface, a capacitance, C_1, is still defined between the dielectric surface and inner surface of the outer conductor. If the outer conductor and dielectric are initially charge neutral, the initial charge on these two surfaces must be equal in magnitude but of opposite sign. As the spacing between the surfaces increase, the voltage drop across this region must increase: for a capacitor of fixed charge $Q = CV$, as the spacing increases, C decreases, and V must increase.

Assuming the charge separation between the outer conductor and dielectric occurs instantaneously, a voltage pulse must appear across the load. The load is shown as the resistor R_L. The capacitance C_2 represents the *local* fixed capacitance between the surface of the dielectric and inner conductor. These variables are defined in Figure 2.8. As time progresses, some of the positive charge on the outer conductor will pass through R_L; in other words, negatively charged electrons will pass from the inner conductor to the outer conductor, neutralizing some of the positive charge. Eventually, the current through R_L drops to zero once the *final* potentials of the inner conductor, V_2, and outer conductor, V_1, are the same:

$$-V_1 + 0 + V_2 = 0 \implies V_1 = V_2$$

The final charges on C_1 and C_2 are easily obtained. Assume that the total available charge is given by $\rho_S A$ where ρ_S is the initial uniform surface charge and A is the local area where this surface charge exists. The final charge on the two capacitors must equal this charge since charge is conserved:

$$Q_1 + Q_2 = \rho_S A$$

In addition, the final voltages across the two capacitors must be equal:

$$V_1 = \frac{Q_1}{C_1} = V_2 = \frac{Q_2}{C_2}$$

Solving for Q_2 from these two equations, the final charge along the surface of the inner conductor is

$$Q_2 = \rho_S A \frac{C_2}{C_1 + C_2}$$

The final charge along the surface of the outer conductor is

$$Q_1 = \frac{C_1}{C_2} Q_2 = \rho_S A \frac{C_1}{C_1 + C_2}$$

The charge on the outer conductor has decreased while the charge on the inner conductor, which was initially zero, has increased. The current that passes through the load resistor during this charge transfer can be obtained by assuming a standard exponential charge decay relationship. The equation describing the charge on the capacitor C_3 is given by

$$q_3(t) = Q_2\, e^{-\frac{t}{\tau}} \quad t \geq 0 \tag{2.2}$$

The time constant for this increase is

$$\tau = RC \approx R_L C_3 \tag{2.3}$$

assuming the sum of the cable and load capacitance, C_3, satisfies $C_3 \gg C_1 C_2 / (C_1 + C_2)$ (and the cable is electrically short at the highest frequency of interest). The inductance and resistance of the cable are ignored. The corresponding current, i, is

$$i = -\frac{dq_3(t)}{dt} = \frac{Q_2}{\tau} e^{-\frac{t}{\tau}} \quad t \geq 0 \tag{2.4}$$

Therefore, the voltage across the load is

$$v_L = iR_L = \rho_S A \frac{C_2}{C_3 (C_1 + C_2)} e^{-\frac{t}{\tau}} \quad t \geq 0 \tag{2.5}$$

The maximum value of this voltage occurs at $t = 0$:

$$v_{L,max} = \rho_S A \frac{C_2}{C_3 (C_1 + C_2)} \tag{2.6}$$

For this maximum voltage to exist across the load resistor, the initial separation must occur much faster than $\tau = R_L C_3$, otherwise, the charge will begin to decay before the final separation distance is obtained. When the air gap closes, the direction of the neutralizing current is in the opposite direction.

The separated charge in the cable generates a current that is a source of electrical noise. The maximum voltage across the load from this current, which is also equal to the initial voltage across the air gap, will be limited by the breakdown strength of the air gap. The maximum voltage, as expected, is proportional to the surface area of the tribocharged region and the charge generated. It is inversely proportional to C_3. The smaller the characteristic impedance of the cable (for a fixed cable inductance), the larger the cable capacitance. Also, the longer the cable, the larger the cable capacitance (assuming the cable is electrically short). Finally, the maximum voltage is also inversely proportional to C_1, the capacitance of the air gap (the smaller the air gap, the larger C_1). To help reduce the maximum voltage, the air gap should be kept small. Although the previous analysis covered only one-half cycle of a typical charge-separation sequence, when a cable is repeatedly flexed or stressed, the results seem similar to random Gaussian distributed noise.

The limited literature in this area focuses on low-frequency (audio) cables connecting high-impedance loads (e.g., a decent amplifier) to high-impedance sources (e.g., a condenser microphone, a piezoelectric force sensor, or an inductive electric guitar pickup). Although, the previous model indicates that the maximum tribogenerated voltage across the resistive load is independent of the load resistance, the rate of decay is a function of R_L via the time constant. As the load resistance increases, the decay time increases. However, when selecting a cable, other factors in addition to the tribogenerated noise must be considered. For example, the frequency response of the system from the input to output of the cable is a factor,

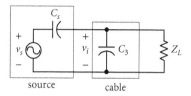

FIGURE 2.9 Capacitive source connected to a load via an electrically-short cable.

especially for high-impedance sources. Also, the magnitude of the current through the cable to the load usually must be considered. For electrically-short cables, the current into the cable from the source decreases as the load and source impedances increase:

$$I_s = \frac{V_{ss}}{Z_s + \left(\dfrac{1}{j\omega C_3}\middle\|Z_L\right)}$$

where Z_s and Z_L are the impedances of the source and load, respectively, and V_{ss} is the phasor form of the supply voltage. (The capacitance of the cable is in parallel with Z_L as is shown in Figure 2.9.) For these low-current measurements involving high source and load impedances, the small current generated by the triboelectric noise can be significant. For very small signal levels, the noise can mask the signal from the source. Filtering the electrical noise once it is generated is a challenging task. Equation (2.6) indicates that a large cable capacitance is desirable. In some product literature, it is claimed that a low cable capacitance is desirable. Actually, several disadvantages of a large cable capacitance are not related to the noise generation. First, a large cable capacitance will reduce the cutoff frequency of the low-pass response of the cable and system. Second, a large cable capacitance will reduce the load voltage. For a source with a capacitance of C_s, the voltage across the input of the cable in the frequency domain, assuming the load impedance is very high, is given by

$$V_{is} \approx V_{ss}\frac{\dfrac{1}{j\omega C_3}}{\dfrac{1}{j\omega C_3}+\dfrac{1}{j\omega C_s}} = V_{ss}\frac{C_s}{C_s+C_3} \quad \text{if } |Z_L| \gg \left|\frac{1}{j\omega C_3}\right|$$

The leakage resistance of the cable is neglected in this model, but for high-impedance sources, it is important that the leakage be small (i.e., high leakage *resistance*). If C_s is small corresponding to a high source impedance, then a small cable capacitance, C_3, is desirable to obtain a large load voltage. Since C_3 is the total cable capacitance, shorter cables will result in lower values for this cable capacitance. *Sometimes* for a given cable geometry (without regard to the characteristic impedance), "softness" of a cable, partially measured by the amount of air in the dielectric (i.e., density), and the ability of the dielectric to stay in contact with conductors, is an indication of a lower capacitance.

The triboelectric noise from a cable is a function of the method of testing. The peak voltage level across a specified load will vary depending on whether the cable is twisted, crushed, or impulsively deformed. On standard cable, voltages more than 50 mV peak can be generated via tribocharging. Cables advertised as low noise or "noise treated" can generate noise levels below 50 μV peak, but the level can vary considerably based on the method of testing.

Designing a cable with a triboelectric noise level below 25 μV peak is more challenging. In the design of low-noise cables, there are four major parameters: the dielectric material, degree of contact between the insulation and conductors, shield or external conductor type, and semiconductive layer's material, position, and geometry. Since the dielectric constant affects the cable's capacitance, characteristic impedance, and velocity of propagation, it is not always feasible to vary the dielectric material used to reduce the level of tribocharging. Also, although the relationship between the insulation and conductor type will affect the degree of tribocharging, the extent of this charging is often unpredictable. Although nonstandard or more exotic conductors can also be used to reduce the level of tribocharging with the dielectric, the additional cost must be considered. Controlling the degree of contact between the insulation and conductors, even for multiconductor cables, reduces the tribocharging effects. This was shown in the analysis through the capacitance C_1. The contact area between the outer conductor and insulation can have an effect on the degree of tribocharging. By using a braided shield rather than a solid or tightly wound spiral shield as the outer conductor, the contact area is reduced. This may have an effect on the degree of tribocharging, but the shielding effectiveness of the cable is also reduced. Semiconductive layers are designed to help in the dissipation or neutralization of any charge that may be generated. The layers are also designed to reduce the actual physical separation so that intimate contact between the dielectric and conductors is maintained. The layer can be an extrusion, a liquid coating, or a textile (which is difficult to work with when dealing with connectors). The layer contains a binder with carbon or silver particles or pigment to increase the binder's conductivity (similar to static-dissipative bags). All of these modifications to reduce the tribocharge-based emissions from a cable also affect the physical properties of the cable such as its flexibility, stripability, and size.

If the cable cannot be replaced with a lower noise cable, or the low-noise cable is not satisfactory, then the mechanical stresses on the cable should be reduced. For example, the coupling of the cable to any vibration source such as a motor or pump should be dampened. The cable itself can be secured, clamped, or tied down to reduce its mechanical motion (or, possibly, even buried in cement assuming no undesirable chemical reaction occurs). Also, because of the thermal stress that can occur, the cable should also be kept away from temperature variations that can cause the cable materials to expand or contract at different rates.

2.5 Voltage and Current Responses

Review the charging expressions for a parallel *RC* circuit connected to a step function voltage and current source. [Haase; Jonassen]

Frequently, when attempting to understand initially the charging and discharging behavior of an object, it is modeled as a parallel *RC* circuit. Although the equations are elementary, it is important to review the behavior of a parallel *RC* circuit connected to a step function voltage and current source. Recall that a decent voltage source generally has a small internal impedance (relative to its load impedance) and, thus, it should provide a relatively constant voltage at its output over a range of load impedances. A decent current source, on the other hand, has a large internal impedance (relative to its load impedance), and it should provide a relatively constant current at its output for a range of load impedances. Ideally, a voltage source has zero internal impedance and a current source has infinite internal impedance. Although engineers are more familiar with voltage sources, in some charging situations, it is better to model the process using a current source.

Using basic circuit analysis, the voltage across a parallel *RC* load when connected to a voltage source with an internal Thévenin resistance of R_s is equal to

$$v_C(t) = \frac{V_s R}{R + R_s}\left(1 - e^{-\frac{t}{\tau}}\right)u(t) \quad \text{where } \tau = \left(R_s\|R\right)C = \frac{R_s R}{R_s + R}C \tag{2.7}$$

A long time after the switch is closed in Figure 2.10, the capacitor appears like an open circuit to the constant, dc voltage source. The voltage across the parallel *RC* is then obtained using voltage division between the two resistors:

$$v_C(\infty) = \frac{V_s R}{R + R_s}$$

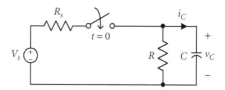

FIGURE 2.10 Charging of an *RC* load by a voltage supply.

The corresponding charge on the capacitor is

$$q(t) = Cv_C(t) = \frac{CV_sR}{R+R_s}\left(1-e^{-\frac{t}{\tau}}\right)u(t) \quad \text{where } \tau = \frac{R_sR}{R_s+R}C \tag{2.8}$$

The charge on the capacitor increases exponentially with time. The charging current through the capacitor is also exponential in nature:

$$i_C(t) = C\frac{dv_C(t)}{dt} = \frac{V_s}{R_s}e^{-\frac{t}{\tau}}u(t) \quad \text{where } \tau = \frac{R_sR}{R_s+R}C \tag{2.9}$$

The charging current decays exponentially to zero.

The expressions for the capacitor voltage and current for a current source with an internal resistance of R_s are

$$v_C(t) = \frac{I_sR_sR}{R+R_s}\left(1-e^{-\frac{t}{\tau}}\right)u(t) \quad \text{where } \tau = (R_s\|R)C = \frac{R_sR}{R_s+R}C \tag{2.10}$$

$$i_C(t) = I_se^{-\frac{t}{\tau}}u(t) \quad \text{where } \tau = \frac{R_sR}{R_s+R}C \tag{2.11}$$

After a long time, the capacitor appears like an open circuit to the constant, dc current source. After a long time, the voltage across the parallel *RC* load shown in Figure 2.11 is

$$v_C(\infty) = \frac{I_sR_sR}{R+R_s} = I_s(R_s\|R)$$

If both R_s and R are neglected (i.e., infinite or very large) and the time constant is infinite (or, at least, large compared to the longest time, *t*, of interest), the voltage across the capacitor can be approximated as

$$v_C(t) \approx \frac{I_sR_sR}{R+R_s}\left[1-\left(1-\frac{t}{\tau}\right)\right]u(t) = \frac{I_st}{C}u(t) \quad \text{if } \tau \gg t \tag{2.12}$$

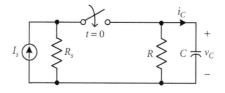

FIGURE 2.11 Charging of an *RC* load by a current supply.

since $e^x \approx 1 + x$ for $x \ll 1$. The charging voltage across the capacitor under these conditions increases approximately linearly with time. Simple examples where the source could be modeled (at least for a period of time) as a current source are the tribocharging of grain flowing into a silo (convection current), tribocharging of an individual walking across a carpet, and tribocharging of a plastic wrap unwinding from its roll. Van de Graaff generators are current sources that can generate large voltages. With these generators, charge generated by corona is sprayed or deposited on a moving belt. This moving charge, which is convection current, accumulates on an (often spherical shape) electrode near the end of the belt. As time increases, this charge on the electrode continues to build up until the surrounding medium breaks down.

As a simple example of current source charging of an *RC* circuit, imagine that a puzzle manufacturer is dropping puzzle pieces through a funnel to a metallic holding bin that is isolated from the ground (i.e., the bin is floating). Unfortunately, the contact of the pieces with the funnel (and each other) causes the pieces to be tribocharged. These falling charged puzzle pieces are a source of convection current in the direction shown in Figure 2.12 (assuming the pieces are negatively charged). An estimate of the voltage of the metallic bin as a function of time can be obtained by assuming quasisteady-state conditions. The puzzle pieces are assumed to be poor insulators and no "trapped" charge is permitted in the bin. The capacitance between the metallic bin and conducting ground plane is C, and the leakage resistance between the bin and ground plane is R. If the charge density per unit volume of the charge on the pieces is ρ_V and the flow rate is U, then the charging current is $\rho_V U$. If the bin is initially uncharged and the current "source" is ideal,[4] a first-order approximation for the voltage of the bin is, using Equation (2.10),

$$v_b(t) = I_s R \left(1 - e^{-\frac{t}{\tau}}\right) u(t) = \rho_V U R \left(1 - e^{-\frac{t}{\tau}}\right) u(t) \quad \text{where } \tau = RC \tag{2.13}$$

Although some of the charge on the bin dissipates through the slab neutralizing some of the charge on the ground plane, the charge on the bin increases according to the expression

$$q(t) = Cv_b(t) = C\rho_V U R \left(1 - e^{-\frac{t}{\tau}}\right) u(t) \quad \text{where } \tau = RC \tag{2.14}$$

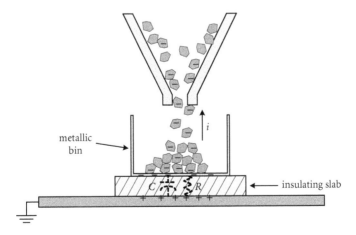

FIGURE 2.12 Tribocharged puzzle pieces collecting in an isolated bin.

---

[4]The flow rate and the charge from the funnel are not affected by the load (i.e., the puzzle pieces in the bin).

If $\rho_V = 10^{-7}$ C/m³, $U = 3 \times 10^{-3}$ m³/sec, $R = 10^{12}$ Ω, and $C = 250$ pF, then

$$v_b(t) = 300\left(1 - e^{-\frac{t}{\tau}}\right)u(t)\,\text{V} \quad \text{where } \tau = 250 \text{ sec}$$

The maximum possible voltage of the bin is 300 V. After one time constant or 250 sec, the voltage of the bin will be about 190 V. Although the voltage is a function of the quantity of puzzle pieces to be stored per bin, it is not likely to take one time constant, or a few minutes, to fill one bin. The maximum possible electrical energy stored in this bin-to-ground system is

$$W = \frac{1}{2}CV^2 \approx 0.01 \text{ mJ}$$

As discussed later, this energy level is unlikely to ignite most vapors, gases, or powders. However, in other tribocharging situations, the voltage can easily be in the kV range, and the electrical energy can be in the mJ range.

2.6   Sources of Current

Discuss conduction, convection, and diffusion current. Provide examples of each type. [Melcher; Haus; Kaiser; Sze; Adamczewski; Thomson]

The importance of charge flow in electroquasistatic phenomena cannot be overemphasized. Although the magnitude of the current is usually very small (otherwise the situation would not be electroquasistatic), the control of the quantity and position (and sometimes type) of charge is important in ESD prevention.

Current is a measure of charge flow. Current does not flow but is the rate of charge flow:

$$i = \frac{dq}{dt} \tag{2.15}$$

where q is the total positive charge passing through a surface. The direction of the positive charge flow is the direction of this current.[5] The volume current density or just current density, $\vec{J}$, is the charge flow per unit area and thus has units of A/m². To most electrical engineers, conduction (or drift) current density is the most familiar. Current in solid conductors and in low-velocity fluid flow is usually dominated by conduction. The conduction current density is defined as

$$\vec{J}_c = b|\rho_V|\vec{E} \tag{2.16}$$

where b is the mobility of the volume charge distribution, ρ_V.[6] The mobility has units of m²/V-s. The mobility, which will be discussed later in greater depth, is a measure of a charged particle's ability to move under an applied electric field. Notice that the current is in the direction of the applied electric field, independent of the sign of the charge distribution. Recall that the force on a particle of charge q in an electric field is

$$\vec{F} = q\vec{E} \tag{2.17}$$

---

[5]Occasionally, the direction of the current is defined as the direction of the negatively charged electrons. This definition has been used in some high school, technology, and military courses.

[6]Sometimes, the conduction current is defined as a function of b, n, and Q, where n is the number of Q charges per unit volume.

In Figure 2.13, positive and negative particles are shown in a right-directed uniform electric field. Although the negatively charged particles are moving to the left, opposite to the direction of the electric field, their current contribution is right directed. The positively charged particle's current contribution is also right directed. The mechanism of conduction in charge transport increases in importance as the conductivity increases. When only one type of charge carrier is conductively carried in a neutral background material, it is

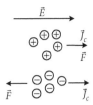

FIGURE 2.13 Conduction current direction for positive and negative charges in an applied electric field.

referred to as unipolar conduction. In this case, there can be a net volume charge at various locations in the material. Most materials have more than one carrier type. A more realistic model for many situations is the bipolar two-charge carrier system. The conductivity σ (1/Ω-m or siemens/m) is defined as

$$\sigma = b_+ |\rho_{V+}| + b_- |\rho_{V-}| \Rightarrow \vec{J}_c = \sigma\vec{E} \qquad (2.18)$$

where b_+ and b_- are the mobilities of the positive and negative charges, respectively. In materials where more than two charge carrier types are present, $\rho_{V\pm}$ are the dominant carriers. A neutral material has equal number of positive and negative charge carriers, and the net free charge is zero, $\rho_{V+} + \rho_{V-} = 0$. Often, materials in their "natural unperturbed" state are neutral. To state that the conductivity is mainly due to either the positive or negative charge carriers for a neutral material implies that the respective mobility of the carrier is much greater than the minor, or minority, carrier. For solids such as copper, the major charge carrier is the electron. Electrons are small negative particles that have high mobilities. Although in metals there are positive atoms, these atoms are fixed in position, and their mobility is zero. Although the atoms do not move, they neutralize the negative electron charge. When the conductivity is mainly due to electron flow, the conduction is referred to as electronic. For solid semiconducting materials such as silicon, the major charged carriers are the electron and hole (a positive "particle"). In many semiconducting materials, both the electron and hole contribute significantly to the conduction process. For liquids and gases, free electrons are rare—they quickly attach themselves to some atom resulting in an ion.[7] When the conductivity is mainly due to ions, the conduction is referred to as ionic. Most liquid conduction is ionic.

Another major method of transporting charge is convection; that is, the medium itself carries any residing charge. The convection current density is defined intuitively as

$$\vec{J}_v = \rho_V\vec{V} \qquad (2.19)$$

where $\vec{V}$ is the material's velocity vector. In Figure 2.14, the direction of the charge flow in a right-directed velocity field is illustrated. Unlike charge particles in an electric field, the velocity of the medium transports both negative and positive charge in the direction of the velocity without regard to the particle's sign. In this case, the current direction is a function of the sign of the charge. The right-directed negative particles correspond to left-directed current. Convection current can be important in fluid flow, especially

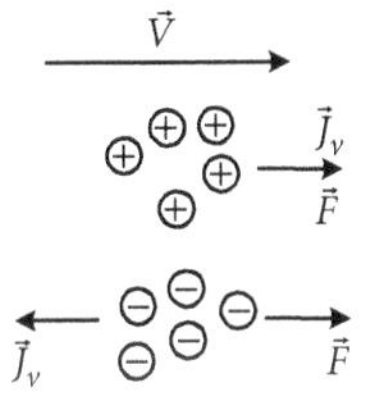

FIGURE 2.14 Convection current direction for positive and negative charges in a velocity field.

[7]Two exceptions to this are liquid metals and liquid ammonia.

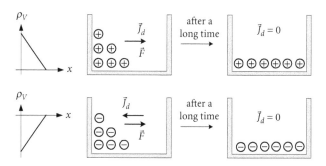

FIGURE 2.15 Diffusion current direction and final charge distribution for positive and negative charges with a nonuniform initial charge distribution.

when the fluid velocity is high and the fluid is a poor conductor. Insulating oils are sometimes pumped around electronic devices (and transformers) for cooling purposes. Convection current can also be a dominant charge flow mechanism for solids. For example, surface charge can be transported on moving conveyers or belts. Although near electrical discharges in air conduction current is often dominant, fans can be used to transport charged ions to help neutralize charge buildup on static-sensitive objects. Gravity can also be used to transport heavier particles. For example, gravity can transport tribocharged cereal flakes from a third-floor processing facility to a main-floor storage bin via a plastic conduit. When it rains, the falling charged water droplets from the clouds are a source of convection current.

The final major means of charge transport is diffusion. Diffusion is the tendency of particles to spread themselves out uniformly. Nature attempts to level out any concentration gradients (including spatial charge concentrations) through random thermal motions. This leveling of gradients occurs even if the electric forces are small.[8] Diffusion current is defined by the relationship

$$\vec{J}_d = -D\nabla\rho_V \tag{2.20}$$

where D is the charge diffusion coefficient, which has units of m²/sec, and $\nabla\rho_V$ is the gradient of the volume charge density. If the charges are uniformly distributed, then this gradient term is zero as expected, and there is no diffusion current. The diffusion coefficient is directly related to the mobility, b, and temperature. As the temperature rises, the charges become more thermally active and the diffusion current increases. Figure 2.15 illustrates the diffusion process for both negative and positive particles in "enclosed" boxes. Regardless of the sign of the particles, the charge will diffuse until an equilibrium zero-gradient condition is reached. The gradient is merely a single derivative in this one-dimensional case corresponding to the slope of the charge-vs.-position graph shown in Figure 2.15. The current is thus right directed in the positive charge case (a negative slope negated by the minus sign in $-D$), and the current is left directed in the negative charge case (a positive slope multiplied by $-D$). Note that both of these results are intuitively consistent with the previous charge transport examples. The relative strength of diffusion with respect to conduction and convection is usually small. However, there are cases when diffusion can be important. For example, if a charged fluid is stationary and the electric fields are weak, then near electrodes and other boundaries, diffusion can play a major part in the conduction process. The diffusion term is also extremely important in solid state devices such as along the interface of a *pn* junction. The ratio of the diffusion to conduction current is

$$\frac{\vec{J}_d}{\vec{J}_c} = \frac{-D\nabla\rho_V}{b|\rho_V|\vec{E}} \tag{2.21}$$

---

[8]For example, certain recognizable fetid vapors have a tendency to spread quickly throughout a room.

Einstein's relation links the mobility to the diffusion constant for ideal gases and liquids:

$$\frac{D}{b} = \frac{\kappa T}{q} \quad (\approx 0.0266 \text{ V } @ \, T = 20°\text{C and when } q = \text{charge of an electron}) \tag{2.22}$$

where κ is Boltzmann's constant ($\approx 1.381 \times 10^{-23}$ J/K), T is the temperature (K), and q is the particle's charge. Substituting Equation (2.22) into Equation (2.21) and taking the magnitude of both sides yields

$$\frac{\left|\vec{J}_d\right|}{\left|\vec{J}_c\right|} = \frac{\kappa T \nabla \rho_V}{q |\rho_V| \|\vec{E}|} \tag{2.23}$$

If the volume charge density does not vary by more than ρ_V over some characteristic length, l_{th}, of interest, the maximum gradient of the charge density can be approximated as

$$\nabla \rho_V \approx \frac{\rho_V}{l_{th}} \tag{2.24}$$

resulting in the expression

$$\frac{\left|\vec{J}_d\right|}{\left|\vec{J}_c\right|} \approx \frac{\kappa T}{q l_{th} |\vec{E}|} \tag{2.25}$$

The relative magnitude of diffusion current relative to conduction current can be demonstrated by examining two cases. Assuming $\kappa T/q = 0.0266$ and the magnitude of the electric field, $|\vec{E}|$, is roughly 1 V/m, then the diffusion term is strongly dominant for lengths less than about 0.27 cm:

$$\frac{\left|\vec{J}_d\right|}{\left|\vec{J}_c\right|} \approx \frac{0.027}{l_{th}} > 10 \quad \Rightarrow l_{th} < 0.0027 \text{ m}$$

This electric field is typical of values inside conducting liquids where the potential is roughly constant and the electric field is low. If the study of some process (e.g., the production of a charged conducting jet) focuses on distances on the order of 2.7 cm, or less, and $|\vec{E}| \approx 1$ V/m, then the diffusion current should not be neglected. At 2.7 cm, the magnitude of Equation (2.25) is about one. If the magnitude of the electric field is roughly 10 kV/m, then the diffusion term is about the same strength as the conduction term for lengths near 2.7 μm. The diffusion term can be a significant fraction of the current in regions of small voltage drop and very near interfaces or boundaries.

The total current density is the sum of the conduction, convection, and diffusion terms. Thus, for each charge species, the current density contribution can be described by the equation

$$\vec{J} = \vec{J}_c + \vec{J}_v + \vec{J}_d = b |\rho_V| \vec{E} + \rho_V \vec{V} - D \nabla \rho_V \tag{2.26}$$

Measurement of the conductivity is relatively easy (but not always repeatable) unless the material is highly insulative. For unipolar conduction, σ is a function of two variables: the mobility and charge density. The mobility is an important parameter in the rate of charge decay. The ability of a charged particle to move in an applied electric field is a measure of its mobility:

$$\vec{v} = b\vec{E} \tag{2.27}$$

where $\vec{v}$ is the velocity of the particle. The mobility is defined as positive. For many solids and semiconductors, the electron and hole mobilities are well established. For many liquids, unfortunately, the mobilities are not well established. The convection process and possible nonlinear ion production are just two of many phenomena responsible for the increased complexity in measuring the mobility of particles in a liquid. There are a few expressions available that provide an estimate of the charge mobility. First, in many liquids the mobilities of the negative and positive charges are approximately equal. Second, for larger particles that are spherical in shape and obey Stoke's law, the mobility is given by

$$b = \frac{q}{6\pi\mu a} \tag{2.28}$$

where μ is the absolute viscosity of the surrounding gas or liquid (in kg/m-s), a is the particle's radius, and q is the particle's charge. According to (2.28), for a given charge, the mobility of a particle decreases with increasing radius. If the spherical particle is ion-impact charged to saturation, this expression can be written in terms of the magnitude of the electric field acting on the particle, $|\vec{E}|$:

$$b \approx \frac{2\varepsilon_o a |\vec{E}|}{\mu} \tag{2.29}$$

In air under standard conditions, this expression is valid for particles greater than about 0.5 μm. The unipolar charge decay expression to be derived assumes that the mobility is constant and not a function of the electric field. Obviously, in this unipolar decay expression, some sort of "average" mobility based on the expected electric field must be used. Approximate relationships for the mobility of positive and negative ions in highly insulating liquids are

$$b_+ \approx \frac{1.5 \times 10^{-11}}{\mu}, \; b_- \approx \frac{3.0 \times 10^{-11}}{\mu} \tag{2.30}$$

These equations are correct within an order of magnitude for most situations. These empirically derived equations are quite interesting in that they relate the viscosity of a liquid, a macroscopic mechanical property, to the mobility of a residing charge particle, a microscopic electrical property. The absolute viscosity of water is about 0.001 kg/m-s. For highly insulating liquids of the same viscosity of water (which is typically not insulating in its "natural" state), the corresponding mobilities are $b_+ \approx 1.5 \times 10^{-8}$ m^2/V-s and $b_- \approx 3 \times 10^{-8}$ m^2/V-s. The absolute viscosity of baby oil is about 0.03 kg/m-s, and the corresponding mobilities are $b_+ \approx 5 \times 10^{-10}$ m^2/V-s and $b_- \approx 1 \times 10^{-9}$ m^2/V-s. The absolute viscosity of transmission oil is about 0.07 kg/m-s, and the corresponding mobilities are $b_+ \approx 2.1 \times 10^{-10}$ m^2/V-s and $b_- \approx 4.3 \times 10^{-10}$ m^2/V-s. For ions in gas at atmospheric pressure and 20°C, the mobilities are much larger. For example, in dry air, $b_+ \approx 1.4 \times 10^{-4}$ m^2/V-s and $b_- \approx 2.1 \times 10^{-4}$ m^2/V-s, and in CO$_2$, $b_+ \approx 0.84 \times 10^{-4}$ m^2/V-s and $b_- \approx 0.98 \times 10^{-4}$ m^2/V-s.

For semiconductors, the mobilities are generally much higher. For example, in silicon the electron mobility is 0.15 m²/V-s and hole mobility is 0.045 m²/V-s. For metals, the electron mobility is in the range of 0.001–0.01 m²/V-s. It would initially seem that the electron mobility in metals should be much greater than in semiconductors. However, the conductivity is a function of the mobility and number density (or volume charge density) of free charge carriers. For metals, the charge density is extremely high. This high density is the reason a metal's conductivity is so high, not an electron's exceptional mobility in the metal. The difference in the conductivity of various metals is determined mainly by the difference in their electron mobilities.

2.7   Rate of Charge Decay

Derive the expressions for the rate of volume charge decay in ohmic and nonohmic materials. When does the total current into a node not equal to the total current out of the node? [Melcher; Zahn]

The rate of charge decay in or on an object is an important parameter of interest in ESD since charge is a source of electric field. The manner that charge decays is based on the charge conservation equation.

The insight that the charge conservation equation provides is worthy of a few pages in this book. The very fundamental conservation of charge (or continuity) equation in differential form is[9]

$$\nabla \cdot \vec{J} + \frac{\partial \rho_V}{\partial t} = 0 \qquad (2.31)$$

where $\vec{J}$ is the current density in A/m^2 and ρ_V is the volume charge density in C/m^3. The charge conservation equation states that if the charge density is increasing with time or positive ($\partial \rho_V / \partial t > 0$), then the divergence of the current density must be negative ($\nabla \cdot \vec{J} < 0$). Often, the charge conservation equation is more easily understood in its integral form. Integrating both terms over some volume, V,

$$\iiint_V \left(\nabla \cdot \vec{J} + \frac{\partial \rho_V}{\partial t} \right) dv = \iiint_V (\nabla \cdot \vec{J}) \, dv + \iiint_V \frac{\partial \rho_V}{\partial t} \, dv = 0$$

and then using the divergence theorem to convert the volume integral involving the divergence of $\vec{J}$ to a surface integral involving just $\vec{J}$,

$$\oiint_S \vec{J} \cdot d\vec{s} + \iiint_V \frac{\partial \rho_V}{\partial t} \, dv = 0 \qquad (2.32)$$

The closed or complete surface, S, corresponds to the volume V. The integral in (2.32) involving $\vec{J}$ is the total current (not current density) *leaving* the region of interest through the surface S. The second integral involving $\partial \rho_V / \partial t$ is the rate of change of the total charge inside the region of interest. As is expected, if the rate of change of total charge in the volume region V is increasing, then a net current must be passing into (not out of) this region since this current is the source of this additional charge. Charge is neither created nor destroyed but conserved. In this particular example, the first integral in (2.32) is negative and the second integral is of equal magnitude but opposite sign.

Kirchoff's current law (KCL) is obtained from the charge conservation expression. One of the first concepts learned in elementary circuit courses is that the total current into (or out of) a node must sum to zero. Although it is probably not discussed, since it might confuse the novice, this assumes steady-state conditions with respect to the charge. Imagine that the volume of interest, V, is around a node where several current-carrying conductors meet. In steady-state conditions, the rate of charge increase or decrease inside this node is zero ($\partial \rho_V / \partial t = 0$). Therefore,

$$\oiint_S \vec{J} \cdot d\vec{s} = 0 \quad \text{in steady state} \qquad (2.33)$$

The total current leaving the node is zero (i.e., KCL) in steady-state conditions. After a few circuits courses, a student may state, "Of course, the current into a node must always equal the current out of a node!" This, though, is only true when the charge (not current) is in steady state. Next, imagine that current is injected into a "sponge" (e.g., shoes tribocharged from walking on a carpet). The total current into the sponge is positive *not zero* over some period of time, and the charge on the "sponge" is increasing with time. Of course, eventually the charge level on the "sponge" will be so great that the surrounding air will break down and the charge will leave the "sponge" through some mechanism. This charge that is leaving the "sponge" is analogous to current leaving a node. The charge conservation equation indicates

---

[9]This expression is easily derived from fundamental principles or by using two of Maxwell's equations: $\nabla \times \vec{H} = \vec{J} + \partial \vec{D}/\partial t$ and $\nabla \cdot \vec{D} = \rho_V$. The divergence of both sides of $\nabla \times \vec{H} = \vec{J} + \partial \vec{D}/\partial t$ is taken and the vector identity $\nabla \cdot \nabla \times \vec{A} = 0$ is used: $0 = \nabla \cdot \vec{J} + \nabla \cdot (\partial \vec{D}/\partial t) = \nabla \cdot \vec{J} + \partial(\nabla \cdot \vec{D})/\partial t = \nabla \cdot \vec{J} + \partial \rho_V/\partial t$.

that the total current leaving a region may not be zero if not in steady state: charge sources or sinks can exist in the region of interest.

Assume that the volume charge inside a solid spherical ball constructed of ohmic material is given by

$$\rho_V(t) = \rho(0)e^{-\frac{t}{\tau}}$$

The charge is uniformly distributed in the sphere of radius R, with an initial value of $\rho(0)$, and it is decreasing exponentially with time with a time constant of τ. Substituting into the integral form of the conservation equation,

$$\oiint_S \vec{J} \cdot d\vec{s} = -\iiint_V \frac{\partial \rho_V}{\partial t} dv = -\iiint_V \frac{d}{dt}\left[\rho(0)e^{-\frac{t}{\tau}}\right]dv = \int_0^{2\pi}\int_0^{\pi}\int_0^{R} \frac{\rho(0)e^{-\frac{t}{\tau}}}{\tau} r^2 \sin\theta\, dr\, d\theta\, d\phi$$

$$= \frac{\rho(0)e^{-\frac{t}{\tau}}}{\tau}\frac{4\pi R^3}{3}$$

Since the charge is uniformly distributed, the total charge at any time within the sphere is merely $\rho_V(t)$ multiplied by the volume of the sphere. As stated, the surface integral involving the current density is the total current leaving the volume. Therefore, the total current leaving, although positive, is also decreasing exponentially with time. The current density at the surface, which is entirely in the radial direction, is easily obtained for this spherical surface:

$$\int_0^{2\pi}\int_0^{\pi} J_r \hat{a}_r \cdot r^2 \sin\theta\, d\theta\, d\phi\, \hat{a}_r\Big|_{r=R} = J_r 4\pi R^2 = \frac{\rho(0)e^{-\frac{t}{\tau}}}{\tau}\frac{4\pi R^3}{3} \quad \Rightarrow J_r = \frac{\rho(0)e^{-\frac{t}{\tau}}}{\tau}\frac{R}{3}$$

Obviously, if free space surrounds the spherical ball and there are no paths for the charge to leave the sphere, the charge must be building up along its surface. This surface charge is shown in Figure 2.16 with "+'s" and is labeled $\rho_S(t)$. The current density immediately outside the sphere in free space is zero.

As seen in the previous example, surface charge can build up in transient conditions along boundaries. As with electric and magnetic fields, there is a very intuitive boundary condition involving the normal components of the current density at both sides of a boundary or interface:

$$J_{n2} - J_{n1} = -\frac{d\rho_S}{dt} \tag{2.34}$$

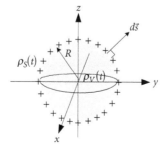

FIGURE 2.16 Volume charge decaying to the spherical surface.

This expression assumes that there are no surface currents along the interface. Although this expression will not be derived, it is obtained directly from the charge conservation equation when the region of interest is taken about an interface between two materials as shown in Figure 2.17. The volume region about the interface has the shape of a very thin disk. The disk is so thin that zero or negligible current leaves through the sides of the disk. (This

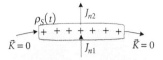

FIGURE 2.17   Normal component of the current density and the surface charge along an interface.

boundary condition assumes zero surface current. If surface current was present, then current could enter and leave the disk along its thin sides.) In transient conditions, if the current density in region 2 is zero (e.g., free space with no arcing or other charge loss mechanisms),

$$0 - J_{n1} = -\frac{d\rho_S}{dt} \quad \Rightarrow J_{n1} = \frac{d\rho_S}{dt} \tag{2.35}$$

In other words, the current into the interface is equal to the rate of change of the surface charge at the interface. In steady-state conditions, $d\rho_S/dt = 0$, and the normal component of the current density entering the interface must equal the normal component of the current density leaving the interface. According to Equation (2.35), the rate of surface charge buildup on the interface is equal to the current density into the interface from region 1. By integrating both sides of Equation (2.34) over the area, A, of the interface, another form of the boundary condition is obtained:

$$\iint_A (J_{n2} - J_{n1})\,ds = -\iint_A \frac{d\rho_S}{dt}\,ds \quad \Rightarrow I_{n2} - I_{n1} = -\frac{dq}{dt} \tag{2.36}$$

where q is the total charge along the interface and the I's are the currents corresponding to the current densities. For example, if $I_{n2} = 1$ A and $I_{n1} = 3$ A, then the total charge, q, along the interface is increasing with respect to time or $dq/dt = 2$ C/sec > 0. The current into the interface is greater than the current out of the interface so charge must be building up along it.

The charge decay equations can now be derived. For ohmic materials where the current is entirely due to (or dominated by) conduction, $\vec{J} = \sigma\vec{E}$. Substituting into the charge conservation equation, (2.31),

$$\nabla \cdot \sigma\vec{E} + \frac{\partial \rho_V}{\partial t} = \sigma\nabla \cdot \vec{E} + \frac{\partial \rho_V}{\partial t} = 0 \tag{2.37}$$

where it was assumed that the conductivity, σ, does not vary with position (so that it could be taken past the ∇ operator). Next, Maxwell's equation that relates the electric flux density to the volume charge density

$$\nabla \cdot \vec{D} = \nabla \cdot \varepsilon\vec{E} = \rho_V \quad \Rightarrow \nabla \cdot \vec{E} = \frac{\rho_V}{\varepsilon} \tag{2.38}$$

is used. It was also assumed in Equation (2.38) that the permittivity, $\varepsilon = \varepsilon_r\varepsilon_o$, does not vary with position. Substituting Equation (2.38) into (2.37),

$$\sigma\frac{\rho_V}{\varepsilon} + \frac{\partial \rho_V}{\partial t} = 0$$

The solution to this simple differential equation is quickly obtained. It has a decaying exponential solution:

$$\int_{\rho_V(0)}^{\rho_V(t)} \frac{d\rho_V}{\rho_V} = -\int_0^t \frac{\sigma}{\varepsilon_r \varepsilon_o} dt \quad \Rightarrow \ln\left[\frac{\rho_V(t)}{\rho_V(0)}\right] = -\frac{\sigma}{\varepsilon_r \varepsilon_o} t$$

$$\rho_V(t) = \rho_V(0)e^{-\frac{\sigma}{\varepsilon_r \varepsilon_o}t} = \rho_V(0)e^{-\frac{t}{\tau}} \quad \text{where } \tau = \frac{\varepsilon_r \varepsilon_o}{\sigma} \tag{2.39}$$

where ε_r is the relative permittivity or dielectric constant of the medium and τ is the time constant or relaxation time. Equation (2.39) is an important and a classical result. For an ohmic material with uniform conductivity and permittivity, any volume charge in the material will decay in an exponential fashion with a time constant equal to τ. This result is remarkable. Regardless of any external fields such as from charged electrodes *and* regardless of the geometry and spatial distribution of the volume charge, the volume charge decays exponentially with respect to time! No boundary conditions were used to obtain this simple result, and it does not assume electrostatic conditions. As will be seen in a later discussion, this simple exponential relationship does not usually apply to surface charge only to volume charge. Because of the similarity of Equation (2.39) to the current and voltage response of an *RC* circuit to a step voltage or current source, frequently an ohmic object (and even a nonohmic object) with uniform electric properties is modeled as an *R* and a *C*. As the time constant increases, the time required for the charge to decay to some level increases. The time constant decreases as the σ increases. Good conductors dissipate charge faster than poor (ohmic) conductors.

The simple exponential relationship for the charge decay of volume charge can sometimes perplex even the best students. Assume that the initial volume charge density inside a solid spherical ball of radius *R* is given by

$$\rho_V(r, t=0) = \begin{cases} 2r^2 & 0 < r < \dfrac{R}{2} \\ 0 & \dfrac{R}{2} < r < R \end{cases}$$

Free space surrounds the ball. Notice that the volume charge density is zero from *R*/2 to *R*, and there is spatial variation in the volume charge from 0 to *R*/2. If the permittivity and conductivity of the ohmic material are ε and σ, respectively, throughout the entire sphere, the charge in the sphere will decay according to the expressions

$$\rho_V(r, t) = \begin{cases} 2r^2 e^{-\frac{\sigma}{\varepsilon_r \varepsilon_o}t} & 0 < r < \dfrac{R}{2} \\ 0 & \dfrac{R}{2} < r < R \end{cases}$$

The *shape* of the charge distribution at every position inside the sphere does not change with time. The charge does not spread or diffuse outward for this ohmic material. The volume charge density remains zero in the outer half of the sphere. The charge everywhere is decaying at the same rate, and the charge is building up on the surface of the sphere. The surface charge on the sphere is given by

$$\rho_S(R, t) = \frac{\displaystyle\int_0^{2\pi}\int_0^{\pi}\int_0^{\frac{R}{2}} 2r^2\left(1 - e^{-\frac{\sigma}{\varepsilon_r \varepsilon_o}t}\right) r^2 \sin\theta \, dr \, d\theta \, d\phi}{4\pi R^2} = \frac{1}{80}\left(1 - e^{-\frac{\sigma}{\varepsilon_r \varepsilon_o}t}\right) R^3$$

At $t = 0$, there is no surface charge. After several time constants, the exponential term is small compared to one, and the surface charge is essentially constant:

$$\rho_s(R,t) \approx \frac{1}{80}R^3 \quad t > 3\tau \text{ to } 5\tau$$

What is initially perplexing is that the portion of the sphere from $R/2$ to R remains charge neutral. The question inquisitive students ask is, "How can the charge pass from the inner half of the sphere to the surface without passing through and thereby charging the outer half of the sphere?" For every charge ejected from the inner volume of the sphere at $r = R/2$ to the neutral half of the sphere, a charge is ejected from the outer volume of the sphere at $r = R$ (to the sphere's outer surface). Thus, the outer volume always remains charge neutral. As will be seen in a later discussion, the shape of the surface and any external fields will affect the distribution of the surface (not volume) charge. Because of the repeated use and misuse of the exponential decay relationship, the assumptions in its deviation are summarized:

1. the permittivity of the region is constant and unaffected by the perturbing charge and any electric field
2. the conductivity of the region is constant and unaffected by the perturbing charge and any electric field
3. the current density is described by Ohm's law
4. the perturbing charge is of the same type as the intrinsic background charge.

The classical exponential decay expression was derived assuming the medium was ohmic and the current density could be described by $\vec{J} = \sigma\vec{E}$ (i.e., Ohm's law). The conductivity was assumed constant and independent of $\vec{E}$. Many materials such as plastics and other insulators are not ohmic and do not obey Ohm's law. For these materials, the "σ" can change with $\vec{E}$, so that $\vec{J} = \sigma(E)\vec{E}$. For these materials, the charge decay is not exponential with time. Before these nonohmic materials can be modeled, conductivity must be understood. Recall that the conductivity is a measure of the number density (n), mobility (b), and charge (q) of the free carriers of charge in a medium:

$$\sigma = nb|q| \tag{2.40}$$

The mobility of a charged particle (e.g., electron or ion) is a measure of the ability of the particle to move in an applied field. (The larger the mobility, the greater the average velocity of the particle when placed in an electric field.) As is well known for ohmic materials, the greater the conductivity, the greater the current density for a given applied voltage. It is important to distinguish between the conductivity of a medium and the net charge in the medium. A nonzero conductivity does *not* imply a net charge. In layman terms, the conductivity is a measure of the number, size, and potential speed of "electricity carriers" in a medium. With ohmic material, when a voltage is applied across the material, the number of injected electrons from the voltage source has a negligible effect on n, the number of free intrinsic carriers available for conduction. With nonohmic materials, these injected electrons from the source can dramatically affect the intrinsic n of the material. Furthermore, under certain conditions when the applied voltage or electric field is sufficiently strong, the field itself can generate additional free carriers via ionization. Sometimes, the current density in nonohmic materials, when only charge of one type is present (i.e., unipolar conduction), can be described by the expression

$$\vec{J} = \rho_V\vec{v} = \rho_V b\vec{E} \tag{2.41}$$

where $\vec{v}$ is the velocity of the volume charge present in the medium. This expression assumes a single species or type of positive charge. The current density in this nonohmic model is a function of the product of the charges (per unit volume) present and their velocity, which has the appropriate units of $(C/m^3)(m/sec) = C/(m^2sec) = A/m^2$. The velocity of the volume charge is equal to the product of their mobility and the applied electric field. Volume charge has little effect on the conductivity of ohmic materials. However, the volume charge in a nonohmic material is a source of "free carriers." The expression for the rate of decay of charge in nonohmic materials is easily obtained by substituting Equation (2.41) into the charge conservation equation:

$$\nabla \cdot (\rho_V b \vec{E}) + \frac{\partial \rho_V}{\partial t} = 0$$

The mobility will be assumed constant (although it can be a function of several factors including the electric field). The volume charge density will initially be allowed to vary over the volume of the material:

$$b \nabla \cdot (\rho_V \vec{E}) + \frac{\partial \rho_V}{\partial t} = 0$$

The ∇ operator, which involves differentiation, is operating on the product of a scalar function and vector function. Similar to the product rule in calculus, the divergence of the product is given by

$$\underbrace{b \underbrace{(\nabla \rho_V)}_{\text{vector}} \cdot \overbrace{\vec{E}}^{\text{scalar}}}_{\text{vector}} + \underbrace{b \rho_V \underbrace{\nabla \cdot \vec{E}}_{\text{scalar}}}_{\text{scalar}} + \overbrace{\frac{\partial \rho_V}{\partial t}}^{\text{scalar}} = 0$$

(The ∇ of a scalar function, referred to as the gradient of the function, is a vector. The ∇ dotted with a vector function, referred to as the divergence of the vector, is a scalar.) Using one of Maxwell's equations and assuming a constant permittivity,

$$\nabla \cdot \vec{D} = \nabla \cdot \varepsilon \vec{E} = \rho_V \quad \Rightarrow \nabla \cdot \vec{E} = \frac{\rho_V}{\varepsilon}$$

$$b(\nabla \rho_V) \cdot \vec{E} + b \rho_V \frac{\rho_V}{\varepsilon} + \frac{\partial \rho_V}{\partial t} = 0$$

The final assumption is that the first term in this expression is negligible:

$$b \rho_V \frac{\rho_V}{\varepsilon} + \frac{\partial \rho_V}{\partial t} \approx 0 \quad \text{if} \left| b \rho_V \frac{\rho_V}{\varepsilon} + \frac{\partial \rho_V}{\partial t} \right| \gg \left| (\nabla \rho_V) \cdot b \vec{E} \right| = \left| (\nabla \rho_V) \cdot \vec{v} \right| \qquad (2.42)$$

If the space variation in the volume charge density is zero or small ($\nabla \rho_V \approx 0$), then the condition given in Equation (2.42) is reasonable. The solution of this differential equation is obtained in the traditional way:

$$\int_{\rho_V(0)}^{\rho_V(t)} \frac{d\rho_V}{\rho_V^2} \approx - \int_0^t \frac{b \, dt}{\varepsilon_r \varepsilon_o} \quad \Rightarrow \frac{-1}{\rho_V} \bigg|_{\rho_V(0)}^{\rho_V(t)} = -\frac{bt}{\varepsilon_r \varepsilon_o} \bigg|_0^t$$

$$\frac{-1}{\rho_V(t)} + \frac{1}{\rho_V(0)} = -\frac{bt}{\varepsilon_r \varepsilon_o} \quad \Rightarrow \rho_V(t) = \frac{\rho_V(0)}{1 + \dfrac{t}{\tau}} \quad \text{where } \tau = \frac{\varepsilon_r \varepsilon_o}{\rho_V(0)b} \qquad (2.43)$$

For nonohmic materials satisfying Equation (2.42), the volume charge density is approximately inversely proportional to t. As expected, the rate of decay increases as the mobility, b, of the charge increases. The time constant for this unipolar charge relaxation, τ, is also a function of the initial charge distribution.

The time constant decreases as the initial charge density increases. For times much greater than the time constant, the volume charge is approximately independent of the initial charge distribution:[10]

$$\rho_V(t) \approx \frac{\rho_V(0)}{\dfrac{t}{\tau}} = \frac{\varepsilon_r \varepsilon_o}{bt} \quad t \gg \tau \tag{2.44}$$

(The inequality $t \gg \tau$ was used instead of $t > 3\tau$ to 5τ, which is common with ohmic decay, since the decay rate in (2.43) is not exponential.) To summarize, the assumptions in the deviation of the unipolar charge decay relationship are

1. the permittivity of the region is constant and unaffected by the perturbing charge and any electric field
2. the mobility of the particles is constant and unaffected by the perturbing charge and any electric field
3. the charge density is approximately uniform (and the inequality in (2.42) is satisfied)
4. the perturbing charge density is of one sign and much larger in magnitude than any intrinsic "free" charge carriers.

2.8 Maximum Surface Charge before Breakdown

What is the maximum charge that can exist on a surface? [Crowley, '86; Davidson; Cross; Felici; Dascalescu; Taylor, '94]

Charge on a surface generates an electric field. If the field is sufficiently large, the surrounding air (or other medium) can break down, and charge transfer from the surface can occur. This breakdown limits the maximum charge that can exist on a surface. There are essentially two criteria for breakdown due to ionization. One is based on the voltage and the other on the electric field. Generally, both criteria must be satisfied for breakdown to occur. For "large" gaps in standard conditions, the electric field must exceed about 3 MV/m, while for "small" gaps, the voltage must exceed about 320 V. These numbers are based on the breakdown between two parallel plate electrodes, which are clean and smooth. For a corona discharge, the required electric field is less.

The maximum surface charge on an isolated spherical object is determined first. The maximum electric field from a uniformly charged isolated sphere is easily obtained using Laplace's equation. To obtain this field, the potential between two concentric spheres of radii a and b is initially determined. The voltage of the inner electrode is V, and the voltage of the outer electrode is zero. For this symmetrical geometry shown in Figure 2.18, Laplace's equation in spherical coordinates reduces to

FIGURE 2.18   Two concentric spherical electrodes.

$$\nabla^2 \Phi = \frac{1}{r^2} \frac{d}{dr}\left(r^2 \frac{d\Phi}{dr}\right) = 0$$

Integrating twice, results in two constants of integration:

$$r^2 \frac{d\Phi}{dr} = C_1 \quad \Rightarrow \frac{d\Phi}{dr} = \frac{C_1}{r^2} \quad \Rightarrow \Phi = -\frac{C_1}{r} + C_2$$

---

[10]It might be helpful to view the larger initial charge density as having a greater self "spreading" field or force.

The potential is known at both $r = b$ and $r = a$, which allows these constants to be determined:

$$V = -\frac{C_1}{b} + C_2, \quad 0 = -\frac{C_1}{a} + C_2 \quad \Rightarrow C_1 = \frac{ab}{b-a}V, \quad C_2 = \frac{b}{b-a}V$$

The potential between the electrodes is then

$$\Phi = \frac{abV}{a-b}\left(\frac{1}{r} - \frac{1}{a}\right)$$

The corresponding electric field between the electrodes is obtained from the gradient operator:

$$\vec{E} = -\nabla\Phi = -\frac{d}{dr}\left[\frac{abV}{a-b}\left(\frac{1}{r} - \frac{1}{a}\right)\right]\hat{a}_r = \frac{ab}{(a-b)}\frac{V}{r^2}\hat{a}_r$$

The maximum field occurs at the inner electrode at $r = b$:

$$E_{r,max} = \frac{ab}{(a-b)}\frac{V}{b^2} = \frac{a}{(a-b)}\frac{V}{b} \tag{2.45}$$

The total charge on the inner or outer conductor can be determined via the capacitance since $Q = CV$. Instead, in order to introduce a simple but important concept, the boundary condition

$$D_{n2} - D_{n1} = \rho_S$$

will be used along the inner conductor's spherical surface. Assuming the inner electrode is a good conductor, both the $\vec{E}$ and $\vec{D}$ fields inside the conductor must be small (or ideally zero). These internal fields are small since the conductivity, σ, is very high for good conductors. Therefore, for the conduction current density, $\vec{J} = \sigma\vec{E}$, to be finite in a good conductor, $\vec{E}$ must be zero (or at least very small). Otherwise, the current density would be infinite (or at least very large). Referring to Figure 2.19, since the electric field inside the perfect conductor is zero,

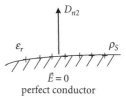

FIGURE 2.19 Electric field and flux density are entirely normal to the surface of a perfect conductor.

$$D_{n2} - 0 = \varepsilon_r\varepsilon_o E_{n2} = \rho_S \quad \Rightarrow E_{n2} = \frac{\rho_S}{\varepsilon_r\varepsilon_o} \tag{2.46}$$

The total charge along the surface of the inner electrode is the surface charge multiplied by the area of the spherical surface:

$$Q = 4\pi b^2 \rho_S = 4\pi b^2 \varepsilon_r\varepsilon_o E_{n2} \tag{2.47}$$

Using Equation (2.45), the total charge can also be written in terms of the voltage:

$$Q = 4\pi b^2 \varepsilon_r\varepsilon_o \frac{a}{(a-b)}\frac{V}{b} = 4\pi b\varepsilon_r\varepsilon_o \frac{a}{a-b}V \tag{2.48}$$

If $a \gg b$, the charge on an "isolated" spherical conductor can be obtained as a function of the electric field at its surface or its potential relative to infinity:

$$Q = 4\pi b^2 \varepsilon_r \varepsilon_o E_{n2} \approx 4\pi b \varepsilon_r \varepsilon_o V \tag{2.49}$$

For this "isolated" charged electrode, the outer spherical electrode is very far from the inner spherical electrode. Actually, the shape of the outer electrode in determining fields near the inner electrode is not important as along as the outer electrode is far away.

Although the previous voltage and field limits were given for two parallel electrodes, they can be used to estimate the voltage and field limits for other shapes. For small spherical objects (i.e., small b), the 320 V value limits the maximum charge on the spherical object. The corresponding charge is referred to as the voltage-limited charge. For large spherical objects, the 3 MV/m limits the maximum charge on the object. The corresponding charge is referred to as the field-limited charge. Using Equation (2.49), the critical radius is given as

$$b = \frac{V_c}{E_c} \approx \frac{320}{3 \times 10^6} \approx 100 \,\mu\text{m} \tag{2.50}$$

Above this radius, the charge is limited by the electric field. Below this radius, the charge is limited by the voltage. Therefore, for larger isolated particles greater than about 100 μm in air, the maximum surface charge is

$$\rho_{S,max} = \varepsilon_r \varepsilon_o E_c \approx (8.854 \times 10^{-12})(3 \times 10^6) \approx 27 \,\mu\text{C/m}^2 \tag{2.51}$$

(A value of 26 μC/m² is also seen in the literature.) This value is actually the maximum magnitude of the surface charge since the charge can also be negative. This maximum surface charge limit can also be used for flat conductors, since the same boundary condition, (2.46), is still valid. (It will be seen, however, that the surface charge can exceed this value for some situations.) The charge given in (2.51) is also referred to as the Gaussian surface charge limit. Again, for corona discharge, the surface charge limit is generally smaller since the field required is less than 3 MV/m. For smaller particles, less than about 100 μm, the electric field can be much greater than 3 MV/m before a discharge occurs. For example, for a particle radius of $b = 50$ μm in air, the maximum charge on the particle is

$$Q \approx 4\pi (50 \times 10^{-6})(8.854 \times 10^{-12})(320) \approx 1.8 \,\text{pC}$$

and the corresponding maximum field is

$$E_{max} = \frac{Q}{4\pi b^2 \varepsilon_o} = \frac{4\pi b \varepsilon_o V_c}{4\pi b^2 \varepsilon_o} = \frac{V_c}{b} \approx 6 \,\text{MV/m}$$

The field increases with decreasing particle size. Again, the 320 V and 3 MV/m criteria are actually for a uniform field. The field from a charged isolated spherical electrode is obviously nonuniform, and the breakdown field is greater than 3 MV/m for smaller particles. Felici estimated that for large spherical conductors with b greater than about 100 mm, the field is about 3 MV/m. For b about 10 mm, the field is about 6 MV/m. For b about 0.1 mm = 100 μm, the field is about 25 MV/m. For powders, usually the electric field from a bulk of particles is of interest rather than the field due to a single particle, if such a single particle can be isolated.

Even though the previous charge expressions were derived for a conducting spherical electrode, the maximum charge for an insulating spherical object is the same. Via Gauss' law, the electric fields outside a uniformly charged conducting and insulating spherical surface are identical. However, for a flat sheet

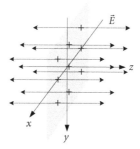

FIGURE 2.20   Electric field from a large sheet of surface charge.

of surface charge not residing on a conducting surface, the maximum charge density is greater. The electric field from the infinite sheet of uniformly distributed surface charge shown in Figure 2.20 is

$$\vec{E} = \begin{cases} \dfrac{\rho_S}{2\varepsilon_o}\hat{a}_z & \text{if } z > 0 \\[2mm] -\dfrac{\rho_S}{2\varepsilon_o}\hat{a}_z & \text{if } z < 0 \end{cases} \tag{2.52}$$

The magnitude of the field strength is identical on both sides of the sheet. The electric field near a uniformly charged flat sheet of plastic is equal to Equation (2.52). The maximum surface charge (magnitude), assuming a breakdown strength of 3 MV/m and air on both sides of the sheet, is

$$\rho_{S,max} = 2\varepsilon_r\varepsilon_o E_{max} \approx 2(8.854\times10^{-12})(3\times10^6) \approx 53\ \mu\text{C/m}^2$$

where ε_r is the relative permittivity of the surrounding medium, which for air is equal to one. This value is twice the value of the maximum surface charge on a conductor. If this infinite surface charge sheet were in direct contact with an insulating dielectric sheet, the maximum field magnitude outside of the dielectric sheet (in free space) would be the same. (Outside the large, flat dielectric sheet, the electric field from the bound charges in the dielectric cancels.) This is possibly why some individuals state that field lines are "transparent" to insulating materials (or insulating materials are "transparent" to electric fields).[11]

When surface charges of opposite sign are located on both sides of an insulating or a dielectric slab, the surface charge can be much greater than the Gaussian limit of 26 μC/m². This larger surface charge is the source for the high-energy propagating brush discharge discussed later. Referring to Figure 2.21,

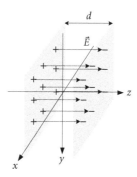

FIGURE 2.21   Electric field from two large sheets of surface charge of equal magnitude but opposite sign.

---

[11]This statement is generally not true. For example, a spherical ball of insulating material with a dielectric constant greater than one when placed in a uniform field will distort the uniform field.

the electric field distribution for two large parallel sheets of surface charge of equal magnitude but opposite sign is given by

$$\vec{E} = \begin{cases} \dfrac{\rho_S}{\varepsilon_o}\hat{a}_z & \text{if } |z| < \dfrac{d}{2} \\ 0 & \text{if } |z| > \dfrac{d}{2} \end{cases}$$

where d is the distance between the charged sheets, which are equally spaced from the xy plane. Free space exists between the sheets of charge. The electric field outside the two sheets of charge is zero. If the two charge sheets have different magnitudes, the field outside the sheets would not be zero. Superposition along with Equation (2.52) can be used to determine the fields in this case. When an insulating slab is present between these surface charges, the electric field inside the slab (between the surface charges) is given by[12]

$$\vec{E} = \begin{cases} \dfrac{\rho_S}{\varepsilon_r\varepsilon_o}\hat{a}_z & \text{if } |z| < \dfrac{d}{2} \\ 0 & \text{if } |z| > \dfrac{d}{2} \end{cases} \tag{2.53}$$

where ε_r is the relative permittivity or dielectric constant of the slab. This distribution is referred to as a double layer of charge. The electric field between the charges has been reduced by the addition of the dielectric insulating slab: fields are generated by the induced bound charges in the slab. If opposite sign charge exists on opposite sides of an insulator, the surface charge can be much greater than 26 μC/m². Although the electric field in the slab can exceed 3 MV/m, the breakdown strength of the insulating slab can be much greater than 3 MV/m, which corresponds to air. For example, imagine that opposite–sign surface charge exists on opposite sides of an insulating flat slab of polystyrene. This material has a dielectric constant of about 2.6 and a breakdown strength of about 30 MV/m. The maximum surface charge magnitude that can exist on the slab is then

$$\rho_{S,max} = \varepsilon_r\varepsilon_o E_{max} = 2.6\,(8.854\times10^{-12})(30\times10^6) \approx 690\ \mu\text{C/m}^2$$

In an industrial setting, a charge density of this magnitude is difficult to obtain or maintain.

The given electric field expression is for two infinite flat sheets of surface charge on both sides of an infinite flat slab. However, it can also be used to approximate the field from the surface charge along curved surfaces when the distance between the surface charges is small compared to the radius of curvature of the surface(s). Surface charge can build up along the inner surface of an insulating tube when insulating liquids and powders flow down the tube. If opposite–sign surface charge also exists along the outer surface of the tube, then the surface charge can become quite large. It is important to recognize that the source of the outer surface charge can be induced charge from a grounded conductor in direct contact with the outer surface. The grounded conductor can be a human hand or water. This double layer of charge can also be used to represent the tribocharging that can occur on both sides of a moving plastic sheet when in direct contact with a roller on each side of the sheet.

The maximum possible surface charge on an object is also a function of its environment. As an example of the influence of external conductors on the maximum surface charge limit, the electric field from a flat sheet of surface charge on a dielectric slab placed between two flat grounded electrodes will be analyzed.

---

[12]This expression is easily obtained by applying the boundary condition $D_{n2} - D_{n1} = \rho_S$ at either sheet of charge. Since the field outside the slab is zero, $D_{n2} - 0 = \varepsilon_r\varepsilon_o E_{n2} = \rho_S$.

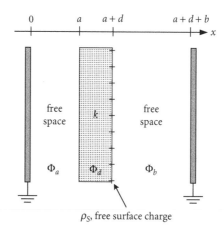

FIGURE 2.22 Surface charge on a dielectric slab located between two grounded plates.

The slab shown in Figure 2.22 is perfectly insulating with a dielectric constant of k. If the distance $a + d + b$ is small compared to the other dimensions, the potentials are one dimensional. Since no volume charges are present, Laplace's equation can be used to determine the general solution for the potential in each of the three regions:

$$\nabla^2 \Phi_a = \frac{d^2\Phi_a}{dx^2} = 0 \quad \Rightarrow \Phi_a = C_1 x + C_2 \quad \Rightarrow \vec{E}_a = -\nabla\Phi_a = -C_1 \hat{a}_x$$

$$\nabla^2 \Phi_d = \frac{d^2\Phi_d}{dx^2} = 0 \quad \Rightarrow \Phi_d = C_3 x + C_4 \quad \Rightarrow \vec{E}_d = -\nabla\Phi_d = -C_3 \hat{a}_x$$

$$\nabla^2 \Phi_b = \frac{d^2\Phi_b}{dx^2} = 0 \quad \Rightarrow \Phi_b = C_5 x + C_6 \quad \Rightarrow \vec{E}_b = -\nabla\Phi_b = -C_5 \hat{a}_x$$

Since there are six constants of integration, six boundary conditions are required. Four of the boundary conditions are obtained immediately since the potential is 0 V at each of the grounded electrodes and the potential is continuous at the dielectric-air interfaces:

$$C_1(0) + C_2 = 0, \quad C_5(a+d+b) + C_6 = 0$$

$$C_1(a) + C_2 = C_3(a) + C_4, \quad C_3(a+d) + C_4 = C_5(a+d) + C_6 \tag{2.54}$$

The remaining two boundary conditions involve the normal flux densities:

$$D_{na} = \varepsilon_o E_{na} = -\varepsilon_o C_1, \quad D_{nd} = k\varepsilon_o E_{nd} = -k\varepsilon_o C_3, \quad D_{nb} = \varepsilon_o E_{nb} = -\varepsilon_o C_5$$

At $x = a$, the $\vec{D}$ fields are continuous since no surface charge is present. However, at $x = a + d$, the $\vec{D}$ field jumps in value from one side of the interface to the other since surface charge is present:

$$-\varepsilon_o C_1 = -k\varepsilon_o C_3 \quad \text{valid} \ @ \ x = a$$

$$D_{nb} - D_{nd} = -\varepsilon_o C_5 - (-k\varepsilon_o C_3) = \rho_S \quad \text{valid} \ @ \ x = a + d \tag{2.55}$$

Using a symbolic math program, these six constants are easily solved for from Equations (2.54) and (2.55):

$$C_1 = bk \frac{\rho_S}{\varepsilon_o[d+k(a+b)]}, \quad C_2 = 0$$

$$C_3 = b \frac{\rho_S}{\varepsilon_o[d+k(a+b)]}, \quad C_4 = ab(k-1)\frac{\rho_S}{\varepsilon_o[d+k(a+b)]}$$

$$C_5 = -(d+ka)\frac{\rho_S}{\varepsilon_o[d+k(a+b)]}, \quad C_6 = (d+ka)(a+b+d)\frac{\rho_S}{\varepsilon_o[d+k(a+b)]}$$

The electric field in the air on both sides of the dielectric slab is of interest in this discussion (since the breakdown strength of the dielectric slab is likely higher than the surrounding air):

$$\vec{E}_a = -C_1\hat{a}_x = -bk\frac{\rho_S}{\varepsilon_o[d+k(a+b)]}\hat{a}_x = -\frac{\rho_S}{\varepsilon_o}\frac{k\dfrac{b}{d}}{1+k\left(\dfrac{a+b}{d}\right)}\hat{a}_x$$

$$(2.56)$$

$$\vec{E}_b = -C_5\hat{a}_x = (d+ka)\frac{\rho_S}{\varepsilon_o[d+k(a+b)]}\hat{a}_x = \frac{\rho_S}{\varepsilon_o}\frac{1+k\dfrac{a}{d}}{1+k\left(\dfrac{a+b}{d}\right)}\hat{a}_x$$

These results are obviously different from the electric field $E_n = \rho_S/(2\varepsilon_o)$ from an infinite sheet of surface charge or the electric field $E_n = \rho_S/\varepsilon_o$ from surface charge on a conductor. As a simple check, if both electrodes are far away from the slab,

$$\vec{E}_a \approx -\frac{\rho_S}{2\varepsilon_o}\hat{a}_x, \quad \vec{E}_b \approx \frac{\rho_S}{2\varepsilon_o}\hat{a}_x \quad \text{if } a \approx b \gg d$$

However, if only the grounded electrode at $x = a + d + b$ is far away,

$$\vec{E}_a \approx -\frac{\rho_S}{\varepsilon_o}\hat{a}_x, \quad \vec{E}_b \approx \frac{\rho_S}{\varepsilon_o}\frac{1+k\dfrac{a}{d}}{k\dfrac{b}{d}}\hat{a}_x \quad \text{if } b \gg d, b \gg a \qquad (2.57)$$

The electric field between $x = a + d$ and $x = a + d + b$ is substantially less than that from an isolated surface charge sheet:

$$\frac{\rho_S}{\varepsilon_o}\frac{1+k\dfrac{a}{d}}{k\dfrac{b}{d}} \ll \frac{\rho_S}{2\varepsilon_o} \quad \text{or} \quad d \ll \frac{1}{2}k(b-2a) \approx \frac{kb}{2} \qquad (2.58)$$

The reason the field is substantially less in this region is that the induced surface charge on the electrode at $x = 0$ is of the opposite sign to ρ_S. This induced charge on this nearby electrode generates an electric field of equal magnitude but opposite sign to the field generated by the surface charge on the slab. Thus, like the electric fields external to a charged parallel plate capacitor, the parallel surface charges of equal magnitude but opposite sign generate a small field for $x > a + d$. Although the magnitude of the surface charge on the slab can be larger than an isolated sheet before breaking down the air for $x > a + d$, the electric field for $x < a$ will break down first since its field is greater than $\rho_S/(2\varepsilon_o)$. If the dielectric slab is in direct contact with the grounded electrode at $x = 0$, then the electric field in the air region is

$$\vec{E}_b = \frac{\rho_S}{\varepsilon_o} \frac{1}{1+k\dfrac{b}{d}} \hat{a}_x \quad \text{if } a = 0 \tag{2.59}$$

Now, if the other electrode is far away,

$$\vec{E}_b = \frac{\rho_S}{\varepsilon_o} \frac{d}{kb} \hat{a}_x \quad \text{if } a = 0, b \gg d \tag{2.60}$$

Thus, as is stated in one of the references, "The maximum surface charge density can be increased by placing an earthed sheet behind the insulating plane." For very large surface charge densities, nonideal effects such as leakage along the "dirty" surface of the insulator limits the surface charge density. The modeling of this leakage current along the surface of a nonideal insulator is done later when discussing static-dissipative work surfaces.

2.9   Grounded Conducting Objects and Charged Insulating Surfaces

How do nearby conducting objects affect the likelihood of an electrostatic discharge from charged insulating surfaces? [Asano; Smythe; Weber]

Explicit analytical expressions for the potential and electric field for charge distributions near conducting surfaces are not plentiful. For some special situations, such as a point charge above an infinitely large conducting plane or an infinitely long line charge parallel to an infinitely long cylinder, analytical expressions are available. The classical method of obtaining the electric field in these cases is via the method of images.

Charge near a ground plane induces charge of the opposite sign along the plane's surface. This charge distribution along the plane's surface is nonuniform. Rather than working with a ground plane with a nonuniform distribution of surface charge, it is sometimes easier to replace the ground plane with an equivalent image charge distribution. Essentially, the effect of a grounded conducting object on the field from a charge distribution is obtained by replacing the grounded object with an appropriately placed image charge distribution. For example, for a point or line charge above a flat plane, the image charge distribution is of the opposite sign and equal distance from the flat conductor as shown in Figure 2.23. In the region above the ground plane, the total electric field from the actual charge distribution and image charge distribution (without the ground plane present) is identical to the electric field from the actual charge distribution with the ground plane. Obviously, the electric fields are different for the two situations below the conducting plane since the electric field is zero in a perfect conductor.

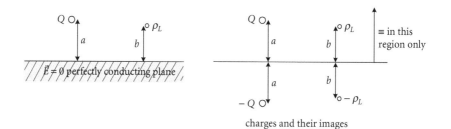

charges and their images

FIGURE 2.23   Point and line charge above a perfectly conducting plane and their image charge equivalent (without the ground plane).

FIGURE 2.24 Tangential components of the electric fields from both the original real charge and image charge are equal in magnitude but in opposite directions along the interface corresponding to the surface of the conductor.

Whether the interface involves a conductor or dielectric, the standard boundary conditions can be checked along the interface using the fields generated by the original and image charges. For perfect conductors, the tangential electric field everywhere along the conductor's surface must be zero. This is illustrated in Figure 2.24 for a single "ray" of electric field emanating from the original and image charges. For dielectric interfaces, a few simple checks can be performed. The most obvious check is to eliminate the interface by allowing the dielectric constants of the two regions to be equal. In this case, the image charges should vanish. A second, not so obvious check, is to allow the permittivity of one region to approach infinity. A charge-neutral dielectric with infinite permittivity simulates a perfectly conducting *uncharged* conductor (but without the ability to conduct charge). For example, referring to Table 2.3, the equivalent image system for a point charge in a dielectric with a permittivity of ε_1 at a distance a from an infinite flat interface where the permittivity changes to ε_2 is shown in Figure 2.25. Notice that the image charge when determining the field in the upper region is equal to $-Q$ when ε_2 is infinitely large:

$$\lim_{\varepsilon_2\to\infty}\left[-Q\left(\frac{\varepsilon_2-\varepsilon_1}{\varepsilon_2+\varepsilon_1}\right)\right]=-Q$$

For a perfect conductor, $-Q$ is also the image charge. Furthermore, when determining the field in the lower region where the permittivity is infinite, the charge has a value of zero:

$$\lim_{\varepsilon_2\to\infty}\left[Q\left(\frac{2\varepsilon_1}{\varepsilon_2+\varepsilon_1}\right)\right]=0$$

Since the charge is zero, no electric field is produced in the lower region of infinite permittivity. The electric field inside a perfect conductor is zero. It is also insightful to examine the consequence of allowing the permittivity, ε_2, approach infinity for a dielectric cylinder, with an outside, parallel line charge. Again, this situation is simulating an uncharged cylindrical conductor. Referring to the image equivalent system shown in Figure 2.26, it is fairly obvious that in the limit as the cylinder's permittivity approaches infinity, the image consists of two line charges: one image exists along the axis with a value of ρ_L and the other image exists at R^2/a from the axis with a value of $-\rho_L$. What may be initially perplexing is that another entry in Table 2.3 indicates that only one image charge exists for a line charge outside and parallel to a *grounded* conducting cylinder. The image, in this case, has a value of $-\rho_L$ at a distance of R^2/a from the axis. However, the grounded conducting cylinder has a net charge on its surface, induced from the outside line charge. The simulated conductor, on the other hand, is uncharged or charge neutral: its total image charge is $\rho_L+(-\rho_L)=0$. Finally, note that the value of the image charge does not have to be equal but opposite to the original charge. For example, for a point charge outside a grounded sphere, the image charge has a value of $-Q(R/a)$ not just $-Q$. The electric field lines from the real charge Q can terminate on the grounded sphere or the other ground located at infinity. If the real charge is very far from the grounded sphere, then $a\gg R$ and the image charge is small—most of the field lines are terminating on

TABLE 2.3   Charge near a Conductor or Dielectric Interface and its Equivalent Image without the Conductor or Dielectric Interface

Actual Configuration	Equivalent Image Configuration

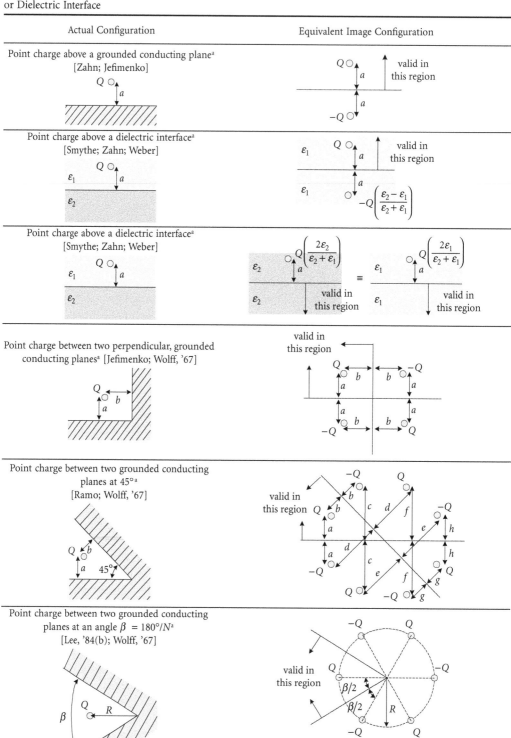

Point charge above a grounded conducting plane[a] [Zahn; Jefimenko]

Point charge above a dielectric interface[a] [Smythe; Zahn; Weber]

Point charge above a dielectric interface[a] [Smythe; Zahn; Weber]

Point charge between two perpendicular, grounded conducting planes[a] [Jefimenko; Wolff, '67]

Point charge between two grounded conducting planes at 45°[a] [Ramo; Wolff, '67]

Point charge between two grounded conducting planes at an angle $\beta = 180°/N$[a] [Lee, '84(b); Wolff, '67]

where N is an integer

a total of $(2N - 1)$ alternating sign image charges equally spaced along a circle of radius R

TABLE 2.3 Charge near a Conductor or Dielectric Interface and its Equivalent Image without the Conductor or Dielectric Interface (Continued)

Actual Configuration	Equivalent Image Configuration
Point charge outside a grounded, thin conducting sphere [Zahn; Jefimenko] 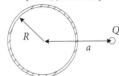	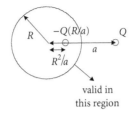 valid in this region
Point charge inside a grounded, thin conducting sphere [Zahn; Jefimenko]	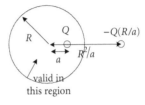 valid in this region
Infinitely long line charge outside and parallel to a grounded, thin conducting cylinder [Zahn; Jefimenko] 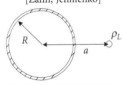	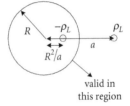 valid in this region
Infinitely long line charge inside and parallel to a grounded, thin conducting cylinder [Zahn; Jefimenko]	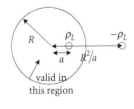 valid in this region
Infinitely long line charge outside and parallel to a cylindrical dielectric interface [Smythe; Zahn] 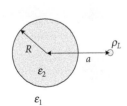	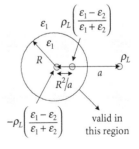 valid in this region

TABLE 2.3   Charge near a Conductor or Dielectric Interface and Its Equivalent Image without the Conductor or Dielectric Interface (Continued)

Actual Configuration	Equivalent Image Configuration

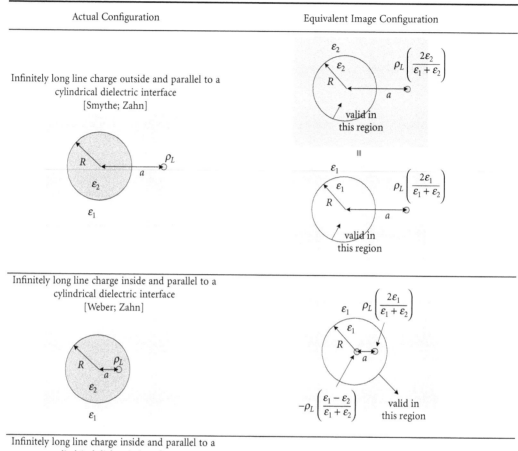

Infinitely long line charge outside and parallel to a cylindrical dielectric interface
[Smythe; Zahn]

Infinitely long line charge inside and parallel to a cylindrical dielectric interface
[Weber; Zahn]

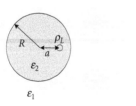

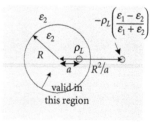

Infinitely long line charge inside and parallel to a cylindrical dielectric interface
[Weber; Zahn]

[a]This image equivalent configuration also applies to an infinitely long line charge parallel to the interface(s) by letting $Q = \rho_L$.

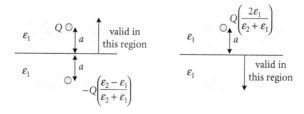

FIGURE 2.25   Image equivalent for a point charge Q a distance a from a dielectric interface.

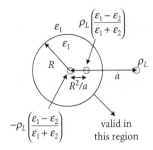

FIGURE 2.26   Image line charges for a line charge outside of a dielectric rod.

the ground at infinity. However, when the real charge is very close to the grounded sphere, then $a \approx R$ and the image charge is about $-Q$ — most of the field lines are terminating on the grounded sphere.[13]

With the method of images, both the location and strength of the image charge can be a function of the distance of the charge from the surface. Also, a single "real" charge can generate more than one image charge. To expand the size of Table 2.3, superposition can be applied to generate other charge distributions such as a charged ring or disk above a ground plane. In this table, the flat conducting planes (and dielectric interfaces) are theoretically infinite in width and length, and the cylindrical surfaces are theoretically infinite in length. However, the expressions can be used with excellent results if the distance between the charge distribution and surface is small compared to the width and length of the plane(s) and the length of the cylinder. For those configurations involving conductors, the expressions were derived assuming the conductors were perfectly conducting; however, the results are usually quite adequate for good conductors. Also, it is not necessary that the conducting planes be infinite in thickness. The method of images for electric charges can also be applied to thin good-conducting sheets. In Table 2.3, the point charges are represented by Q's and line charges by ρ_L's.

In many situations, the voltage across two objects is fixed. As one object approaches the other, the capacitance relationship, $Q = CV$, can be used if the capacitance is known as a function of position. As two objects approach each other, C increases and, for a fixed V, the charge on the objects must increase. If there are only two objects involved, and since charge is conserved, charge is transferred between the objects with help from the voltage supply. However, if the Q on each object is fixed but not V, as the distance between the objects decrease, C increases and V must decrease.

Although the method of images can be used in some numerical work involving realistically shaped charged objects, the distribution of the charge on these objects is required (or must be adjusted to meet the necessary boundary conditions). The charge on the objects as they approach a conducting plane, for example, will redistribute itself. For example, for an isolated sphere, the charge may be initially uniformly distributed on the sphere. When this sphere approaches a conducting plane, the surface charge tends to move toward the point where the sphere is closest to the plane.

The previous work involving surface charge on a flat dielectric slab will be used to show that the potential and electric field can change as a charged object approaches a conducting plane. The flat surface charge could be representing a tribocharged printed circuit board or the bottom of a plastic cup. Since this is a one-dimensional problem, the charge on the slab does not redistribute itself.

---

[13]The magnitude of the image charge for an infinite-length line charge parallel to an infinite-length grounded cylinder is not a function of position. The length of the grounded cylinder is infinite while the radius of the grounded sphere is finite.

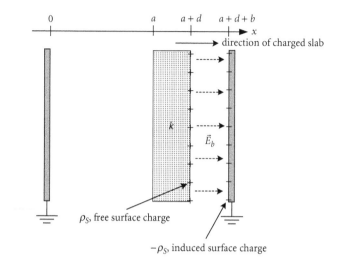

FIGURE 2.27 Dielectric slab with a charged surface approaching a grounded plane.

The electric field between the grounded electrode and surface charge shown in Figure 2.27 is from Equation (2.56)

$$\vec{E}_b = \frac{\rho_S}{\varepsilon_o} \frac{1+k\dfrac{a}{d}}{1+k\left(\dfrac{a+b}{d}\right)}\hat{a}_x$$

The potential of the slab at the location of the surface charge is

$$\Phi_d(x=a+d)=C_3(a+d)+C_4$$

$$=b(a+d)\frac{\rho_S}{\varepsilon_o[d+k(a+b)]}+ab(k-1)\frac{\rho_S}{\varepsilon_o[d+k(a+b)]}$$

When the surface charge is centered between the two electrodes, the electric field is

$$\vec{E}_b = \frac{\rho_S}{\varepsilon_o} \frac{1+k\dfrac{a}{d}}{1+k\left(\dfrac{2a+d}{d}\right)}\hat{a}_x < \frac{\rho_S}{\varepsilon_o}\hat{a}_x \quad \text{if } b=a+d$$

When the electrode at $x = a + d + b$ is close to the surface charge, the expressions for the field and potential are

$$\vec{E}_b \approx \frac{\rho_S}{\varepsilon_o}\hat{a}_x, \quad \Phi_d(a+d)\approx b\frac{\rho_S}{\varepsilon_o} \quad \text{if } a \gg d,\, a \gg b \qquad (2.61)$$

As the sheet approaches the grounded conductor, the electric field increases to the maximum possible value for a conducting surface. Since an electrostatic discharge between two objects is a strong function

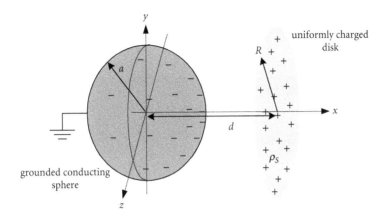

FIGURE 2.28   Uniformly charged disk a distance of d from the center of a grounded conducting sphere.

of the electric field, this increases the likelihood of a discharge. The potential of the "floating" slab approaches 0 V as b approaches zero. Because the charge is fixed on the slab and the voltage is not, the electric field can remain constant while the potential decreases.

A manageable expression for the electric field directly between a uniformly charged insulating disk and a grounded spherical conductor exists. This expression can be used to estimate the strength of the electric field between a grounded object of a given curvature as it approaches a charged insulating surface. This information is also helpful in estimating the likelihood of an electrostatic discharge. The method of images is used to determine the integral expression for the potential distribution at any point (x, y, z) between the sphere and disk shown in Figure 2.28. For those students fortunate enough to work the related problem, the integral expression for the potential is given by

$$\Phi = \frac{\rho_S}{4\pi\varepsilon_o}\int_0^{2\pi}\int_0^{R}\frac{\rho'd\rho'd\phi'}{\sqrt{(x-d)^2+(y-\rho'\sin\phi')^2+(z-\rho'\cos\phi')^2}}$$

$$-\frac{\rho_S}{4\pi\varepsilon_o}\int_0^{2\pi}\int_0^{R}\frac{a}{\sqrt{d^2+(\rho')^2}}\frac{\rho'd\rho'd\phi'}{\sqrt{\left(x-\frac{a^2 d}{d^2+(\rho')^2}\right)^2+\left(y-\frac{a^2\rho'\sin\phi'}{d^2+(\rho')^2}\right)^2+\left(z-\frac{a^2\rho'\cos\phi'}{d^2+(\rho')^2}\right)^2}} \qquad (2.62)$$

where a is the radius of the grounded conducting spherical object, R is the radius of the disk of surface charge, and d is the distance from the center of the sphere to the center of the disk. The first integral is the potential due to the surface charge, ρ_S, on the charged disk. The second integral is the potential due to the induced charge on the grounded sphere (also referred to as the image charge). The disk is centered about the x axis and the sphere is centered about the origin. Being a surface charge, the disk is thin. As approximately illustrated in Figure 2.28, the induced surface charge concentration on the sphere is greatest near the charged disk. The largest value of the electric field *along the surface of the sphere* is also at the nearest point to the disk (i.e., along the x axis at $x = a$). The electric field is the negative gradient of the potential, which involves differentiation with respect to x, y, and z. In general, it is necessary to determine the potential everywhere near the point(s) of interest if the electric field is to be determined from the potential distribution.[14] However, in this particular configuration, if the electric field is only desired along the x axis, the electric field can be obtained from the potential distribution along the x axis. It is not necessary to determine the potential everywhere around the x axis. Since the electric field is entirely in the x direction along the x axis for this configuration, the electric field can be obtained from the negative gradient of the potential distribution along the x axis: $\vec{E} = -\nabla\Phi = (-d\Phi/dx)\hat{a}_x$. Fortunately, the integrals for the potential distribution can be analytically evaluated along the x axis. Along the x axis ($y = 0$, $z = 0$), these integrals reduce to $(\cos^2\phi' + \sin^2\phi' = 1)$

$$\Phi = \frac{\rho_S}{4\pi\varepsilon_o}\int_0^{2\pi}\int_0^R \frac{\rho'd\rho'd\phi'}{\sqrt{(x-d)^2+(\rho')^2}}$$

$$-\frac{\rho_S}{4\pi\varepsilon_o}\int_0^{2\pi}\int_0^R \frac{a}{\sqrt{d^2+(\rho')^2}} \frac{\rho'd\rho'd\phi'}{\sqrt{\left(x-\dfrac{a^2d}{d^2+(\rho')^2}\right)^2 + \left(\dfrac{a^2\rho'}{d^2+(\rho')^2}\right)^2}} \qquad (2.63)$$

$$= \frac{\rho_S}{4\pi\varepsilon_o}\int_0^{2\pi}\int_0^R \frac{\rho'd\rho'd\phi'}{\sqrt{(x-d)^2+(\rho')^2}} - \frac{\rho_S}{4\pi\varepsilon_o}\int_0^{2\pi}\int_0^R \frac{a\rho'd\rho'd\phi'}{\sqrt{(xd-a^2)^2+(x\rho')^2}}$$

These integrals can be evaluated using an integral table, a classical method learned in freshman calculus, or a symbolic manipulation package:

$$\Phi = \frac{\rho_S}{2\varepsilon_o}\left[\sqrt{(x-d)^2+R^2}+x-d-\frac{a\sqrt{(xd-a^2)^2+R^2x^2}}{x^2}+\frac{ad}{x}-\frac{a^3}{x^2}\right] \quad a\leq x\leq d \qquad (2.64)$$

The potential along the x axis between the sphere and disk can be determined from this function. As a simple check, at the surface of the grounded sphere, the potential is zero:

$$\Phi(x=a) = \frac{\rho_S}{2\varepsilon_o}\left[\sqrt{(a-d)^2+R^2}+a-d-\frac{a\sqrt{(ad-a^2)^2+R^2a^2}}{a^2}+\frac{ad}{a}-\frac{a^3}{a^2}\right] = 0$$

The potential is zero inside the sphere since no charge is present inside the equipotential, conducting, closed spherical surface. The electric field along the x axis is given by the negative gradient of the potential:

$$E_x = -\frac{d\Phi}{dx}$$

$$= -\frac{\rho_S}{2\varepsilon_o}\left\{\begin{array}{l}\dfrac{x-d}{\sqrt{(x-d)^2+R^2}}+1\\[2ex]+\dfrac{-ax^2[(dx-a^2)d+R^2x]+2ax[(dx-a^2)^2+R^2x^2]}{x^4\sqrt{(dx-a^2)^2+R^2x^2}}\\[2ex]-\dfrac{ad}{x^2}+\dfrac{2a^3}{x^3}\end{array}\right\} \quad a\leq x\leq d \qquad (2.65)$$

---

[14]The derivative(s) of a function should be taken before specific values are substituted into the function.

The maximum value of the electric field along the surface of the grounded sphere is at $x = a$. At this point, Equation (2.65) can be simplified to

$$E_x(x = a) = -\frac{\rho_S}{2\varepsilon_o}\left[3 - \frac{d}{a} + \frac{(d-a)\left(\frac{d}{a}-3\right)+\frac{R^2}{a}}{\sqrt{(a-d)^2 + R^2}}\right] \tag{2.66}$$

If the sphere is very close to the charged disk, the field at the spherical surface is

$$E_x(x = a) \approx -\frac{\rho_S}{2\varepsilon_o}\left(2 + \frac{R}{a}\right) \quad \text{if } a \approx d \tag{2.67}$$

For positive surface charge, the electric field is in the negative x direction, directed from the positively charged disk to the negatively induced-charged sphere. While, if the sphere is very far from the charged disk,

$$E_x(x = a) \approx -\frac{\rho_S}{2\varepsilon_o}\left(-\frac{d}{a} + \frac{d^2 + R^2}{a\sqrt{d^2 + R^2}}\right) = -\frac{\rho_S}{2\varepsilon_o}\left(\frac{\sqrt{d^2 + R^2} - d}{a}\right) \quad \text{if } d \gg a \tag{2.68}$$

As expected, when $d \gg R$ and $d \gg a$, the electric field approaches zero since the disk is small and far away. What is possibly unexpected is that when the disk (radius) is large compared to a and d, the electric field between the grounded sphere and disk along the x axis is a function of R:

$$E_x \approx -\frac{\rho_S}{2\varepsilon_o}\frac{aR}{x^2} \quad \text{if } R \gg a, R \gg d, R \gg x \tag{2.69}$$

Recall that the magnitude of the electric field from an infinite plane of surface charge is $|\rho_S|/2\varepsilon_o$, directed normally from the surface. As seen by this last expression, at the center of the surface of the disk, the electric field is not equal to this value. The induced charge on the sphere is affecting the electric field from the disk. For that matter, unlike the infinite surface charge where all grounds are at infinity, one ground for this configuration is the nearby spherical conductor. The electric fields emanating from the surface charge on the disk (assuming $\rho_S > 0$) must terminate on the grounded spherical conductor or the ground located at infinity. The total charge on the sphere is given by

$$Q_s = -\left(\pi R^2 \rho_S\right)\frac{2a(\sqrt{d^2 + R^2} - d)}{R^2} \tag{2.70}$$

where $\pi R^2 \rho_S$ is the total charge on the disk. For disks close to the sphere, the charge induced on the sphere is directly related to the sphere's radius:

$$Q_s \approx -\left(\pi R^2 \rho_S\right)\frac{2a}{R} \quad \text{if } R \gg d \tag{2.71}$$

When $R > 2a$, the charge on the sphere is less than on the disk. The remaining induced charge (for the electric fields from the disk to terminate on), must be on the other ground located at infinity. The electric field between two conducting objects with a fixed potential difference across them varies in a different manner than a conducting grounded object and a "floating" charged insulating object. If the separation distance is large compared to the disk radius, however, the charge on the sphere is directly a function of the ratio of the sphere's radius to the separation distance:

$$Q_s = -\left(\pi R^2 \rho_S\right) \frac{2a\left[d\sqrt{1+\left(\dfrac{R}{d}\right)^2}-d\right]}{R^2}$$

$$\approx -\left(\pi R^2 \rho_S\right) \frac{2a\left\{d\left[1+\dfrac{1}{2}\left(\dfrac{R}{d}\right)^2\right]-d\right\}}{R^2} = -\left(\pi R^2 \rho_S\right)\frac{a}{d} \quad \text{if } d \gg R$$

(2.72)

(The binomial expansion was used to obtain this approximation.) Furthermore, when the disk is small compared to the sphere and it is close to the sphere ($a \approx d \gg R$), the magnitude of the induced charge on the grounded sphere is nearly the same as the charge on the disk. In this case, almost all of the electric field from the disk charge terminates (if the charge is positive) on the grounded sphere.

In Mathcad 2.1, the electric field variation along the x axis is plotted for various sphere radii. The distance between the disk and nearest part of the sphere's surface is kept constant. In the second plot in Mathcad 2.1, the radius of the disk is varied while the other parameters are kept constant. The electric field increases everywhere along the x axis as the total charge on the disk increases. When the sphere radius is small compared to the radius of the disk, the maximum electric field along the x axis is at the sphere's surface. However, when the sphere's radius is not small compared to the radius of the disk, the maximum electric field can occur at the disk. When the disk is small compared to the sphere, the fields from the disk "spread out" to the induced charge on the large spherical surface. Finally, in Mathcad 2.2 the electric field is plotted for a constant sphere and disk radius for several spacings. As expected, as the distance between the sphere and disk decreases, the electric field increases. It was previously shown via Equation (2.69) that the electric field along the x axis is not a function of the spacing, d, for large disks. This is numerically verified. The electric field is a function of the radius of the grounded sphere. The smaller the grounded sphere, the smaller the electric field.

A potential contour plot is given in Figure 2.29. It is obtained from Equation (2.63), the integral expression for the potential anywhere outside the sphere. The electric field is normal to the potential contours, and it is greatest where the contours are the closest. Thus, the electric field is clearly the strongest where the grounded spherical contour and charged insulating disk are the closest. The surface of the grounded spherical conductor, although not labeled, is along the 0 V contour and nearly tangential to the 500 V contour. The insulating surface, however, is not along an equipotential contour. Equipotentials are actually intersecting the charged disk in this figure. Generally, insulating surfaces are not equipotentials, although they can be for certain symmetrical configurations.

As a charged object approaches a conducting plane (and other shapes), charge is induced on the plane. The resultant charge flow along the conducting plane is a real current. Charge can also be induced on imperfect insulating planes (and other shapes). If the speed of the approaching object is fast compared to the charge response time of the imperfect plane (often measured via the time constant or equivalent time constant of the plane), then the induced charge on the plane when the object is nearby should be initially small (compared to the charge of the approaching object). The relationship between the speed of the approaching object and charge response time is important in the design of static-dissipative work surfaces and bags. The charge on resistive or insulative objects is sluggish. To begin and sustain an arc, a minimum current is required. Insulating surfaces naturally limit the charge flow or current required to begin and sustain an arc.

$$\rho_S := 10^{-7} \quad \varepsilon_o := 8.854 \cdot 10^{-12} \quad a := 0.02 \quad d := 0.1 \quad R := 0.01$$

$$x := 0.001, 0.002 .. 0.1$$

$$E_{xp}(a, d, R, x) := \frac{-\rho_S}{2 \cdot \varepsilon_o} \cdot \left[\frac{x - d}{\sqrt{(x - d)^2 + R^2}} + 1 \dots \right.$$

$$\left. + \frac{\left[-a \cdot x^2 \cdot \left[\left(d \cdot x - a^2 \right) \cdot d + R^2 \cdot x \right] \right] + 2 \cdot a \cdot x \cdot \left[\left(d \cdot x - a^2 \right)^2 + R^2 \cdot x^2 \right]}{x^4 \cdot \sqrt{\left(d \cdot x - a^2 \right)^2 + R^2 \cdot x^2}} - \frac{a \cdot d}{x^2} + \frac{2 \cdot a^3}{x^3} \right]$$

$$E_x\left(a_t, d_t, R_t, x\right) := if\left(a_t \le x \le d_t, E_{xp}\left(a_t, d_t, R_t, x\right), 0\right)$$

distance from spherical surface to disk constant

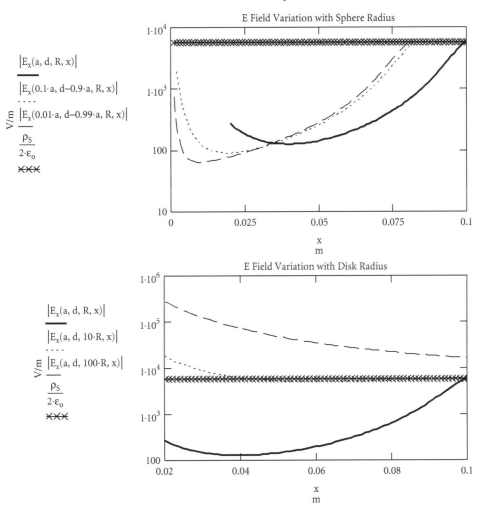

MATHCAD 2.1 Electric field between the grounded sphere and charged disk as a function of the sphere radius and disk radius.

$$\rho_S := 10^{-7} \quad \varepsilon_o := 8.854 \cdot 10^{-12} \quad a := 0.02 \quad d := 0.1 \quad R := 0.01$$

$$x := 0.001, 0.002 .. 0.1$$

$$E_{xp}(a, d, R, x) := \frac{-\rho_S}{2 \cdot \varepsilon_o} \cdot \left[\frac{x - d}{\sqrt{(x-d)^2 + R^2}} + 1 \ldots \right.$$

$$\left. + \frac{\left[-a \cdot x^2 \cdot \left[\left(d \cdot x - a^2\right) \cdot d + R^2 \cdot x\right]\right] + 2 \cdot a \cdot x \cdot \left[\left(d \cdot x - a^2\right)^2 + R^2 \cdot x^2\right]}{x^4 \cdot \sqrt{\left(d \cdot x - a^2\right)^2 + R^2 \cdot x^2}} - \frac{a \cdot d}{x^2} + \frac{2 \cdot a^3}{x^3} \right]$$

$$E_x(a_t, d_t, R_t, x) := \text{if}(a_t \le x \le d_t, \ E_{xp}(a_t, d_t, R_t, x), \ 0)$$

E Field Variation with Decreasing Spacing

$$\left| E_x(a, d, R, x) \right|$$

$$\left| E_x(a, 0.8 \cdot d, R, x) \right|$$
- - - -
$$\frac{V}{m} \quad \left| E_x(a, 0.4 \cdot d, R, x) \right|$$

$$\frac{\rho_S}{2 \cdot \varepsilon_o}$$
× × ×

E Field Variation with Decreasing Spacing

$$\left| E_x\left(a, d, 10^2 \cdot R, x\right) \right|$$

$$\left| E_x\left(a, 0.8 \cdot d, 10^2 \cdot R, x\right) \right|$$
+ + +
$$\frac{V}{m} \quad \left| E_x\left(a, 0.4 \cdot d, 10^2 \cdot R, x\right) \right|$$

$$\frac{\rho_S}{2 \cdot \varepsilon_o}$$
× × ×

MATHCAD 2.2 Electric field between the grounded sphere and charged disk as a function of the spacing between the sphere and disk.

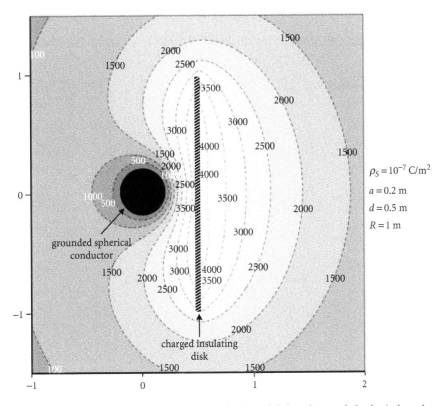

$\rho_S = 10^{-7}$ C/m^2

$a = 0.2$ m

$d = 0.5$ m

$R = 1$ m

FIGURE 2.29 Equipotential contours between a positively charged disk and grounded spherical conductor.

2.10   Charge Accumulation along Interfaces

Why does charge tend to accumulate along interfaces where the conductivity or permittivity changes? When does this charge not accumulate along these interfaces? How can the charge be dissipated or removed? [Haus; Zahn]

Since it is generally desirable to dissipate or neutralize charge on objects to reduce the probability of electrostatic discharge, it is important to understand where charge can accumulate. The conductivity and permittivity of a material and its surroundings determine the rate of discharge. Conductivity is partially a measure of the number of free charges available for "carrying electricity" in a material. Thus, it should not be too surprising that when the conductivity changes in a smooth fashion, the charge can build up smoothly as a volume charge or that when the conductivity changes abruptly, the charge can build up abruptly as a surface charge. Although the charges associated with the permittivity of a material are not free but bound, bound charges can also accumulate. When the permittivity varies smoothly, a volume bound charge can build up, and when the permittivity varies abruptly, a surface bound charge can build up. Charge along the surface of a perfect conductor in free space is a simple example of a surface charge along an interface where there is an abrupt change in the conductivity.

To add some mathematical rigor to this discussion, Gauss' law and the charge conservation law are used. It is assumed that both the conductivity, σ, and permittivity, ε, can vary with position. Gauss' law in differential form is

$$\nabla \cdot \vec{D} = \nabla \cdot (\varepsilon \vec{E}) = \rho_V$$

where ρ_V is the volume charge. Because the permittivity can vary with position, it cannot be pulled directly past the divergence operation. But, using a vector property similar to the product rule in calculus (see the discussion on charge decay for a similar operation), this equation can be written as

$$(\nabla \varepsilon) \cdot \vec{E} + \varepsilon \nabla \cdot \vec{E} = \rho_V \quad \Rightarrow \nabla \cdot \vec{E} = \frac{\rho_V}{\varepsilon} - \frac{(\nabla \varepsilon) \cdot \vec{E}}{\varepsilon} \tag{2.73}$$

Recall that the charge conservation equation is

$$\nabla \cdot \vec{J} + \frac{\partial \rho_V}{\partial t} = 0$$

Assuming that the current density is adequately described by Ohms law, $\vec{J} = \sigma \vec{E}$, where the conductivity can vary with position,

$$\nabla \cdot (\sigma \vec{E}) + \frac{\partial \rho_V}{\partial t} = (\nabla \sigma) \cdot \vec{E} + \sigma \nabla \cdot \vec{E} + \frac{\partial \rho_V}{\partial t} = 0$$

Substituting Equation (2.73),

$$(\nabla \sigma) \cdot \vec{E} + \sigma \left[\frac{\rho_V}{\varepsilon} - \frac{(\nabla \varepsilon) \cdot \vec{E}}{\varepsilon} \right] + \frac{\partial \rho_V}{\partial t} = 0$$

or

$$\frac{\partial \rho_V}{\partial t} + \frac{\sigma}{\varepsilon} \rho_V = \left[\frac{\sigma}{\varepsilon} (\nabla \varepsilon) - (\nabla \sigma) \right] \cdot \vec{E} \tag{2.74}$$

Although it is probably not obvious, this partial differential equation indicates that a volume charge can exist (at least temporarily) where the permittivity and/or conductivity varies in the direction of the electric field. (A permittivity variation is noted by a nonzero value in $\nabla \varepsilon$, and a conductivity variation is noted by a nonzero value in $\nabla \sigma$.) For dc steady-state situations, this relationship is simpler:

$$\rho_V = \left[(\nabla \varepsilon) - \frac{\varepsilon}{\sigma} (\nabla \sigma) \right] \cdot \vec{E} \quad \text{if } \frac{\partial \rho_V}{\partial t} = 0 \tag{2.75}$$

Of course, for the special situation where $\sigma \nabla \varepsilon = \varepsilon \nabla \sigma$ there is no volume charge accumulation. In Figure 2.30, variation in the permittivity between two charged parallel plates is indicated via the gradation in the tone. The permittivity varies continuously in the y direction for the left region and in the x direction for the right region. In the left region, volume charge will exist due to the permittivity variation since the electric field is in the same direction as $\nabla \varepsilon (y)$. However, in the right region, the electric field and $\nabla \varepsilon (x)$ are perpendicular, and their dot product is zero. Hence, no volume charge will exist in this right region due to permittivity variation. The conductivity, of course, can also vary with position. For example, in high-voltage cables, the conductivity of the oil insulation between power conductors can vary with position because of temperature variations between the conductors.

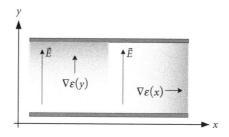

FIGURE 2.30 Spatial variation of the permittivity between two charged plates.

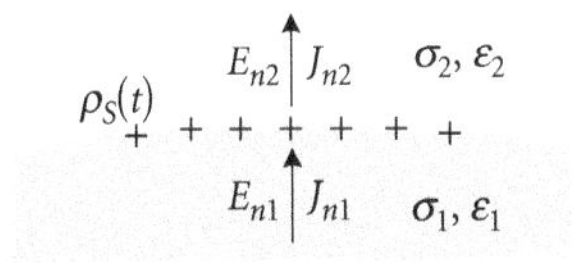

$$\rho_S(t) \quad \begin{array}{c} E_{n2} \uparrow J_{n2} \quad \sigma_2, \varepsilon_2 \\ + \ + \ + \ + \ + \ + \\ E_{n1} \uparrow J_{n1} \quad \sigma_1, \varepsilon_1 \end{array}$$

FIGURE 2.31 Interface where both the conductivity and permittivity could abruptly change.

In Figure 2.30, the permittivity was shown varying in a continuous fashion in the two individual halves. Probably, a step or sudden change in the permittivity or conductivity is more familiar to students. The boundary conditions based on the discontinuity of the normal flux density and on the charge conservation equation are

$$D_{n2} - D_{n1} = \rho_S, \quad J_{n2} - J_{n1} = -\frac{d\rho_S}{dt} \tag{2.76}$$

Referring to Figure 2.31, using Equation (2.76), and assuming that $\vec{J} = \sigma \vec{E}$ applies for both sides of the interface,

$$J_{n2} - J_{n1} = -\frac{d}{dt}(D_{n2} - D_{n1}) \tag{2.77}$$

$$\sigma_2 E_{n2} - \sigma_1 E_{n1} = -\frac{d}{dt}(\varepsilon_2 E_{n2} - \varepsilon_1 E_{n1}) \tag{2.78}$$

This boundary condition can be applied along an interface that has a sudden jump in both the conductivity and permittivity. As was shown, volume charge can exist where the conductivity and/or permittivity varies continuously. As a special case of this time-varying boundary condition, for dc steady-state conditions (or when the rate of change of the charge has negligible influence on the current), the current density across an interface is continuous:

$$J_{n2} - J_{n1} = 0 \quad \Rightarrow \sigma_2 E_{n2} = \sigma_1 E_{n1} \tag{2.79}$$

The surface charge along the interface can be determined from the normal flux densities:

$$\rho_S = D_{n2} - D_{n1} = \varepsilon_2 E_{n2} - \varepsilon_1 E_{n1} = \varepsilon_2 E_{n2} - \varepsilon_1 \frac{\sigma_2}{\sigma_1} E_{n2} = \left(\varepsilon_2 - \varepsilon_1 \frac{\sigma_2}{\sigma_1} \right) E_{n2}$$

$$= \left(1 - \frac{\varepsilon_1 \sigma_2}{\varepsilon_2 \sigma_1} \right) \varepsilon_2 E_{n2} \quad \text{for dc steady-state conditions} \tag{2.80}$$

$$= \left(\frac{\varepsilon_2 \sigma_1}{\varepsilon_1 \sigma_2} - 1 \right) \varepsilon_1 E_{n1} \quad \text{for dc steady-state conditions}$$

An equivalent form of this condition is

$$\rho_S = \left(1 - \frac{\varepsilon_1 \sigma_2}{\varepsilon_2 \sigma_1} \right) \varepsilon_2 \hat{a}_n \cdot \vec{E}_2 \quad \text{for dc steady-state conditions} \tag{2.81}$$

where $\hat{a}_n$ is the unit vector normal or perpendicular to the interface directed from region 1 to region 2. As a simple use of this dc steady-state boundary condition, assume that region 2 is a perfect insulator so that $\sigma_2 = 0$, or region 1 is a perfect conductor so that $\sigma_1 \to \infty$, or $\sigma_2 \ll \sigma_1$. In all three cases, assuming $\varepsilon_1/\varepsilon_2$ is not extreme,

$$\rho_S = \varepsilon_2 E_{n2} \tag{2.82}$$

This equation is a boundary condition for perfect conductors. For sinusoidal steady-state conditions, a simple relationship also exists for the surface charge. The boundary condition (2.78) in the frequency domain is

$$\sigma_2 E_{n2s} - \sigma_1 E_{n1s} = -j\omega\left(\varepsilon_2 E_{n2s} - \varepsilon_1 E_{n1s}\right) \tag{2.83}$$

Solving for the electric field in region 1,

$$E_{n1s} = \frac{\sigma_2 + j\omega\varepsilon_2}{\sigma_1 + j\omega\varepsilon_1} E_{n2s} \tag{2.84}$$

The surface charge in the frequency domain is then

$$\rho_{Ss} = \varepsilon_2 E_{n2s} - \varepsilon_1 E_{n1s} = \varepsilon_2 E_{n2s} - \varepsilon_1 \frac{\sigma_2 + j\omega\varepsilon_2}{\sigma_1 + j\omega\varepsilon_1} E_{n2s} = \left(\frac{\varepsilon_2\sigma_1 - \varepsilon_1\sigma_2}{\sigma_1 + j\omega\varepsilon_1}\right) E_{n2s} \tag{2.85}$$

The ratio of the electric fields, converting to polar form, is

$$\frac{E_{n1s}}{E_{n2s}} = \frac{\sqrt{\sigma_2^2 + \omega^2\varepsilon_2^2}\,\angle\tan^{-1}\left(\dfrac{\omega\varepsilon_2}{\sigma_2}\right)}{\sqrt{\sigma_1^2 + \omega^2\varepsilon_1^2}\,\angle\tan^{-1}\left(\dfrac{\omega\varepsilon_1}{\sigma_1}\right)} = \sqrt{\frac{\sigma_2^2 + \omega^2\varepsilon_2^2}{\sigma_1^2 + \omega^2\varepsilon_1^2}}\,\angle\left[\tan^{-1}\left(\frac{\omega\varepsilon_2}{\sigma_2}\right) - \tan^{-1}\left(\frac{\omega\varepsilon_1}{\sigma_1}\right)\right]$$

$$= \frac{\sigma_2}{\sigma_1}\sqrt{\frac{1 + \left(\dfrac{\omega}{\dfrac{\sigma_2}{\varepsilon_2}}\right)^2}{1 + \left(\dfrac{\omega}{\dfrac{\sigma_1}{\varepsilon_1}}\right)^2}}\,\angle\left[\tan^{-1}\left(\frac{\omega}{\dfrac{\sigma_2}{\varepsilon_2}}\right) - \tan^{-1}\left(\frac{\omega}{\dfrac{\sigma_1}{\varepsilon_1}}\right)\right] \tag{2.86}$$

For low frequencies, the magnitude of this field ratio reduces to the ratio of the conductivities:

$$\left|\frac{E_{n1s}}{E_{n2s}}\right| \approx \frac{\sigma_2}{\sigma_1} \quad \text{if } \omega \ll \frac{\sigma_2}{\varepsilon_2},\ \omega \ll \frac{\sigma_1}{\varepsilon_1} \tag{2.87}$$

As will be seen, in sinusoidal steady-state conditions at low frequencies, the materials can be modeled as resistors, which are a function of the conductivity. At high frequencies, the magnitude of this field ratio reduces to the ratio of permittivities:

$$\left|\frac{E_{n1s}}{E_{n2s}}\right| \approx \frac{\varepsilon_2}{\varepsilon_1} \quad \text{if } \omega \gg \frac{\sigma_2}{\varepsilon_2},\ \omega \gg \frac{\sigma_1}{\varepsilon_1} \tag{2.88}$$

As will be seen, at high frequencies both materials can be modeled as capacitors, which are a function of the permittivity.

To understand the use and implications of the time-varying boundary condition for an abrupt change in electrical properties, it will be used to determine the surface charge at $y = a$ for a two-layer lossy capacitor. This lossy capacitor, shown in Figure 2.32, is known as Maxwell's capacitor. The distance $a + b$ is assumed small compared to the other dimensions so that the fields between the plates can be

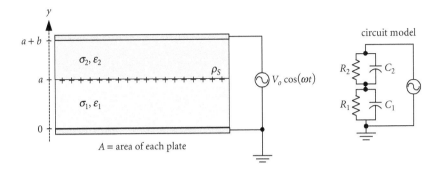

FIGURE 2.32 Two-layer lossy capacitor and its equivalent circuit model.

assumed one dimensional. This capacitor can be modeled as the series combination of two parallel *RC* circuits as shown where

$$R_1 = \frac{a}{\sigma_1 A}, \ C_1 = \frac{\varepsilon_1 A}{a}, \ R_2 = \frac{b}{\sigma_2 A}, \ C_2 = \frac{\varepsilon_2 A}{b} \tag{2.89}$$

These standard expressions for the resistance and capacitance between two parallel plates could be used since the electric field and current density are uniform. Notice that the *RC* time constants for each region are equal to the material time constants for the corresponding medium:

$$\tau_1 = R_1 C_1 = \frac{\varepsilon_1}{\sigma_1}, \ \tau_2 = R_2 C_2 = \frac{\varepsilon_2}{\sigma_2} \tag{2.90}$$

The sinusoidal steady-state results for the fields and interface surface charge will be determined. It is therefore advantageous to use phasors. The relationship between the time-domain and frequency-domain representation for the voltage source is

$$V_o \cos(\omega t) \ \Leftrightarrow \ V_o \angle 0°$$

For this one-dimensional problem where the conductivity and permittivity are uniform over each of the two regions, the electric field is constant in both the lower and upper regions. (The initial volume charge density is zero for each region.) In the frequency domain, these electric fields are given by

$$\vec{E}_{1s} = E_{1s}\hat{a}_y \ \text{ if } 0 < y < a, \ \vec{E}_{2s} = E_{2s}\hat{a}_y \ \text{ if } a < y < a+b$$

where the subscript *s* corresponds to complex or frequency-domain quantities. Since the voltage is fixed across the two regions, the basic definition for voltage in terms of the electric field can be used:

$$V_o = -\int_0^{a+b} \vec{E}_s \cdot d\vec{L} = -\int_0^a \vec{E}_{1s} \cdot dy\hat{a}_y - \int_a^{a+b} \vec{E}_{2s} \cdot dy\hat{a}_y = -E_{1s}\int_0^a dy - E_{2s}\int_a^{a+b} dy = -aE_{1s} - bE_{2s}$$

$$\Rightarrow E_{1s} = -\frac{V_o}{a} - \frac{b}{a}E_{2s}$$

The next equation is obtained using the sinusoidal steady-state version of the boundary condition along the *y = a* interface:

$$\sigma_2 E_{2s} - \sigma_1 E_{1s} = -j\omega(\varepsilon_2 E_{2s} - \varepsilon_1 E_{1s})$$

where d/dt corresponded to $j\omega$. Substituting the expression given for E_{1s} and then solving for E_{2s} yields

$$\sigma_2 E_{2s} - \sigma_1 \left(-\frac{V_o}{a} - \frac{b}{a} E_{2s} \right) = -j\omega \left[\varepsilon_2 E_{2s} - \varepsilon_1 \left(-\frac{V_o}{a} - \frac{b}{a} E_{2s} \right) \right]$$

$$\Rightarrow E_{2s} = -\frac{(\sigma_1 + j\omega\varepsilon_1)V_o}{a(\sigma_2 + j\omega\varepsilon_2) + b(\sigma_1 + j\omega\varepsilon_1)} \tag{2.91}$$

$$\Rightarrow E_{1s} = -\frac{V_o}{a} - \frac{b}{a} E_{2s} = -\frac{(\sigma_2 + j\omega\varepsilon_2)V_o}{a(\sigma_2 + j\omega\varepsilon_2) + b(\sigma_1 + j\omega\varepsilon_1)}$$

Now that the electric field is known in each region in the frequency domain, the surface charge along the interface can be determined:

$$\rho_{Ss} = D_{n2s} - D_{n1s} = \varepsilon_2 E_{2s} - \varepsilon_1 E_{1s}$$

$$= -\frac{\varepsilon_2(\sigma_1 + j\omega\varepsilon_1)V_o}{a(\sigma_2 + j\omega\varepsilon_2) + b(\sigma_1 + j\omega\varepsilon_1)} + \frac{\varepsilon_1(\sigma_2 + j\omega\varepsilon_2)V_o}{a(\sigma_2 + j\omega\varepsilon_2) + b(\sigma_1 + j\omega\varepsilon_1)} \tag{2.92}$$

$$= \frac{(\varepsilon_1\sigma_2 - \varepsilon_2\sigma_1)V_o}{a(\sigma_2 + j\omega\varepsilon_2) + b(\sigma_1 + j\omega\varepsilon_1)}$$

This equation for the charge is in the frequency domain. It can be easily written in standard Bode magnitude plotting form:

$$\rho_{Ss} = \frac{(\varepsilon_1\sigma_2 - \varepsilon_2\sigma_1)V_o}{(a\sigma_2 + b\sigma_1)\left(1 + \dfrac{j\omega}{\dfrac{a\sigma_2 + b\sigma_1}{a\varepsilon_2 + b\varepsilon_1}} \right)} \tag{2.93}$$

Thus,

$$\rho_{Ss} \approx \begin{cases} \dfrac{(\varepsilon_1\sigma_2 - \varepsilon_2\sigma_1)V_o}{(a\sigma_2 + b\sigma_1)} & \text{if } \omega \ll \dfrac{a\sigma_2 + b\sigma_1}{a\varepsilon_2 + b\varepsilon_1} \\[4mm] \dfrac{(\varepsilon_1\sigma_2 - \varepsilon_2\sigma_1)V_o}{j\omega(a\varepsilon_2 + b\varepsilon_1)} & \text{if } \omega \gg \dfrac{a\sigma_2 + b\sigma_1}{a\varepsilon_2 + b\varepsilon_1} \end{cases} \tag{2.94}$$

For dc and very low frequencies, charge accumulates or exits at the interface (but still varies in an oscillatory manner). For higher frequencies above the cutoff frequency (but not so high that the electrostatic definition $\vec{E} = -\nabla\Phi$ cannot be applied), the surface charge along the interface decreases and eventually becomes negligible. This is one reason some measurements are performed at higher frequencies — charge does not have time to accumulate along the interfaces.[15] For example, when measuring the resistance between two locations along the earth, the test signals are usually ac not dc. Notice that the surface charge is always zero for the special case of $\varepsilon_1\sigma_2 = \varepsilon_2\sigma_1$. The cutoff or break frequency can also be written in terms of the

---

[15]If charge was present along a surface or an interface prior to the application of the high-frequency voltage, then it can remain on the surface.

material time constants for the two layers:

$$\omega_c = \frac{a\sigma_2 + b\sigma_1}{a\varepsilon_2 + b\varepsilon_1} = \frac{\dfrac{a}{\varepsilon_1}\dfrac{\sigma_2}{\varepsilon_2} + \dfrac{b}{\varepsilon_2}\dfrac{\sigma_1}{\varepsilon_1}}{\dfrac{a}{\varepsilon_1} + \dfrac{b}{\varepsilon_2}} = \frac{\dfrac{a}{\varepsilon_1}\dfrac{1}{\tau_2} + \dfrac{b}{\varepsilon_2}\dfrac{1}{\tau_1}}{\dfrac{a}{\varepsilon_1} + \dfrac{b}{\varepsilon_2}} \tag{2.95}$$

Although the break frequency is also a function of the dimensions of Maxwell's capacitor, for fixed dimensions, as the time constant of either layer decreases (e.g., conductivity increases), the cutoff frequency increases. The expression for the surface charge in the time domain is obtained in the standard manner:

$$\rho_S(t) = \mathrm{Re}\left(\rho_{Ss}e^{j\omega t}\right) = \mathrm{Re}\left[\frac{(\varepsilon_1\sigma_2 - \varepsilon_2\sigma_1)V_o e^{j\omega t}}{(a\sigma_2 + b\sigma_1)\sqrt{1^2 + \left(\dfrac{\omega}{\dfrac{a\sigma_2 + b\sigma_1}{a\varepsilon_2 + b\varepsilon_1}}\right)^2} \angle\tan^{-1}\left(\dfrac{\omega}{\dfrac{a\sigma_2 + b\sigma_1}{a\varepsilon_2 + b\varepsilon_1}}\right)}\right]$$

$$\rho_S(t) = \frac{(\varepsilon_1\sigma_2 - \varepsilon_2\sigma_1)V_o}{(a\sigma_2 + b\sigma_1)\sqrt{1 + \left(\dfrac{\omega}{\dfrac{a\sigma_2 + b\sigma_1}{a\varepsilon_2 + b\varepsilon_1}}\right)^2}} \cos\left[\omega t - \tan^{-1}\left(\dfrac{\omega}{\dfrac{a\sigma_2 + b\sigma_1}{a\varepsilon_2 + b\varepsilon_1}}\right)\right] \tag{2.96}$$

In Maxwell's capacitor, the material interface is perpendicular to a uniform electric field. The surface charge, although time varying, is uniform or constant along the width of the capacitor. Surface charge can accumulate along any interface where the field is in direction of the change in the conductivity and/or permittivity. For example, a spherical ball of material with a conductivity and/or permittivity different from its surroundings will have a surface charge present along its surface when placed in an external uniform or nonuniform electric field. Surface charge can also vary along the dimensions of the interface. This variation is a function of the shape of the interface, variation of the electrical properties along the interface, and uniformity of the applied field.

In most cases, excess charge will find a path to ground or some other nearby object. If the object is conducting, then often the simplest method of dissipating the charge is by connecting the object to a good ground. Even for insulating objects, charge will eventually find a path to ground or some other nearby object. There are several ways that charge can dissipate or decay. It can dissipate by conduction along the surface of the charged body. It can dissipate by conduction through the volume of the charged body. It can dissipate by conduction through leakage paths of surrounding insulators (including air), since all real insulators have a finite resistivity. The charge, if it and the field strength are sufficiently high, can dissipate through corona or another discharge process. Finally, assuming the sign and magnitude of the charge are known, the charge can be (partially) neutralized through the use of an ionizer.

2.11   Convection Charge Flow

Derive several of the basic one-dimensional expressions for the charge distribution where both conduction and convection current are important. Provide several examples where convection current is important. [Melcher; Zahn; Crowley, '86; Cross]

   Although conduction current is emphasized in this book, in some applications convection current can be extremely important. With conduction current, both external and self-generated fields are responsible for charge movement. With convection current, the medium, such as the surrounding air or liquid, is responsible for the movement of the charge. Charge can also be transported on solid bodies such as on a belt in a Van de Graaff generator.

   For the first situation shown in Figure 2.33, a voltage source is applied to an electrode that produces a constant volume charge density of $\rho_V(0)$ at $x = 0$. The medium at and beyond the electrode is moving at a constant velocity of U in the x direction. (Imagine that the electrode is perforated to allow the moving fluid to pass through it.) The conductivity and permittivity of this moving medium are σ and ε, respectively. The steady-state volume charge density is desired. The current density is assumed conductive and convective:

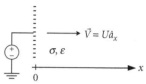

FIGURE 2.33   Charge density generated at $x = 0$ transported in the x direction at a constant velocity. The medium itself is also conductive.

$\vec{J} = \sigma\vec{E} + \rho_V\vec{V}$. Furthermore, the conductivity of the medium is assumed constant and not affected by ρ_V. Beginning with the charge continuity equation,

$$\nabla \cdot \vec{J} + \frac{\partial \rho_V}{\partial t} = \nabla \cdot \left(\sigma\vec{E} + \rho_V\vec{V}\right) + \frac{\partial \rho_V}{\partial t} = \sigma\nabla \cdot \vec{E} + \nabla \cdot \left(\rho_V\vec{V}\right) + \frac{\partial \rho_V}{\partial t} = 0$$

The divergence of the electric field can be written in terms of the volume charge density by using Gauss' law:

$$\nabla \cdot \vec{D} = \nabla \cdot (\varepsilon\vec{E}) = \varepsilon\nabla \cdot \vec{E} = \rho_V$$

Substituting into the charge continuity equation,

$$\sigma\frac{\rho_V}{\varepsilon} + \nabla \cdot \left(\rho_V\vec{V}\right) + \frac{\partial \rho_V}{\partial t} = 0$$

As was previously done, the divergence operation can be applied to both terms of the scalar and vector product (similar to the product rule in calculus):

$$\frac{\sigma\rho_V}{\varepsilon} + (\nabla\rho_V)\cdot\vec{V} + \rho_V(\nabla\cdot\vec{V}) + \frac{\partial \rho_V}{\partial t} = 0 \tag{2.97}$$

This rather general expression does not assume the situation is electrostatic. It assumes that the current is the sum of a conductive term (describable via Ohm's law) and convective term. For the particular situation shown in Figure 2.33, the velocity is constant so its divergence, which involves space differentiation, is zero. Since dc steady-state results are desired, the time-change term in (2.97) is zero. Finally, the problem is assumed one dimensional so that the charge density only varies in the x direction. Thus, the expression reduces to

$$\frac{\sigma\rho_V}{\varepsilon} + (\nabla\rho_V)\cdot U\hat{a}_x = \frac{\sigma\rho_V}{\varepsilon} + \left(\frac{d\rho_V}{dx}\hat{a}_x\right)\cdot U\hat{a}_x = \frac{\sigma\rho_V}{\varepsilon} + U\frac{d\rho_V}{dx} = 0 \tag{2.98}$$

This differential equation can be easily integrated:

$$\int_{\rho_V(0)}^{\rho_V(x)} \frac{d\rho_V}{\rho_V} = \int_0^x -\frac{\sigma dx}{\varepsilon U} \Rightarrow \ln \rho_V \Big|_{\rho_V(0)}^{\rho_V(x)} = -\frac{\sigma x}{\varepsilon U}\Big|_0^x \Rightarrow \ln\left[\frac{\rho_V(x)}{\rho_V(0)}\right] = -\frac{\sigma x}{\varepsilon U}$$

Therefore,

$$\rho_V(x) = \rho_V(0)e^{-\frac{\sigma}{\varepsilon U}x} = \rho_V(0)e^{-\frac{x}{l_m}} \quad \text{where } l_m = \frac{\varepsilon U}{\sigma} \tag{2.99}$$

The charge density decays exponentially with a spatial charge "skin depth" or decay length equal to $\varepsilon U/\sigma$. After a few of these spatial decay lengths, the volume charge density is small since an exponential function decreases rapidly. Although perhaps not immediately apparent, if the volume charge density is to appear within a few centimeters or more from the charge source at $x = 0$, the velocity must be large and dc conductivity of the medium must be small. Recall that the permittivity is a very small number since $\varepsilon = k\varepsilon_o = k(8.854 \times 10^{-12})$ where k is the dielectric constant that often ranges from 1 to 10. If l_m is to be greater than 10 cm, then

$$\frac{\varepsilon U}{\sigma} = \frac{k(8.854 \times 10^{-12})U}{\sigma} > 0.1\,\text{m} \tag{2.100}$$

Conductivities of some oils can be quite small (e.g., 10^{-10} 1/Ω-m), which would allow for reasonable velocities. The conductivity of air, of course, is also very small. However, it is dubious whether Ohm's law can be applied to low-conductivity mediums such as air or insulating oils. The conductivity is likely not constant and is easily influenced by the injected charge density. For conductive mediums, *extremely large* velocities would be required to produce spatial lengths even in the cm range.

Equation (2.99) has also been used to describe the charge density variation on the belt of a Van de Graaff generator and the charge along a conveyer belt. The variables σ and ε are the conductivity and permittivity, respectively, of the belt. Although frequently the charge on a belt is considered a surface charge, far from a charged belt, the electric field from surface charge on and volume charge in the belt are similar. The initial charge at $x = 0$ can be generated via corona from a high-voltage sharp electrode as shown in Figure 2.34. It can also be unintentionally generated via tribocharging between the insulating belt and a conducting roller. With the Van de Graaff generator, the charge is collected at the well-known metallic dome. Another interesting application of this exponential charge variation with distance is in the noncontact resistivity monitoring of moving materials. By introducing charge on the material, the electric field from the charge at a specific position x along the material is an indication of its conductivity.

In the previous situation, imagine that a parallel grounded electrode is added downstream to collect the charge as shown in Figure 2.35. In this case, the applied voltage from $x = 0$ to $x = d$ is V_{dc}. Because the volume charge distribution between the electrodes is still one dimensional, the electric field is also

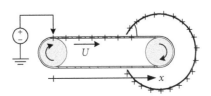

FIGURE 2.34 Simplified model of a Van de Graaff generator.

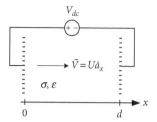

FIGURE 2.35 Fixed voltage applied across a conductive medium with a known velocity.

one dimensional. Using Gauss' law and Equation (2.99),

$$\nabla \cdot \vec{E} = \frac{dE_x}{dx} = \frac{\rho_V(x)}{\varepsilon} = \frac{\rho_V(0)e^{-\frac{x}{l_m}}}{\varepsilon} \quad \text{where } l_m = \frac{\varepsilon U}{\sigma}$$

Integrating both sides,

$$E_x(x) = -\frac{l_m \rho_V(0)e^{-\frac{x}{l_m}}}{\varepsilon} + C$$

The constant, C, can be determined since the voltage between the electrode is V_{dc}, and it must be equal to the line integral of this electric field between the electrodes:

$$V_{dc} = -\int_d^0 E_x(x)\,dx = -\int_d^0 \left(-\frac{l_m \rho_V(0)e^{-\frac{x}{l_m}}}{\varepsilon} + C \right) dx = \left(-\frac{l_m^2 \rho_V(0)e^{-\frac{x}{l_m}}}{\varepsilon} - Cx \right)\Bigg|_d^0$$

$$= \frac{l_m^2 \rho_V(0)\left(e^{-\frac{d}{l_m}} - 1 \right)}{\varepsilon} + Cd$$

Solving for C, the electric field between the electrodes is given by

$$E_x(x) = -\frac{l_m \rho_V(0)e^{-\frac{x}{l_m}}}{\varepsilon} + \frac{V_{dc}}{d} + \frac{l_m^2 \rho_V(0)}{d\varepsilon}\left(1 - e^{-\frac{d}{l_m}} \right) \tag{2.101}$$

Now that both the electric field and velocity are known, the current density is

$$J_x = \sigma E_x(x) + \rho_V(x)U$$

$$= -\frac{l_m \sigma \rho_V(0)e^{-\frac{x}{l_m}}}{\varepsilon} + \frac{\sigma V_{dc}}{d} + \frac{l_m^2 \sigma \rho_V(0)}{d\varepsilon}\left(1 - e^{-\frac{d}{l_m}} \right) + U\rho_V(0)e^{-\frac{x}{l_m}} \tag{2.102}$$

$$= \frac{\sigma V_{dc}}{d} + \frac{\varepsilon U^2 \rho_V(0)}{d\sigma}\left(1 - e^{-\frac{d}{l_m}} \right)$$

If the cross-sectional area of the electrodes is A, the current is

$$I = \frac{A\sigma V_{dc}}{d} + \frac{A\varepsilon U^2 \rho_V(0)}{d\sigma}\left(1 - e^{-\frac{d}{l_m}}\right)$$

$$= \frac{V_{dc}}{R} + \frac{\varepsilon U^2 \rho_V(0)}{R\sigma^2}\left(1 - e^{-\frac{d}{l_m}}\right) \quad \text{where } R = \frac{d}{\sigma A} \tag{2.103}$$

As is expected, the current is not a function of x. Furthermore, if the velocity, U, of the medium is zero, the current between the electrodes is entirely conductive. The conduction current is entirely determined by V_{dc} and the resistance of the ohmic material between the electrodes. If the applied voltage is zero (and the volume charge can still be maintained at $x = 0$ by some other external means), then the current is entirely due to convection. The limiting value of the convection current after a few l_m is

$$I \approx \frac{\varepsilon U^2 \rho_V(0)}{R\sigma^2} = \left(\frac{l_m}{d}\right)UA\rho_V(0) \quad \text{if } d > 3l_m \text{ to } 5l_m, \ V_{dc} = 0 \tag{2.104}$$

The importance of the medium's velocity and the spatial decay length relative to the electrode spacing is apparent.

In this next example, the conduction current is determined by the volume charge density. (In the previous examples, the conductivity was assumed so large that the volume charge density had little effect on it.) The current density is given by $\vec{J} = b\rho_V\vec{E} + \rho_V\vec{V}$, where b is the mobility of the volume charge, and is referred to as unipolar conduction and convection. The velocity is assumed constant, uniform, and in the x direction as is shown in Figure 2.36. The charge density at $x = 0$ is $\rho_V(0)$. In this case, a load can be placed across the output of this system at V_o. The steady-state distributions for the volume charge density, electric field, and potential are desired. For a cross-sectional area of A and a one-dimensional model, $I/A = b\rho_V E_x + \rho_V U$, where I is the current in one electrode and out the other. Solving for the electric field,

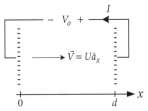

FIGURE 2.36 Unipolar conduction and convection between two electrodes.

$$E_x = \frac{I}{Ab\rho_V} - \frac{U}{b}$$

This electric field is then substituted into Gauss' law to obtain a differential equation involving the volume charge density:

$$\nabla \cdot \vec{D} = \varepsilon\nabla \cdot (E_x\hat{a}_x) = \varepsilon\nabla \cdot \left[\left(\frac{I}{Ab\rho_V} - \frac{U}{b}\right)\hat{a}_x\right] = \rho_V \tag{2.105}$$

Since the velocity is assumed constant and uniform, its spatial derivative is zero, and (2.105) reduces to

$$\nabla \cdot \left(\frac{1}{\rho_V}\hat{a}_x\right) = \frac{Ab\rho_V}{\varepsilon I}$$

For this one-dimensional model, the charge density varies only in the x direction so

$$\frac{d}{dx}\left[\rho_V^{-1}(x)\right] = \frac{Ab\rho_V(x)}{\varepsilon I}$$

Using the chain rule,

$$-\rho_V^{-2}(x)\frac{d\rho_V(x)}{dx} = \frac{Ab\rho_V(x)}{\varepsilon I} \quad \Rightarrow -\frac{d\rho_V(x)}{\rho_V^3(x)} = \frac{Abdx}{\varepsilon I}$$

Integrating both sides,

$$-\int_{\rho_V(0)}^{\rho_V(x)}\frac{d\rho_V(x)}{\rho_V^3(x)} = \int_0^x \frac{Abdx}{\varepsilon I} \quad \Rightarrow \frac{1}{2\rho_V^2(x)}\Bigg|_{\rho_V(0)}^{\rho_V(x)} = \frac{Abx}{\varepsilon I}\Bigg|_0^x$$

and solving for the volume charge density:

$$\rho_V(x) = \frac{\rho_V(0)}{\sqrt{1+\dfrac{2Abx\rho_V^2(0)}{\varepsilon I}}} \tag{2.106}$$

For very large x values, the charge density is not even a function of the initial volume charge density:

$$\rho_V(x) \approx \sqrt{\frac{\varepsilon I}{2Abx}} \quad \text{if } x \gg \frac{\varepsilon I}{2Ab\rho_V^2(0)} \tag{2.107}$$

Although the electric field is a function of the velocity, U, the charge density is not. From the current density expression, the electric field distribution between the electrodes can be determined:

$$E_x(x) = \frac{I}{Ab\rho_V(x)} - \frac{U}{b} = \frac{I\sqrt{1+\dfrac{2Ab\rho_V^2(0)x}{\varepsilon I}}}{Ab\rho_V(0)} - \frac{U}{b} \tag{2.108}$$

The potential distribution is obtained by integrating Equation (2.108):

$$\Phi(x) = -\int_0^x E_x(x)dx = -\int_0^x \left[\frac{I\sqrt{1+\dfrac{2Ab\rho_V^2(0)x}{\varepsilon I}}}{Ab\rho_V(0)} - \frac{U}{b}\right]dx$$

$$= -\left[\frac{\varepsilon I^2\left(1+\dfrac{2Ab\rho_V^2(0)x}{\varepsilon I}\right)^{\frac{3}{2}}}{3A^2b^2\rho_V^3(0)} - \frac{U}{b}x\right]\Bigg|_0^x \tag{2.109}$$

$$= \frac{\varepsilon I^2}{3A^2b^2\rho_V^3(0)}\left[1-\left(1+\frac{2Ab\rho_V^2(0)x}{\varepsilon I}\right)^{\frac{3}{2}}\right] + \frac{U}{b}x$$

The electrical output power is given by $P = \Phi(d)I$. Depending on factors such as the velocity of the medium and the size of the volume charge density at $x = 0$, this power can be positive, negative, or zero. When the power is positive, this device is acting like a generator. It can supply power to a load connected across the voltage terminals. The mechanical energy, which is moving the medium, is supplying this power. When

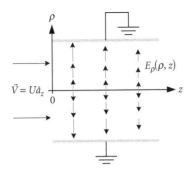

FIGURE 2.37 Charge flowing down a cylindrical grounded pipe. Although the medium velocity is mostly in the z direction, the electric field is strongest in the radial direction.

the power is negative, this device is acting like a pump. The voltage supply (and the electric field associated with the volume charge density) is supplying the electrical energy to pump the fluid.

In this last example, the major components of the electric field and velocity are in different directions. Imagine that charges are flowing down a cylindrical grounded pipe as shown in Figure 2.37. The steady-state volume charge and electric field distributions are desired. The only source of electric field is the self field from the volume charges themselves. The variation in the charge density in the z direction down the length of the pipe is considered small. Hence, the electric field in the z direction is considered small compared to the electric field in the radial direction toward the walls of the grounded pipe. (The charge variation in the radial direction is assumed zero.) The air or medium velocity in the z direction, however, is considered constant and large compared to the velocity in the ρ direction toward the walls. Therefore, the current density is given by

$$\vec{J} = b\rho_V E_\rho \hat{a}_\rho + \rho_V U \hat{a}_z \tag{2.110}$$

Gauss' law can be used to determine the radial component of the electric field in the cylindrical coordinate system (where ρ is the coordinate variable not the volume charge density):

$$\nabla \cdot \vec{D} = \nabla \cdot (\varepsilon_o \vec{E}) = \varepsilon_o \frac{1}{\rho} \frac{\partial}{\partial \rho}(\rho E_\rho) + \varepsilon_o \frac{1}{\rho} \frac{\partial E_\phi}{\partial \phi} + \varepsilon_o \frac{\partial E_z}{\partial z} = \varepsilon_o \frac{1}{\rho} \frac{\partial}{\partial \rho}(\rho E_\rho) = \rho_V(z) \tag{2.111}$$

$$\Rightarrow E_\rho(\rho, z) = \frac{\rho_V(z)}{2\varepsilon_o} \rho$$

where $\varepsilon = \varepsilon_o$ was assumed. (In order to have a finite value of the field at $\rho = 0$, the constant of integration must be zero.) The charge density can vary in the z direction even though its field in this direction is small compared to the radial field component. The electric field is a function of both ρ and z. Next, expression (2.110) can be used in the charge conservation equation for steady-state operation:

$$\nabla \cdot \vec{J} + \frac{\partial \rho_V}{\partial t} = \nabla \cdot [b\rho_V(z)\vec{E} + \rho_V(z)\vec{V}] + 0 = \nabla \cdot [b\rho_V(z)E_\rho(\rho, z)\hat{a}_\rho + \rho_V(z)U\hat{a}_z] = 0$$

Using the product-rule equivalent of the gradient operator,

$$b\frac{d\rho_V(z)}{dz}\hat{a}_z \cdot E_\rho(\rho, z)\hat{a}_\rho + b\rho_V(z)\nabla \cdot [E_\rho(\rho, z)\hat{a}_\rho] + \frac{d\rho_V(z)}{dz}\hat{a}_z \cdot U\hat{a}_z + \rho_V(z)\nabla \cdot (U\hat{a}_z) = 0$$

This reduces to

$$b\rho_V(z)\frac{1}{\rho}\frac{\partial}{\partial \rho}[\rho E_\rho(\rho, z)] + U\frac{d\rho_V(z)}{dz} = 0$$

Substituting (2.111),

$$bp_V(z)\frac{1}{\rho}\frac{\partial}{\partial\rho}\left[\rho^2\frac{\rho_V(z)}{2\varepsilon_o}\right]+U\frac{d\rho_V(z)}{dz}=\frac{b}{\varepsilon_o}\rho_V^2(z)+U\frac{d\rho_V(z)}{dz}=0$$

The volume charge density is obtained by solving this differential equation:

$$-\int_{\rho_V(0)}^{\rho_V(z)}\frac{d\rho_V(z)}{\rho_V^2(z)}=\int_0^z\frac{bdz}{\varepsilon_oU} \quad\Rightarrow\quad \frac{1}{\rho_V(z)}\Big|_{\rho_V(0)}^{\rho_V(z)}=\frac{bz}{\varepsilon_oU}\Big|_0^z \quad\Rightarrow \rho_V(z)=\frac{\rho_V(0)}{1+\dfrac{b\rho_V(0)z}{\varepsilon_oU}} \qquad (2.112)$$

After a great distance, this charge density reduces to

$$\rho_V(z)\approx\frac{\varepsilon_oU}{bz} \quad\text{if } z\gg\frac{\varepsilon_oU}{b\rho_V(0)} \qquad (2.113)$$

The charge density is then independent of the initial charge density at $z = 0$. This may at first surprise callow individuals. But, as the initial charge density is increased, the self field from these charges also increases. This stronger space-charge field is a greater driving force on the charge toward the grounded walls of the cylinder. If a larger charge density is needed at some location z, the velocity can be increased. Static charge buildup on objects is sometimes neutralized by the use of ionizers. Ions are generated by various methods (e.g., breakdown of the air near a high-voltage sharp electrode), and then the ions are often blown onto the object to be neutralized. According to Equation (2.113), once the ions are generated, larger air velocities are the key to increasing the charge density at some distance from the output of an ionizer nozzle. Since the charge density is known, the electric field inside the grounded nozzle is

$$E_\rho(\rho,z)=\frac{\rho_V(0)}{1+\dfrac{b\rho_V(0)z}{\varepsilon_oU}}\frac{\rho}{2\varepsilon_o} \qquad (2.114)$$

The charged particles are propelled by both the electric field and medium (via the air flow). The strength of the electric field in the radial direction, however, weakens as the particles travel down the length of the nozzle (i.e., as z increases). The strength of the electric field is greatest at $z = 0$.

2.12   Potential of an Insulator's Surface

Can the potential along an insulating surface be measured directly? [Jonassen]

Analytically or numerically, the potential along a charged *insulator's* surface (or in a charged insulator) can be determined if the charge distribution is known. In practice, the potential along (or in) a charged insulator cannot be measured directly. (The potential can sometimes be measured indirectly through the measurement of the electric field.) When a potential probe is connected to the surface of a charged insulator, the probe drains some of the charge from the surface. This, of course, affects the measurement since the probe is loading the insulating "circuit." If there were large quantities of charge along the surface, the loss of charge into the voltage instrument would be less important. For very large quantities of charge, however, the insulating surface would appear more like a conducting sheet.

There is another issue that should be considered when attempting to measure the potential along an insulating surface. Unlike charged good-conducting objects where the potential is essentially the same everywhere along and in the conducting object, the potential along and in insulating objects usually vary.

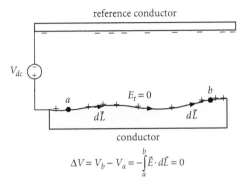

FIGURE 2.38 Sketch of the charge distribution on two conductors.

Recall that the voltage difference between two locations a and b is given by

$$V = -\int_a^b \vec{E} \cdot d\vec{L} \qquad (2.115)$$

where $\vec{E}$ is the electric field and $d\vec{L}$ is the differential path taken from a to b. If a and b are along the surface of the same perfect conductor, and $d\vec{L}$ is taken along this surface, the voltage difference must be zero since the electric field tangential to perfect conductors, E_t, is zero. That is, the potential at locations a and b along the same surface must be equal. The charge distribution along a conducting surface is crudely illustrated in Figure 2.38. As seen in this figure, the density or distribution of the surface charge along a conductor's surface can vary. The charge can also vary in sign along the same surface. However, the tangential electric field is always zero along a perfectly conducting surface.

Along charged or uncharged insulating surfaces, there is no general rule that any component of the electric field along its surface must be zero. Therefore, the potential difference along a charged or an uncharged insulating surface is not necessarily zero: the potential can vary along the surface of the insulator. The electric field will depend on the distribution of free charge along the insulator's surface. Therefore, the concept of surface potential along insulating surfaces should generally be avoided and instead the charge distribution or local electric field be used. There are some cases involving high degrees of symmetry where the electric field tangential to a charged insulating surface is zero. One example is shown in Figure 2.39. The free charges along one side of the insulating slab are uniformly distributed. The free charges along the conducting surfaces are not shown. The electric field is entirely normal to the insulating surface. Even if the free charge were not present along the insulator's surface, the electric field would still be entirely perpendicular to the insulator's surface. Although not shown, near the ends, the tangential electric field is not zero due to fringing. Another example where the tangential electric field

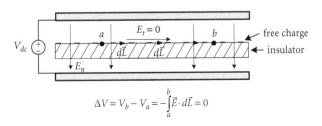

FIGURE 2.39 Situation with a high degree of symmetry where the tangential electric field along an insulating surface is zero.

along an insulating surface is zero is a concentric, cylindrical insulating slab between two charged, conducting concentric cylinders. Assuming the free charge distribution along one or both sides of the insulating slab is uniform or zero, the electric field is entirely normal to the slab.

2.13   Electric Field from Simple Charge Distributions

Tabulate the expressions for the electric field from simple point, line, surface, and volume charge distributions. How can this table be used to generate other distributions? [Hayt]

Knowledge of the electric field is very useful in ESD (and other) work, and the source of the electric field is charge. Because voltage is a commonly used and useful engineering quantity, engineers generally prefer to work with voltage rather than charge. For example, if 10 kV is applied to the electrode of an electrostatic liquid sprayer, the voltage everywhere along the good-conducting electrode of the sprayer is known to be about 10 kV. An analytical description of the charge distribution along the surface of the electrode is not usually known unless its shape (and its environment) is simple and a great deal of symmetry is present.[16] There are situations, though, where it is preferable to know the charge distribution rather than the voltage.

Before the tabulated results are given, the meaning of point, line, surface, and volume charge will be discussed. With the exception of the volume charge, these charge distributions are idealizations, but they all have practical applications and are quite useful in analytical analysis.

A point charge of strength Q coulombs (C) is a charge that exists over zero volume. Its total charge over this zero volume is Q. Of course, a point charge cannot actually exist. However, point charges are commonly used because they can sometimes dramatically simplify analysis. Imagine that charge is distributed, in any manner, over or in a soccer ball (surrounded by free space). The total charge on the ball is equal to Q. Near to the soccer ball, the way that the charge is distributed can strongly influence the electric field in the vicinity of the ball. However, 100 yards from the soccer ball, the soccer ball appears like a point charge and the way that the charge is distributed on the ball is irrelevant. At this great distance, the electric field from this soccer ball is approximately that of a point charge and equal to

$$\vec{E} = \frac{Q}{4\pi\varepsilon_o r^2}\hat{a}_r$$

where ε_o is the free-space permittivity and r is the distance between the soccer ball and far point of interest. The electric field is directed radially from the ball in the $\hat{a}_r$ direction. This field is referred to as the point charge or Coulomb field. The point charge is often used to represent any real charge distribution when the largest dimension of the charge distribution is small compared to the distance to where the electric field is desired.[17] It is also used as a test charge to determine the electric field in a given region (assuming the electric field from the test charge has negligible effect on the overall external field). When using a point charge to represent a real charge distribution, it is assumed that the distance between any charge in the real charge distribution and the point of interest is approximately the same, irrelevant of the actual charge selected. A physical analogy may be helpful. Several miles from a skyscraper along the ground, the distance between the top of the skyscraper and this distant location is approximately the same as the distance between the bottom of the skyscraper and this distant location. The skyscraper appears like a physical point object at this large distance.

The second idealized charge distribution is the line or filament charge. It is frequently described using the variable ρ_L with units of C/m. A line charge has one nonzero dimension — its length. Although a

---

[16]The total charge, Q, on an object (not its distribution) can be determined from the applied voltage if the capacitance between the object and voltage reference electrode is known via $Q = CV$.

[17]For a uniformly charged disk of radius a, for example, the electric field from the disk (along its axis) is essentially equal to that from a point charge at a distance of $10a$ or greater from the disk.

line charge has a length, it has zero cross-sectional area. If the line charge is constant over its length, the total charge is equal to ρ_L multiplied by the length. If ρ_L is varying with position, it can be integrated along its length, L, to determine the total charge:

$$Q_{total} = \int_L \rho_L \, dL \qquad (2.116)$$

A line charge does not need to be straight. The beam of a cathode ray tube (CRT) is often modeled as a line charge. The charge along parallel transmission lines is frequently modeled as two line charges. Even the charge along a rectangular trace on a printed circuit board is frequently modeled as line charge. When the cross-sectional dimensions of a line charge distribution are small compared to the distance to the point of interest, the charge is frequently modeled as a line charge. The expression for the electric field from a straight, infinite, uniformly distributed line charge is commonly used:

$$\bar{E} = \frac{\rho_L}{2\pi\varepsilon_o\rho}\hat{a}_\rho$$

where the ρ is the smallest distance from the point of interest to the line charge. The infinite line charge is also referred to as a long, thin line charge. The electric field is directed radially from the line charge in the $\hat{a}_\rho$ direction. An infinite line charge has infinite length and charge. The expression for the electric field for an infinite-length line charge is frequently used for a *finite*-length line charge when the smallest distance between the point of interest and line charge is small compared to the actual length of the finite-length charge: when the point of interest is close to the line charge, the line charge appears very long. Also, the point of interest should be far from either end of the finite-line charge to avoid fringing fields.

   A surface charge has two nonzero dimensions — its width and length. It is frequently described using the variable ρ_S with units of C/m^2. Surface charge has zero depth. Again, if the surface charge is not varying with position, the total charge can be determined by merely multiplying ρ_S by the area of the surface. If it is varying with position, the surface charge can be integrated over its surface:

$$Q_{total} = \iint_S \rho_S \, ds \qquad (2.117)$$

Surface charge does not have to lie in a flat plane. Surface charge is frequently used to represent the charge distribution along conductors in electrostatic situations and along conductors at higher frequencies where most of the charge is contained within about a skin depth of the conductor's surface. When the depth of the charge distribution is small compared to the distance to the point of interest, it is frequently modeled as a surface charge. It is also used when the depth of the charge distribution is small compared to the other dimensions of the charge distribution. For example, the charges along the belt in a Van de Graaff generator and along a surface sprayed with charged powder are commonly modeled as surface charges. It is also used to describe the double layer of charge common when two materials are placed in contact (e.g., tribocharging). Frequently, the expression for the electric field from a flat, uniform surface charge of infinite width and length is used:

$$\bar{E} = \frac{\rho_S}{2\varepsilon_o}\hat{a}_n$$

The electric field is directed normally from the surface charge in the $\hat{a}_n$ direction. This expression for the field from an infinite surface charge is used for flat, finite, uniform surface charge distributions when the smallest distance between the point of interest and surface charge plane is small compared to the plane's dimensions. Again, to avoid the fringing fields, the point of interest should be far from the edges of the surface charge plane. Notice that the electric field from an infinite charge sheet is not a function of the distance from this surface charge. A good analogy is the light from many parallel fluorescent tubes in the ceiling. This light source roughly provides the same illumination a foot from the ceiling as on the floor. The illumination from a single small light bulb, however, does vary strongly with position from the bulb. Far from the light bulb, the bulb can be viewed as a point light source.

Finally, a volume charge given as ρ_V with units of C/m³ has three nonzero dimensions. If the volume charge is not varying with position, the total charge can be determined by merely multiplying ρ_V by the volume. If it is varying with position, the volume charge can be integrated over its volume:

$$Q_{total} = \iiint_V \rho_V \, dV \tag{2.118}$$

Poisson's equation is a function of the volume charge. Volume charge is used to model myriad situations, including tribogenerated charge in silos.

The electric field expressions in Appendix A are a function of the charge distribution. Free space is assumed everywhere. The electric field inside a perfect conductor is zero, in electrostatic situations, and charge can only exist along the conductor's surface. Therefore, for the line and surface charge expressions, the shaded regions shown can be assumed to be a conductor (if needed) with the line or surface charge residing along the conductor's surface. Additional external conductors generally will affect the electric field since charge will be induced on these external conductors. (Sometimes, the method of images is used to determine the effect of these external conductors.) When the point, line, surface, and volume charge densities are given in C, C/m, C/m², and C/m³, respectively, and the distances are given in meters, the electric field is in V/m. In the figures in Appendix A, the charge is crudely represented using + symbols. Unless –'s are shown, the electric field sketches are assuming positive charge(s). In Appendix A, sometimes the electric field is specified only along a certain direction (i.e., "along the z axis"), in a given region (i.e., "if $r \gg d$"), or at a specific location (i.e., "at the origin"). An absence of an expression in this table does not imply that the fields are zero in the unspecified region or location. The free-space permittivity is given by $\varepsilon_o = 8.854 \times 10^{-12}$ F/m. In this table, $a, b, d, t, w, A, L, Q, \rho_S, \rho_L$, and ρ_V are constant with respect to position unless specifically stated otherwise. The Cartesian coordinate system variables are x, y, and z, the cylindrical coordinate systems variables are ρ, ϕ, and z, and the spherical coordinate system variables are r, θ, and ϕ. Although Appendix A is not exhaustive, it does contain many of the commonly seen and used expressions for the electric fields from simple charge distributions.

If the charge distribution is not provided in this table, superposition of two or more of the provided distributions can possibly be used to generate the desired distribution. For example, assume that the electric field along the z axis is desired for the bipolar cylindrical surface distribution shown in Figure 2.40. The total length of the cylinder is 2L. The surface charge distribution on the top half of the cylinder is ρ_S ($0 < z < L, \rho = a$) while the surface charge on the bottom half of the cylinder is $-\rho_S$ ($-L < z < 0, \rho = a$). The total charge on the cylinder is zero. The table result for a uniformly charged cylindrical surface of *total length L* can be used:

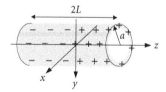

FIGURE 2.40 Bipolar surface charge distribution along a cylinder of length 2L.

$$\vec{E}_{-\frac{L}{2}to\frac{L}{2}}(0,0,z) = \frac{\rho_S a}{2\varepsilon_o}\left[\frac{1}{\sqrt{a^2+\left(z-\frac{L}{2}\right)^2}} - \frac{1}{\sqrt{a^2+\left(z+\frac{L}{2}\right)^2}}\right]\hat{a}_z$$

This electric field is for a cylinder that is centered about the $z = 0$ plane. For a cylinder that exists from $z = 0$ to L, the electric field expression is obtained by a simple shift in the z variable by $L/2$:

$$\vec{E}_{0\,to\,L}(0,0,z) = \frac{\rho_S a}{2\varepsilon_o}\left[\frac{1}{\sqrt{a^2+(z-L)^2}} - \frac{1}{\sqrt{a^2+z^2}}\right]\hat{a}_z$$

For a cylinder with a negative charge distribution of $-\rho_S$ from $z = -L$ to $z = 0$, the electric field is obtained in a similar manner:

$$\vec{E}_{-L\,to\,0}(0,0,z) = -\frac{\rho_S a}{2\varepsilon_o}\left[\frac{1}{\sqrt{a^2+z^2}} - \frac{1}{\sqrt{a^2+(z+L)^2}}\right]\hat{a}_z$$

The electric field along the z axis for the bipolar distribution is the sum of these two electric field expressions:

$$\vec{E}_{bipolar}(0,0,z) = \vec{E}_{0\,to\,L}(0,0,z) + \vec{E}_{-L\,to\,0}(0,0,z)$$

$$= \frac{\rho_S a}{2\varepsilon_o}\left[\frac{1}{\sqrt{a^2+(z-L)^2}} - \frac{1}{\sqrt{a^2+z^2}}\right]\hat{a}_z - \frac{\rho_S a}{2\varepsilon_o}\left[\frac{1}{\sqrt{a^2+z^2}} - \frac{1}{\sqrt{a^2+(z+L)^2}}\right]\hat{a}_z$$

$$= \frac{\rho_S a}{2\varepsilon_o}\left[\frac{1}{\sqrt{a^2+(z-L)^2}} + \frac{1}{\sqrt{a^2+(z+L)^2}} - \frac{2}{\sqrt{a^2+z^2}}\right]\hat{a}_z$$

The distributions can also be subtracted from each other to generate "holes" and other zero-charge regions such as in Figure 2.41. The electric field along the z axis of this uniformly charged infinite sheet with a zero-charge hole of radius a is desired. The field is simply the difference between the field from a uniformly charged infinite sheet and the field from a uniformly charged circular disk:

$$\vec{E}(0,0,z) = \pm\frac{\rho_S}{2\varepsilon_o}\hat{a}_z - \frac{\rho_S}{2\varepsilon_o}\left(\pm 1 - \frac{z}{\sqrt{a^2+z^2}}\right)\hat{a}_z = \frac{\rho_S z}{2\varepsilon_o\sqrt{a^2+z^2}}\hat{a}_z$$

Superposition should be done carefully. The origins, coordinate systems, and field's directions for the various expressions should all be noted.

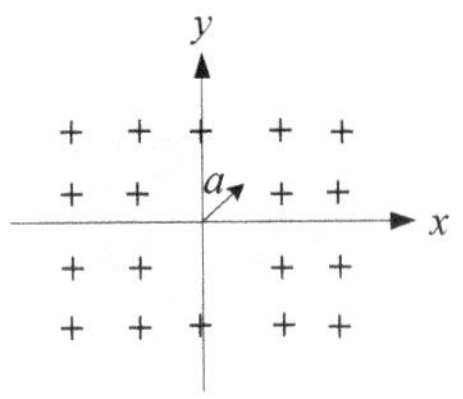

FIGURE 2.41 Uniformly charged sheet with a zero-charge hole.

2.14   Electric Field from Other Charge Distributions

How can the electric field for a general charge distribution be determined? What precautions should be taken when evaluating the integrals in the cylindrical and spherical coordinate systems? [Hayt; Demarest; Haus; Lerner]

For many simple charge distributions possessing a great deal of symmetry, the electric field can be obtained quickly by the use of Gauss' law:

$$\nabla \cdot \vec{D} = \rho_V$$

This equation relates the divergence of the flux density vector, $\vec{D}$, and the volume charge density, ρ_V. The integral form of Gauss' law is obtained by first integrating both sides of this differential or point form over a volume, V:

$$\iiint_V (\nabla \cdot \vec{D})\, dv = \iiint_V \rho_V\, dv$$

The powerful divergence theorem can then be used to convert the volume integral involving $\nabla \cdot \vec{D}$ to a closed surface integral:

$$\oiint_S \vec{D} \cdot d\vec{s} = \iiint_V \rho_V\, dv \tag{2.119}$$

where S is the complete (i.e., closed) surface corresponding to the volume, V.[18] Equation (2.119) is Gauss' law in integral form. It is also seen as

$$\oiint_S \vec{D} \cdot d\vec{s} = Q_{encl} \tag{2.120}$$

where Q_{encl} is the total charge enclosed by the surface of interest. The charge is contained within the volume, V. This important law states that

> "The electric flux passing outward through any closed surface is equal to the total charge enclosed by that surface."

The electric flux through a surface is the total $\vec{D}$ that passes through the given surface. This surface is not necessarily a "real" surface. It is often just an imaginary surface that encloses some charge of interest. For example, the dotted lines in Figure 2.42 are outlining a surface surrounding a collection of charges. The resultant electric field emanating from these charges in all directions, which is not shown, passes in a complex manner in and out of the six surfaces of this volume. If the total charge contained in the corresponding volume were zero, the total outward flux through the six sides of this region would be zero. When the total charge enclosed is zero, some students *incorrectly* believe that the electric field outside the volume must be everywhere zero. (The electric field outside a dipole charge configuration, which has zero net charge, is not zero.) Gauss' law states that the net outward flux passing through all of the surfaces of the volume is zero when the total charge enclosed is zero. When the net charge enclosed by the closed surface is zero, on some parts of the surface the flux is passing outward, and on other parts of the surface the flux is passing inward. If the net charge were positive, the total flux passing out would be greater than that passing in. If the net charge were negative, the total flux passing in would be greater than that passing out.

---

[18]A closed surface integration implies that all sides of the corresponding volume are included in the integration. There should be no holes or other openings in the surface.

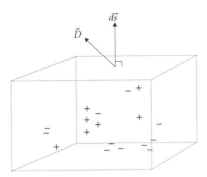

FIGURE 2.42 Imaginary rectangular enclosure containing a charge distribution.

Charge outside the volume of interest, of course, also generates an electric field. Imagine that a positive charge is outside the volume shown in Figure 2.42 and that its field passes through the surfaces of the volume. This charge, because it is not contained in the volume, will generate zero net outward flux through the surfaces of the volume. It will pass in through portions of some surfaces and out through portions of other surfaces. However, the net outward flux generated by this *external* charge through these surfaces is zero. Finally, if the net charge enclosed is the same, the actual shape of the surface around the charge is irrelevant. For example, in Figure 2.42, if a spherical imaginary volume were taken instead of a rectangular prism, the net outward flux would be the same. The shape of the volume and surface is usually selected based on mathematical convenience. For example, a very long cylindrical tube of uniform charge would usually have a cylindrical imaginary surface enclosing it. A spherical uniform surface charge would usually have a spherical imaginary surface enclosing it.

There are many excellent textbooks in electromagnetics that clearly discuss how Gauss' law can sometimes be used to determine the electric field. Generally, to apply easily Gauss' law, a great deal of charge symmetry is required and the proper Gaussian surface (i.e., closed surface surrounding the charge distribution) must be selected. This usually implies that the electric field from the charges is (mostly) normal to the surface selected. Gauss' law is wonderful for obtaining elegant and fast electric-field solutions to many charge distributions. But, unless the reader has a good understanding and "feel" for how the electric fields from a collection of charges add and subtract, it is recommended that Gauss' law not be used to determine specific field information.

In some ways, Gauss' law is analogous to the Thévenin equivalent of a circuit. Like the Thévenin equivalent, Gauss' law does not provide much in terms of the details of how the charge is distributed inside the "box" or Gaussian surface. By measuring the fields everywhere outside the closed surface, the net charge inside the surface can be determined via Gauss' law. Many of the external electric fields from point, line, surface, and volume charges that can be simply obtained using Gauss' law are contained in Appendix A.

For a general charge distribution that does not necessarily possess symmetry or is not simple, the electric field can be determined directly or indirectly (via the potential). For any collection of point, line, surface, and volume charges, the electric field can be determined directly from

$$\vec{E}(r) = \sum_{m=1}^{n} \frac{Q_m}{4\pi\varepsilon_o \left|\vec{r} - \vec{r}_m\right|^2} \hat{a}_{r-r'}$$

$$+ \int_{L'} \frac{\rho_L(r') \, dL'}{4\pi\varepsilon_o \left|\vec{r} - \vec{r}'\right|^2} \hat{a}_{r-r'} + \iint_{S'} \frac{\rho_S(r') \, dS'}{4\pi\varepsilon_o \left|\vec{r} - \vec{r}'\right|^2} \hat{a}_{r-r'} + \iiint_{V'} \frac{\rho_V(r') \, dV'}{4\pi\varepsilon_o \left|\vec{r} - \vec{r}'\right|^2} \hat{a}_{r-r'}$$

$$(2.121)$$

These integrals can be evaluated numerically or, in some special cases, analytically. The primed variables represent the location of the actual charges so that $\vec{r}'$ is the vector from the origin to where the charge(s)

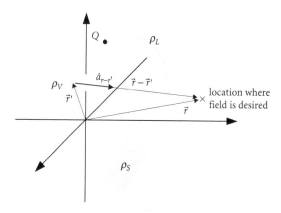

FIGURE 2.43 Variables used in Equation (2.121). The electric field due to the point, line, surface, and volume charge distributions is desired.

is located (the source of the electric field). The unprimed variable $\vec{r}$ is the vector from the origin to where the field is desired. Therefore, $\vec{r} - \vec{r}'$ is the vector from the charge to the location where the field is desired. The quantity $|\vec{r} - \vec{r}'|$ is the distance between the source of the field (i.e., the charge) and location where the field is desired. The vector $\hat{a}_{r-r'}$ is the unit vector from the charge to this point of interest. In Figure 2.43, these vectors are shown for one location inside a volume charge and for one point of interest. The vectors for other locations inside the volume charge and from the point, line, and surface charges are not shown. It is probably fairly obvious that the field at any point of interest is the superposition or sum of the fields from every charge, whether the charge is a point, line, surface, or volume charge. For the point charges, the electric field from each of the charges is merely summed. For line, surface, and volume charges, integration must be performed since the charges are continuous (i.e., not at discrete locations). Initially, these integrals may seem overwhelming, but after a few examples their setup usually becomes mundane.

As a first example, the integral will be set up for the electric field from a square-shaped surface charge located in the *xy* plane. The length of each side of the square is *a*. The surface charge over this square is nonuniformly distributed according to the expression

$$\rho_S(x, y) = Axy \tag{2.122}$$

where *A* is a positive constant. The charge is positive in the first and third quadrants while negative in the second and fourth quadrants. Referring to Figure 2.44, the electric field at the general point (x, y, z) is given

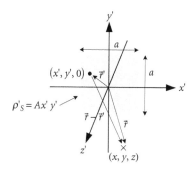

FIGURE 2.44 Variables used to determine the electric field due to the charge distribution in (2.122).

by the one integral (since no point, line, or volume charge is present)

$$\vec{E}(x,y,z) = \int_{-\frac{a}{2}}^{\frac{a}{2}} \int_{-\frac{a}{2}}^{\frac{a}{2}} \frac{A x' y' \, dx' \, dy'}{4\pi\varepsilon_o |(x\hat{a}_x + y\hat{a}_y + z\hat{a}_z) - (x'\hat{a}_x + y'\hat{a}_y)|^2} \frac{(x\hat{a}_x + y\hat{a}_y + z\hat{a}_z) - (x'\hat{a}_x + y'\hat{a}_y)}{|(x\hat{a}_x + y\hat{a}_y + z\hat{a}_z) - (x'\hat{a}_x + y'\hat{a}_y)|}$$

$$= \int_{-\frac{a}{2}}^{\frac{a}{2}} \int_{-\frac{a}{2}}^{\frac{a}{2}} \frac{A x' y' \, dx' \, dy'}{4\pi\varepsilon_o \left[\sqrt{(x-x')^2 + (y-y')^2 + z^2}\right]^2} \frac{(x-x')\,\hat{a}_x + (y-y')\,\hat{a}_y + z\hat{a}_z}{\sqrt{(x-x')^2 + (y-y')^2 + z^2}}$$

$$= \int_{-\frac{a}{2}}^{\frac{a}{2}} \int_{-\frac{a}{2}}^{\frac{a}{2}} \frac{A x' y' [(x-x')\,\hat{a}_x + (y-y')\,\hat{a}_y + z\hat{a}_z] \, dx' \, dy'}{4\pi\varepsilon_o [(x-x')^2 + (y-y')^2 + z^2]^{\frac{3}{2}}}$$

The primed variables represent the location of the surface charge, where $z' = 0$. The integration is taken over the location of this surface charge from $-a/2$ to $a/2$ in both the x' and y' directions. Notice that the surface charge is a function of the primed variables. These integrals can be solved analytically. If the solution is only desired along the z axis where $x = y = 0$, the integration reduces to

$$\vec{E}(0,0,z) = \int_{-\frac{a}{2}}^{\frac{a}{2}} \int_{-\frac{a}{2}}^{\frac{a}{2}} \frac{A x' y' [-x'\hat{a}_x - y'\hat{a}_y + z\hat{a}_z] \, dx' \, dy'}{4\pi\varepsilon_o [(x')^2 + (y')^2 + z^2]^{\frac{3}{2}}} = \int_{-\frac{a}{2}}^{\frac{a}{2}} \int_{-\frac{a}{2}}^{\frac{a}{2}} \frac{-A (x')^2 y' \hat{a}_x \, dx' \, dy'}{4\pi\varepsilon_o [(x')^2 + (y')^2 + z^2]^{\frac{3}{2}}}$$

$$+ \int_{-\frac{a}{2}}^{\frac{a}{2}} \int_{-\frac{a}{2}}^{\frac{a}{2}} \frac{-A x' (y')^2 \hat{a}_y \, dx' \, dy'}{4\pi\varepsilon_o [(x')^2 + (y')^2 + z^2]^{\frac{3}{2}}} + \int_{-\frac{a}{2}}^{\frac{a}{2}} \int_{-\frac{a}{2}}^{\frac{a}{2}} \frac{A x' y' z\hat{a}_z \, dx' \, dy'}{4\pi\varepsilon_o [(x')^2 + (y')^2 + z^2]^{\frac{3}{2}}}$$

One advantage of working in the Cartesian coordinate system is that the unit vectors, $\hat{a}_x$, $\hat{a}_y$, and $\hat{a}_z$, do not change their direction with integration. Therefore, they and other unprimed variables can be pulled outside the integrals (when it is convenient to do so):

$$\vec{E}(0,0,z) = -\frac{A\hat{a}_x}{4\pi\varepsilon_o} \int_{-\frac{a}{2}}^{\frac{a}{2}} \int_{-\frac{a}{2}}^{\frac{a}{2}} \frac{(x')^2 y' \, dx' \, dy'}{[(x')^2 + (y')^2 + z^2]^{\frac{3}{2}}} - \frac{A\hat{a}_y}{4\pi\varepsilon_o} \int_{-\frac{a}{2}}^{\frac{a}{2}} \int_{-\frac{a}{2}}^{\frac{a}{2}} \frac{x' (y')^2 \, dx' \, dy'}{[(x')^2 + (y')^2 + z^2]^{\frac{3}{2}}}$$

$$+ \frac{A z\hat{a}_z}{4\pi\varepsilon_o} \int_{-\frac{a}{2}}^{\frac{a}{2}} \int_{-\frac{a}{2}}^{\frac{a}{2}} \frac{x' y' \, dx' \, dy'}{[(x')^2 + (y')^2 + z^2]^{\frac{3}{2}}}$$

$$= 0$$

Since each integral involves the symmetrical integration of at least one odd function of x' or y', all of these integrals are zero. Therefore, along the z axis, the electric field is zero. The more insightful reader will realize the reasonableness of this result after examining the fields from the odd symmetrical distribution of the charge.

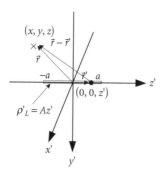

FIGURE 2.45   Variables used to determine the electric field due to the charge distribution in Equation (2.123).

This same procedure can also be applied to line and volume charge distributions. For example, suppose the electric field is desired from a linearly varying line charge from $-a$ to a along the z axis:

$$\rho_L(z) = \begin{cases} Az & \text{if } -a < z < a \\ 0 & \text{if } |z| > a \end{cases} \tag{2.123}$$

where A is a constant. The total charge along this line charge is zero. The setup of the integrals to determine the electric field is determined in the same manner except now the integration is only with respect to one variable, z'. Referring to Figure 2.45,

$$\vec{E}(x,y,z) = \int_{-a}^{a} \frac{Az'\,dz'}{4\pi\varepsilon_o\,|(x\hat{a}_x + y\hat{a}_y + z\hat{a}_z) - (z'\hat{a}_z)|^2} \frac{(x\hat{a}_x + y\hat{a}_y + z\hat{a}_z) - (z'\hat{a}_z)}{|(x\hat{a}_x + y\hat{a}_y + z\hat{a}_z) - (z'\hat{a}_z)|}$$

$$= \int_{-a}^{a} \frac{Az'\,dz'}{4\pi\varepsilon_o\left[\sqrt{x^2 + y^2 + (z-z')^2}\right]^2} \frac{x\hat{a}_x + y\hat{a}_y + (z-z')\hat{a}_z}{\sqrt{x^2 + y^2 + (z-z')^2}}$$

$$= \int_{-a}^{a} \frac{Az'[x\hat{a}_x + y\hat{a}_y + (z-z')\hat{a}_z]\,dz'}{4\pi\varepsilon_o\,[x^2 + y^2 + (z-z')^2]^{\frac{3}{2}}}$$

$$= \frac{Ax\hat{a}_x}{4\pi\varepsilon_o} \int_{-a}^{a} \frac{z'\,dz'}{[x^2 + y^2 + (z-z')^2]^{\frac{3}{2}}} + \frac{Ay\hat{a}_y}{4\pi\varepsilon_o} \int_{-a}^{a} \frac{z'\,dz'}{[x^2 + y^2 + (z-z')^2]^{\frac{3}{2}}}$$

$$+ \frac{A\hat{a}_z}{4\pi\varepsilon_o} \int_{-a}^{a} \frac{z'\,(z-z')\,dz'}{[x^2 + y^2 + (z-z')^2]^{\frac{3}{2}}}$$

Along the $+z$ axis where $x = y = 0$, these integrals reduce to a single integral that can be evaluated in the usual fashion:

$$\vec{E}(0,0,z) = \frac{A\hat{a}_z}{4\pi\varepsilon_o} \int_{-a}^{a} \frac{z'\,(z-z')\,dz'}{[(z-z')^2]^{\frac{3}{2}}} = \frac{A\hat{a}_z}{4\pi\varepsilon_o} \int_{-a}^{a} \frac{z'\,dz'}{(z-z')^2} = \frac{A}{4\pi\varepsilon_o}\left[\ln\left(\frac{z-a}{z+a}\right) + \frac{2az}{z^2 - a^2}\right]\hat{a}_z$$

The electric field can also be determined indirectly by first determining the potential distribution via the expression

$$\Phi(r) = \sum_{m=1}^{n} \frac{Q_m}{4\pi\varepsilon_o |\vec{r} - \vec{r}_m|} + \int_{L'} \frac{\rho_L(r')\,dL'}{4\pi\varepsilon_o |\vec{r} - \vec{r}'|} + \iint_{S'} \frac{\rho_S(r')\,dS'}{4\pi\varepsilon_o |\vec{r} - \vec{r}'|} + \iiint_{V'} \frac{\rho_V(r')\,dV'}{4\pi\varepsilon_o |\vec{r} - \vec{r}'|} \quad (2.124)$$

The variables have the same meaning as in Equation (2.121). Once the potential distribution is known, the electric field can be determined using $\vec{E} = -\nabla\Phi$. The potential reference (i.e., the location of 0 V) is at infinity for Equation (2.124). Therefore, the potential for charge distributions that extend to infinity, such as infinite line charges, cannot be determined using these integrals. One advantage of working with Equation (2.124) is that the integrals are simpler than in Equation (2.121) and do not involve vectors. An obvious disadvantage is that an additional operation, the gradient operation, must be performed to determine the electric field. For comparison, for the previous line charge distribution described by Equation (2.123), the potential anywhere is given by the much simpler, single integral

$$\Phi(x,y,z) = \int_{-a}^{a} \frac{Az'\,dz'}{4\pi\varepsilon_o |(x\hat{a}_x + y\hat{a}_y + z\hat{a}_z) - (z'\hat{a}_z)|} = \int_{-a}^{a} \frac{Az'\,dz'}{4\pi\varepsilon_o \sqrt{x^2 + y^2 + (z-z')^2}}$$

Along the $+z$ axis where $x = y = 0$ and $z > z'$, this potential integral has a rather simple form:

$$\Phi(0,0,z) = \int_{-a}^{a} \frac{Az'\,dz'}{4\pi\varepsilon_o \sqrt{(z-z')^2}} = \frac{A}{4\pi\varepsilon_o} \int_{-a}^{a} \frac{z'\,dz'}{z-z'} = \frac{A}{4\pi\varepsilon_o}[-z' - z\ln(z-z')]\Big|_{-a}^{a}$$

$$= -\frac{A}{4\pi\varepsilon_o}\left[2a + z\ln\left(\frac{z-a}{z+a}\right)\right]$$

By assuming that $z > a$, the electric field along the $+z$ axis is

$$\vec{E}(0,0,z) = -\nabla\Phi = -\frac{d\Phi}{dz}\hat{a}_z = \frac{A}{4\pi\varepsilon_o}\left[\ln\left(\frac{z-a}{z+a}\right) + \frac{2az}{z^2 - a^2}\right]\hat{a}_z$$

The electric field is only a function of the variation of the potential in the z direction. By examining the charge distribution, it is clear that along the z axis there is only a z component of the electric field. In general, unless symmetry arguments are invoked, the potential function should be determined everywhere (or at and near the point of interest if done numerically) if the electric field is desired. The potential is required at more than one location simply because $\vec{E} = -\nabla\Phi$ and the gradient operation involves differentiation with respect to all of the variables in the coordinate system used. When using Equation (2.121), the electric field can be determined at any point by evaluating the integral at the point of interest.

There is one major precaution when evaluating Equation (2.121) (or any other integral-based relation) in other coordinate systems. The unit vectors in other coordinate systems can change direction with respect to the integration. In these situations, these unit vectors should not be pulled past the integrals. This mistake is unfortunately seen in a few books. As a first example, to illustrate the proper method of solving for the electric field in other coordinate systems, the electric field at any point along the z axis for a ring of charge in the xy plane and centered on the z axis is desired. The radius of the ring is a, and the line charge density varies according to the relation

$$\rho_L(\phi) = A\sin\phi \quad (2.125)$$

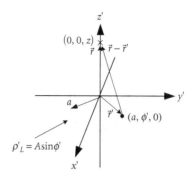

FIGURE 2.46 Variables used to determine the electric field along the z axis due to the charge distribution in Equation (2.125).

The cylindrical coordinate system will be used. Referring to Figure 2.46, the position vectors are given as

$$\vec{r} = z\hat{a}_z, \quad \vec{r}' = a\hat{a}_\rho$$

Immediately notice that $\vec{r}'$ does not *appear* to be a function of the angle ϕ. However, for different positions around the ring, the direction of $\hat{a}_\rho$, the radial unit vector from the z axis, changes. Therefore, $\hat{a}_\rho$ is a "hidden" function of the angle ϕ! The integral expression for the electric field is

$$\vec{E}(0,0,z) = \int_L \frac{\rho'_L \, dL'}{4\pi\varepsilon_o |\vec{r} - \vec{r}'|^2} \hat{a}_{r-r'} = \int_0^{2\pi} \frac{A\sin(\phi') \, a d\phi'}{4\pi\varepsilon_o |z\hat{a}_z - a\hat{a}_\rho|^2} \frac{(z\hat{a}_z - a\hat{a}_\rho)}{|z\hat{a}_z - a\hat{a}_\rho|}$$

$$= \int_0^{2\pi} \frac{A\sin(\phi') \, a d\phi' \, (z\hat{a}_z - a\hat{a}_\rho)}{4\pi\varepsilon_o (z^2 + a^2)^{\frac{3}{2}}}$$

$$= \frac{Aza\hat{a}_z}{4\pi\varepsilon_o (z^2 + a^2)^{\frac{3}{2}}} \int_0^{2\pi} \sin(\phi') \, d\phi' - \frac{Aa^2}{4\pi\varepsilon_o (z^2 + a^2)^{\frac{3}{2}}} \int_0^{2\pi} \sin(\phi') \, d\phi' \, \hat{a}_\rho$$

$$= -\frac{Aa^2}{4\pi\varepsilon_o (z^2 + a^2)^{\frac{3}{2}}} \int_0^{2\pi} \sin(\phi') \, d\phi' \, \hat{a}_\rho$$

Notice that the unit vector $\hat{a}_z$ was taken past the first integral since its direction is always the same. This first integral is also equal to zero since the integration of a sinusoidal function over a full period is zero. The unit vector $\hat{a}_\rho$ cannot be pulled past the second integral. By converting this radial unit vector to the Cartesian coordinate unit vector equivalent, the unit vectors can be pulled past the integrals and the integrals evaluated:

$$\vec{E}(0,0,z) = -\frac{Aa^2}{4\pi\varepsilon_o (z^2 + a^2)^{\frac{3}{2}}} \int_0^{2\pi} \sin(\phi') \, d\phi' \, [\cos(\phi') \, \hat{a}_x + \sin(\phi') \, \hat{a}_y]$$

$$= -\frac{Aa^2 \hat{a}_x}{4\pi\varepsilon_o (z^2 + a^2)^{\frac{3}{2}}} \int_0^{2\pi} \cos(\phi') \sin(\phi') \, d\phi' - \frac{Aa^2 \hat{a}_y}{4\pi\varepsilon_o (z^2 + a^2)^{\frac{3}{2}}} \int_0^{2\pi} \sin^2(\phi') \, d\phi'$$

$$= -\frac{Aa^2 \hat{a}_y}{4\varepsilon_o (z^2 + a^2)^{\frac{3}{2}}}$$

The electric field along the z axis is entirely in the $-y$ direction. By examining the fields due to the positive line charge distribution for $y > 0$ and the negative line charge distribution for $y < 0$, this result might be evident "to even the casual observer."

2.15 Discharges Classified

Discuss the major types of discharge. [Glor; Pratt; Taylor, '94; Jones, '91; Williams; Cross; Lüttgens; Larsen; Heidelberg; Maurer; Tolson]

Electrical breakdown of air (and other medium) is complex, empirical, and phenomenological. It thus should not be surprising that most of the information contained in this discussion is experimentally based. The details of the experiments are not provided, and the information contained in this section should only be used as a guideline. Exceptions can be found to some of these guidelines.

In the field of electrostatics, it is common to categorize the various electrostatic discharges. Classifying helps in the understanding of air breakdown and assists in determining whether a particular situation is potentially an electrostatic hazard. In the discharges to follow, estimates for the expected energy available to ignite a vapor, gas, or liquid are given. In general, it is very difficult to predict analytically the energy in a discharge. The energy is a function of the conductivity and geometry of the charged surfaces, as well as the electrical characteristics and geometry of the surrounding environment.

As will be seen, it is not necessary to have two conducting electrodes for a discharge to occur. Furthermore, grounding a conducting object does not necessarily make it immune to several types of discharge. Also, discharge can occur from an insulating surface to a nearby conducting electrode.

Spark/Capacitive Discharge

The term spark is actually used by nonspecialists to describe many types of breakdown. In the field of electrostatics, however, spark or capacitive discharge is used to describe the breakdown between two isolated electrodes or conducting objects when the potential difference across them is sufficiently large. During capacitive discharge testing, to help avoid other types of discharge, both electrodes should not be too sharp so that the field between them is somewhat uniform. (Often, the edges or ends of the electrodes may need to be smoothed or rounded to help prevent, for example, a corona discharge.) To obtain a somewhat uniform electric field, the distance between the two electrodes should be small compared to the radii of curvature of both electrode surfaces. In Figure 2.47, although not shown, a large voltage is applied across both sets of electrodes. In the left illustration, the smaller radius of curvature of the two electrodes is R_1. The distance between the two electrodes, d, is much greater than R_1. In this case, for region shown, the field is definitely nonuniform between the two electrodes. In the right illustration, the smaller radius of curvature of the two electrodes is R_3. This same distance, d, is now less than R_3. Now, for the region shown, the electric field between the two electrodes is approximately uniform.

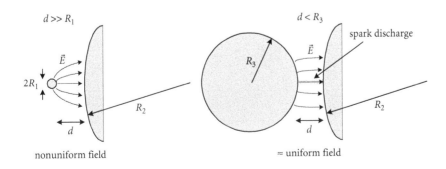

FIGURE 2.47 Electrode pairs with different radii of curvature.

A spark discharge between electrodes is more likely to occur when the distance is small or at least less than the smallest radii of curvature of the electrodes.

For a spark discharge, the discharge is fast over a narrow channel. A spark discharge is thought to be the most common type of discharge in electrostatic hazards. For example, the common shock received when reaching for a metal knob is a spark discharge. In this case, the individual is one "electrode" and the knob is the other electrode. For this type of two-electrode discharge, the maximum possible energy that can be discharged is $CV^2/2$, where C is the capacitance between the two electrodes and V is the voltage difference across them. So that an estimate of the danger or destructive power of charged objects can be determined, it is common to model conducting objects such as shovels, pipes, and people as an isolated capacitor. If the voltage of the conducting object is known, relative to ground or the other electrode of interest, then the *maximum* possible energy of a spark discharge from this object can be determined. This maximum energy is frequently compared to the minimum ignition energy of vapors, gases, and dusts.

Spark discharges can be avoided by eliminating the potential difference between the two conducting objects or electrodes. If both conducting objects are grounded, for example, then the potential difference across them is zero and the capacitive discharge energy is zero. This is probably the basis of the advice to "ground an object" to eliminate (spark) discharges.

Corona Discharge

When the electric field strength near a sharp conducting surface with a radius of curvature less than about 5 mm is sufficiently large, a local breakdown occurs near the surface. (The electric field increases with the sharpness of the electrode for a given potential.) This breakdown is referred to as a corona or silent discharge. The energy density is low, and the region of ionization is over a close space near the sharp electrode. For this reason, it is believed to be too weak to ignite even the most flammable suspended polymer dusts, gases, or vapors (exceptions are hydrogen/air and carbon disulfide/air with minimum ignition energies of 0.011 mJ and 0.009 mJ, respectively).

Wires and rods with diameters less than about 3 mm are used to generate extra ions (i.e., charge) in storage vessels or bins via corona discharge. This charge from corona can be used to neutralize charge in a silo. If the wire diameter is too large, then a much more energetic brush discharge can occur. When draining the charge from bins and other containers, the recommended discharging rod or wire diameter is about 1 to 3 mm.

The voltage required to initiate corona is a function of the polarity of the voltage. Generally, the amplitude of a positive voltage needed to generate corona is greater than the amplitude of a negative voltage needed to generate corona. Corona is further classified as either active or passive. For active corona, the electrode is connected directly to a voltage supply. For passive corona, the electrode is either grounded or floating and the electrode is completely or partially immersed in an externally generated electric field. In Figure 2.48, the electric field to initiate a corona discharge is obtained from a nearby

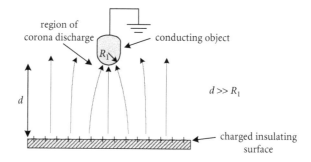

FIGURE 2.48   Generation of passive corona on a grounded object from the field generated from a nearby charged insulating surface.

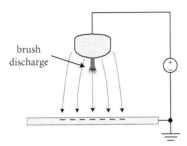

FIGURE 2.49 Brush discharge from an electrode.

charged insulating surface. Even if the conducting object is floating or not connected to ground or any voltage, corona can still occur due to the distortion generated by the conducting object in the electric field. Corona discharge is also referred to as one-electrode discharge (in the field of electrostatics) because this discharge frequently involves one electrode and a charged insulating surface. However, corona discharge can also occur between two electrodes when a sufficiently large potential difference is across them and the field is nonuniform. When the field is approximately uniform, a spark discharge is more likely to occur or, at least, attract the eye.

Brush Discharge

When the electric field near an electrode with a radius or curvature between about 5 mm and 50 mm is sufficiently large (about 500 kV/m), irregular multiple discharge paths are seen that have the look of a brush. This brush-like shape is shown in Figure 2.49. If the electrode is too sharp, a corona discharge will usually occur instead of a brush discharge. Typically, a brush discharge is an electrical breakdown between a grounded conducting electrode and nonconducting surface. Unlike a spark discharge, where a large percentage of the maximum possible stored energy is released per discharge, a much smaller percentage of the maximum energy is released per brush discharge. As with the corona discharge, this breakdown is referred to as a one-electrode discharge (even though it can also occur between two electrodes). Also, two characteristics of a brush discharge are an acoustical "crack" and a burst of current. The energy density is larger than a corona discharge and is thought capable of igniting some flammable gas vapors but not flammable suspended polymer dusts. Common hydrocarbons, however, are easily ignited. Sulfur dusts in an oxygen-enriched atmosphere can occasionally be ignited. About 1 to 3.6 mJ energy is released during a brush discharge, and the surface charge is greater than about 3 $\mu C/m^2$. It has been shown that greater positive surface charge is required to ignite certain gases, such as hydrogen, than negative surface charge via a brush discharge. For some situations such as with a mixture of propane and air, positive surface charge could not even ignite the material while 7.4 $\mu C/m^2$ of negative surface charge could. As the area of the charged surface increases, the strength of the brush discharge generally increases.

When a metal tool or human finger, for example, is brought near charged insulating objects, such as bags or tubes, a brush discharge can occur. It can also occur when conducting objects are inserted into a charged region, such as a bin of charged powder. The electric field to initiate a brush discharge can be generated by charge on the electrode or on a nearby insulating surface. Static eliminators can be used to reduce or neutralize excessive charge on insulating surfaces.

Lichtenberg/Propagating Brush Discharge

For a double layer of charge across an insulating slab or film, the surface charge (magnitude) can be much greater than the Gaussian limit of 26 $\mu C/m^2$. A powerful Lichtenberg discharge, accompanied by a loud cracking, can occur when the charge along an insulating surface exceeds this Gaussian charge limit.

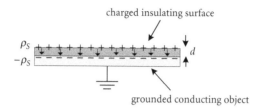

FIGURE 2.50   Double layer of charge on an insulating slab.

If an insulating surface is about 1 to 8 mm thick with a conducting grounded backside,[19] the discharge can have energies greater than 1 J. For this type of discharge to occur, the charge on the surface should be greater than about 230 to 270 μC/m². Also, the breakdown strength of the insulation should be sufficiently large so that film does not break down before this charge level can be reached. Although the actual breakdown voltage is a function of the insulation thickness, for a film thickness of about 20 μm, the voltage is about 4 kV for polyester and polycarbonate materials.

In Figure 2.50, a grounded conducting object is shown in direct contact with an insulating slab. Positive charge is present along the upper surface of the insulator, possibly due to tribocharging with a high-velocity insulating powder. The negative charge on the upper surface of the conducting object is due to the attraction of the electrons in the conductor to this positive charge. For this set of flat, equal-magnitude, parallel, uniform surface charges, the electric field outside the insulator is zero (ignoring fringing). The magnitude of the electric field, $|\vec{E}|$, inside the insulating slab, and potential difference, $\Delta\Phi$, across the insulating slab of thickness d are

$$|\vec{E}| = \frac{\rho_S}{k\varepsilon_o} \quad \Rightarrow \Delta\Phi = \int \vec{E} \cdot d\vec{L} = \frac{\rho_S d}{k\varepsilon_o} \tag{2.126}$$

The energy stored in the electric field between this double layer of charge is

$$E = \frac{1}{2}CV^2 = \frac{1}{2}\left(\frac{k\varepsilon_o A}{d}\right)\left(\frac{\rho_S d}{k\varepsilon_o}\right)^2 = \frac{\rho_S^2 Ad}{2k\varepsilon_o} \tag{2.127}$$

where A is the area of the surface charge layers.[20] Notice that the energy stored increases with the thickness of the insulating slab and decreases with the dielectric constant. (For a fixed total charge, Q, as the capacitance between two objects decreases, the voltage across them must increase.) So, the thicker the insulating slab, the greater the energy available for a Lichtenberg discharge. Because the electrical breakdown strength of many materials is greater than air, the charge density on and electric field in the material can be much larger than if only air was present between the charge layers.

Besides the lightning discharge, this breakdown is the most energetic discharge. This discharge can ignite vapors, gases, and dusts. Insulating linings, pipes, and hoses should be avoided because of the charge that can accumulate along these inner surfaces. Insulating objects where the "other" outer surface is conducting and grounded (via grounding tape, moisture, or hand contact) should especially be avoided. To reduce the likelihood of this type of discharge, it has been recommended that the

---

[19]Actually, it is not necessary that the backside be grounded. As discussed previously, it is only necessary that the backside surface contain surface charge of the opposite sign.

[20]The capacitance is normally defined between two conducting objects. In general, a charged insulating surface is not an equipotential surface. However, for this set of flat, equal-magnitude, parallel, uniform surface charges, the potential of the insulating surface is constant.

insulating lining or material should be less than about 0.25 mm thick for small surface charge densities (less than about 230 to 270 $\mu C/m^2$) and greater than about 8 mm thick for large expected surface charges (greater than about 230 to 270 $\mu C/m^2$).[21] Charge on thin layers of paint and epoxy (and films less than about 10 μm thick) is not normally considered a potential source of propagating brush discharge.

To build up the large surface charge densities required for these high-energy discharges, generally high-speed nonmanual operations are required. For example, hand rubbing a plastic object against another object is not likely to generate the required charge density. However, powder flowing quickly down a plastic tube or high-speed conveyer belt rubbing against a plastic part could generate the necessary high charge density. Sometimes, even a plastic observation window could be an unexpected location for charge to build up.

Clearly, it is best to avoid the charge buildup rather than trying to control the energy of the discharge. To reduce the probability of a Lichtenberg discharge, high-speed operations of the type previously discussed should be avoided. If they cannot be avoided, possibly the key insulating materials can be replaced with more conductive materials so that the charge buildup can be dissipated. In addition, these key insulators could possibly be replaced with insulators with a lower electrical strength (so the material will more likely break down before large charge densities can be reached). If an insulator with a much higher dielectric constant can be located (which is more difficult), then according to Equation (2.127) the energy stored between the double layer will be less. Lichtenberg discharges have been believed to be the source of many serious electrostatic discharge accidents. Pinholes are sometimes an indication of this type of discharge.

In Figure 2.51, high-speed powder is shown flowing down an insulating pipe. The tribocharging that occurs between the powder and pipe generates a surface charge along the inner surface of the pipe. (Volume charge can also exist in the pipe.) Two conducting objects are shown in contact with the pipe: a ground wire and a thumb. Charge of the opposite polarity is induced on the outer surface of the grounding wire and the thumb. The electric fields from the positive and negative surface charges will approximately cancel on both sides of the pipe. Between the charges in the insulating pipe, however, the electric field can be quite large. Many insulating materials can withstand very large electric fields. Eventually, though, the insulating material breaks down, a short circuit occurs between the layers of charge, or the fringing fields

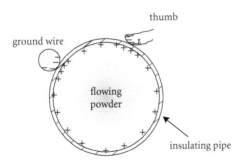

FIGURE 2.51 Induced charge on conductors external to an insulating pipe containing charged powder.

---

[21]The fields outside the double layer of surface charge will cancel if (i) the charge densities are equal but opposite in sign, (ii) the cross-sectional area of the charge layers are very large compared to the distance between the layers, and (iii) the layers are approximately parallel. Therefore, as the thickness increases, the fields are less likely to cancel outside the layers. The charge cannot build up to these very large levels if the fields from the charges do not partially cancel outside the insulation.

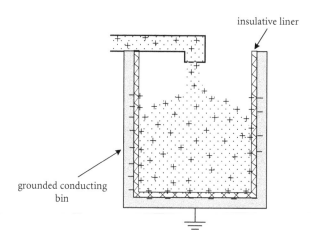

FIGURE 2.52   Charge buildup inside a grounded conducting bin with an insulative liner. Induced charge on the bin is also shown.

break down the nearby air, resulting in a current surge. The energy present in this electric field between the double layer of charge can be quite hazardous.

"Maurer"/Bulking Brush/Cone Discharge

When course particles about 1 to 10 mm in diameter are rapidly transported into a vessel, such as in the filling of a hopper, tribocharging can occur. Larger particles or pellets can produce more energetic discharges than fine powder. The charge in the vessel can accumulate to large values. If the coarse particles have a low conductivity, the charge decay will be slow, especially if there are no contacting grounded surfaces or rods. This charge in the vessel can generate high electric fields. Discharges can be generated when the coarse particles avalanche down the surface of the (cone-shaped) pile in the vessel. The discharges can occur between the surface of the pile and walls of the vessel. This can ignite suspended gases, vapors, and fines, but the evidence is not clear concerning dusts. The energy level has been analytically estimated as 10 mJ, since it is difficult to measure.

In Figure 2.52, a grounded storage vessel or bin, being filled with a charged powder, is lined with an insulator. Thus, the charges cannot easily decay to ground. Some of the charge in the powder is shown "climbing" up the liner. A double layer of charge is also shown, which allows for large surface charges.

Lightning/Lightning-Like Discharge

Large storage vessels (e.g., greater than 3 m in diameter) containing charged particles and a charge dust cloud are believed to be capable of igniting dusts. In nature, for example during volcanic eruptions, the electric field in a cloud can be sufficiently large to break down the air and ignite gases, vapors, and dusts. For comparison to the other discharges, natural lightning can have energies greater than 1 kJ.

2.16   Minimum Ignition Energy

Locate and tabulate the minimum ignition energy (MIE) for many dusts, gases, and vapors. [Jacobson, '61; Jacobson, '62; Jacobson, '64; Hartmann, '48; Hartmann, '51; Nagy, '65; Nagy, '68; Dorsett; Haase; Fenn; Glor; Lüttgens; Pratt; Jones, '91; Cross]

Explosions due to electrostatic discharges have occurred. Electrostatic discharges can ignite vapor-gas mixtures, dusts, and powders resulting in injuries, fatalities, and property loss. Considering the frequency of harmless but annoying human-to-human and human-to-metal electrostatic discharges, most electrostatic discharges do not result in an explosion. For an explosion (or a fire) to occur, a fuel source, sufficient oxygen, and an ignition source must all be present. However, just because all three are present does not imply that an explosion will occur. Most electrostatic discharges do not have sufficient energy to cause an explosion.

For liquid, gaseous, and solid fuels, the combustible mixture must be within an explosive *range*; that is, there is a lower and an upper flammable limit. For gases and vapors, the range is given in terms of the volume concentration of the fuel in air. For solid dusts, the range is given in terms of the mass of the dust per unit volume in air (e.g., kg/m³). For dusts, there is an optimum or best concentration for ignition. For concentrations above or below this level, the energy required for ignition increases above the minimum ignition energy. The oxygen concentration is also a factor in whether an explosion will occur. For many fuels, the minimum required oxygen concentration is about 10% by volume.[22] Often, reducing the oxygen content can dramatically reduce the probability of an explosion. Increasing the oxygen concentration can increase the sensitivity of dust to ignition. Finally, an ignition source is necessary with energy greater than the minimum ignition energy. Examples of ignition sources are an abrasive cutter, a flint lighter, a photoflash, a lightning stroke, an X-ray, or an electrostatic discharge. Because of the frequency of static discharges, it would initially appear that explosions due to electrostatic discharges would be common. However, because of the low-energy level of many electrostatic discharges, their contribution to the overall number of explosions is minor. Nevertheless, they still occur.

The MIE is the smallest energy that can ignite a mixture most easily. A common method of determining this energy level is to place the material in an ignition chamber containing two electrodes of known capacitance, C, and then applying voltage, V, across the plates. When the discharge occurs between the electrodes containing the mixture, the electrical energy contained between the electrodes is given by

$$E = \frac{1}{2}CV^2 \tag{2.128}$$

(Inductor-based discharges can also be used.) The minimum ignition energy is a function of the capacitance between the electrodes. Obviously, energy is lost due to radiation, losses in the dielectric between the plates, which should be very small for air, and the inductance and resistance of the plates and leads. Some energy will also not be discharged and will remain on the plates. Notice that the minimum ignition energy does not consider how the energy is distributed in space or time. For example, there could be a single energy burst over a narrow path between the electrodes or a slow energy release over a larger area.

For many vapors and gases, the minimum ignition energy is frequently given as about 0.25 mJ. By reviewing Table 2.5, it is clear that this guideline is a reasonable "average" value. Hydrogen, acetylene, and carbon disulfide are exceptions with ignition energies of about 0.01 mJ. In other words, energies as low as 0.01 mJ can ignite some sensitive vapors and gases. For dusts (except explosive and reactive dusts such as from Uranium hydride), 10 to 100 mJ is considered the minimum energy range for ignition. But, this energy is a function of factors such as the dust concentration, oxygen level, temperature, particle size distribution, and humidity. Partially, these factors are why it is more difficult to give a minimum ignition energy for dusts or powders compared to gases at a given temperature and pressure. In general, more energy is required to ignite powders and dusts than vapors and gases. Hydrocarbon vapors (e.g., vapors from gasoline and anesthetics) are especially troublesome since their minimum ignition energies are so low and their resistivities are generally so high. The high resistivity implies that the decay time of any charge in the material is very long. Typical energies for specific types of electrostatic discharges were previously given.

The minimum energy data provided in Table 2.4 and Table 2.5 should only be used as a rough guide. The references should be consulted for additional information such as moisture content, maximum oxygen level,

---

[22]The natural atmospheric oxygen content is 21% by volume.

ignition temperature, ignition sensitivity, aging, dust layer energy, particle sizes or distribution of dust, method of ignition, timing of spark, and geometry of test electrodes. In some cases, layers of materials (not clouds) such as "uranium, uranium hydride, and thorium hydride ignited spontaneously at room temperature within a few minutes after exposure in air." For conversion purposes, 1 oz/ft^3 ≈ 1 kg/m^3 = 1 g/l.

MIE = minimum ignition energy
MEC = minimum explosion concentration
§ = no ignition to 8,320 mJ, the highest energy level tried
¥ = no ignition to 2.00 oz/ft^3, the highest concentration tried

2.17   Electrostatic Hazard Case Studies

Produce an abbreviated version of several of the many fascinating electrostatic hazard case studies presented in the Lüttgens and Haase references. Provide a potentially hazardous situation that is commonly encountered by the public. [Lüttgens; Haase; Pidoll; Boxleitner]

Although there is no substitute for personal experience, much may be learned from a study of how experts in the field analyze the facts and hypothecate solutions. As is often the case after the "solution" is provided and understood, the reader will state, "of course, that is obvious." However, when confronted with a new situation without a "book" solution, complete understanding of factors such as the properties of the various electrical discharges, resistances to ground, capacitances between conducting bodies, and minimum ignition energy is required. The references should be consulted for a detailed description of these and many other cases.

Case 1

Organic powder was poured through a funnel into a mixer. The powder was poured from an antistatic polyethylene-lined bag sitting in a plastic drum. An explosion occurred at the funnel. The minimum ignition energy of the powder/air mixture was determined to be less than 1.4 mJ. In addition to the lower resistivity of the antistatic liner, rainwater could have also wetted the outer surface of the bag before it was opened. The capacitance from the somewhat conductive bag through the plastic drum to the nearby ground was estimated to be 100 pF. During a test, when only one-tenth of the bag's contents were poured through the funnel, the bag's potential was greater than 5 kV. This corresponds to an electrostatic energy of about 1.3 mJ. Tribocharging of the powder obviously occurred during the pouring operation. When the bag was near a grounded grating of the mixer, a spark occurred. It was believed that a spark discharge might be the source of the ignition. It was recommended that the antistatic bag be placed in a grounded conducting drum.

Case 2

A solid powder wetted with toluene was delivered in a steel drum for drying. During the hand shoveling of the material from the drum to a dryer, the drum caught on fire. A worker used a wooden handle shovel with a steel blade. The air in the vicinity was well heated and the humidity was low. Although wood can be a decent conductor, its properties are strongly dependent on the humidity and any coatings such as varnish. The charged powder contained in the blade would be slowly dissipated to ground. The path to ground would be through the high resistance from the shovel blade, through the wood handle and worker, to the ground. The resultant time constant of any discharge via this route could be 1 sec or longer. It was believed that a spark discharge between the charged shovel and steel drum might be the source of the ignition. It was recommended that the wood handle be replaced with a metal handle and the worker be properly grounded.

Case 3

During the shredding of a high-resistivity elastomer (a rubber), wetted with hexane, a flame shot from the steel collecting drum. The steel drum was sitting on a wooden stool, the shredder was grounded, and the distance between the shredder and drum was 2 to 10 mm. The resistance and capacitance between

TABLE 2.4   Explosion Characteristics of Dust Clouds[a]

Type of Dust	MIE (mJ)	MEC (oz/ft³)	Type of Dust	MIE (mJ)	MEC (oz/ft³)
Acetal, linear (polyformaldehyde)	20	0.035	Bis-phenol A	15	0.020
Aceto acetanilide	20	0.030	Boron, amorphous, commercial (85% B)	60	<0.100
Acrylamide polymer	30	0.040	Calcium silicide	150	0.060
Acrylonitrile polymer	20	0.025	Carboxy methyl hydroxyethyl cellulose	1,280	0.250
Adipic acid	60	0.035	Carboxy ppolymethylene, regular	§	0.325
			Cashew oil phenolic, soft or hard	25	0.025
Alfalfa meal	370	0.16	Cellucotton	80	0.050
Alkyd molding compound, mineral filler, not self-extinguishing	120	0.155	Cellulose	80	0.055
Alkyd molding compound, mineral filler, self-extinguishing	§	¥	Cellulose acetate	15	0.040
Alkyl nitroso methyl amide	15	0.025	Cellulose acetate butyrate	30	0.035
Allyl alcohol derivative, CR-39, from dust collector	20	0.035	Cellulose flock, fine cut	35	0.055
Aluminum-cobalt alloy (60-40)	100	0.180	Cellulose triacetate	30	0.035
Aluminum	50	0.025	Cellulose, alpha from tunnel walls	40	0.045
Aluminum flake, A422 extra fine lining, polished	10	0.045	Cereal grass, dehydrated	800	0.25
Aluminum-magnesium alloy	80	0.020	Cereal grass, vacuum dried	800	0.20
Aluminum-nickel alloy	80	0.190	Charcoal, hardwood mixture	20	0.140
Aluminum-silicon alloy (12% Si)	60	0.040	Chlorinated paraffin, plant milled, 70% chlorine	§	¥
Aluminum stearate	15	0.015	Chlorinated polyether alcohol	160	0.045
Antimony, milled	1,920	0.420	Chromium, electrolytic, milled	140	0.230
Apricot pit	80	0.035	Cinnamon	30	0.06
Aryl nitroso methyl amide	15	0.050	Citrus peel	100	0.06
Asbestine	§	¥	Citrus peel, natural, dehydrated	45	0.065
Asbestos	§	¥	Coal, Kentucky (bituminous)	30	0.050
Aspirin, fine	25	0.050	Coal, Pennsylvania (experimental mine coal)	60	0.055
Azelaic acid	25	0.025	Cocoa, natural 19% fat	100	0.075
Benzoic acid	20	0.030	Cocoa, natural 22% fat	120	0.065

TABLE 2.4   Explosion Characteristics of Dust Clouds (Continued)

Type of Dust	MIE (mJ)	MEC (oz/ft³)	Type of Dust	MIE (mJ)	MEC (oz/ft³)
Cocoa, prepared, 17% cocoa, 30% whole milk, 51% sugar	140	0.070	Ferrotitanium (19% Ti, 74.1% Fe, 0.06% C)	80	0.140
Coconut shell	60	0.035	Flour, cake, with 10% sugar	60	0.055
Coffee, fully roasted	160	0.085	Flour, cake, with 25% cornmeal	25	0.065
Coffee, raw bean	320	0.15	Flour, cake, with 40% sugar, 10% shortening	80	0.060
Cork	45	0.035	Garlic, dehydrated, from dust collector	240	0.10
Corn (from floor, ledges, and beams)	40	0.055	Gilsonite, from Michigan or Utah	25	0.020
Corncob grit, unused	45	0.045	Grain dust, winter wheat, corn, oats, barley	30	0.055
Cornstarch, commercial product	40	0.055	Grass seed, blue (from thrashing and cleaning operation)	260	0.29
Cornstarch, pelletized	40	0.050	Green base harmon dye	50	0.030
Cotton linters, raw	1,920	0.50	Hemp, hurd	35	0.040
Cottonseed meal from top of bin	120	0.055	Hexamethylene tetramine	10	0.015
Cottonseed meal, before extraction	260	0.055	Hydroxyethyl cellulose	40	0.025
Coumarone-indene, medium or hard	10	0.015	Iron, carbonyl	20	0.105
Cube root, South American	40	0.040	Iron, hydrogen reduced	80	0.120
Dimethyl isophthalate	15	0.025	Isophthalic acid	25	0.035
Egg white	640	0.14	Lactalbumin	50	0.040
Epoxy, no catalyst, modifier, or additives	15	0.020	Lignite, California	30	0.030
Epoxy-bisphenol A mixture	35	0.030	L-Sorbose	80	0.065
Ethyl cellulose molding compound	10	0.025	Lycopodium	40	0.025
Ethyl hydroxyethyl cellulose	30	0.020	Magnesium, milled, grade B	40	0.030
Ethylene oxide polymer	30	0.030	Malt, barley	35	0.055
Ethylene-maleic anhydride copolymer, dry	40	0.095	Melamine formaldehyde, unfilled laminating type, no plasticizer	320	0.085
Ethylene-maleic anhydride copolymer, wet with benzene solvent	10	0.075	Methacrylic acid polymer, modified	100	0.045
Ferromanganese, medium carbon	80	0.130	Methyl methacrylate polymer	20	0.030
Ferrosilicon (88% Si, 9% Fe)	400	0.425	Methyl methacrylate-ethyl acrylate copolymer	10	0.030

(Continued)

TABLE 2.4　　Explosion Characteristics of Dust Clouds (Continued)

Type of Dust	MIE (mJ)	MEC (oz/ft³)	Type of Dust	MIE (mJ)	MEC (oz/ft³)
Mica	§	¥	Polystyrene, clear	120	0.020
Milk, skimmed	50	0.050	Polystyrene, molding compound	40	0.015
Monochlorotrifluoro-ethylene polymer	§	¥	Polyurethane foam (toluene diisocyanate-polyhydroxy with fluorocarbon blowing agent), not fire retardant	20	0.030
Nitrosamine, 100%	60	0.025	Polyurethane foam (toluene diisocyanate-polyhydroxy with fluorocarbon blowing agent), fire retardant	15	0.025
Nitrosamine-oil-silica mixture (80-5-15)	80	0.025	Polyvinyl acetate	160	0.040
Nylon (polyhexamethylene adipamide) polymer	20	0.030	Polyvinyl acetate alcohol	120	0.035
Para nitro chlor benzol ferric sulfonate	§	0.420	Polyvinyl butyral	10	0.020
Pea flour	40	0.050	Polyvinyl chloride (fine or course or copolymer or powdered or binder)	§	¥
Peanut crackling, feed ingredient	370	0.085	Polyvinyl toluene, sulfonated	2,880	1.000
Peat, sphagnum, sun dried	50	0.045	Rayon (viscose) flock, 3.0-denier, 0.030-inch, red	240	0.060
Pentaerythritol	10	0.030	Red dye intermediate	50	0.055
Phenol formaldehyde	15	0.025	Rice, from dust collector, Louisiana	50	0.050
Phenol formaldehyde molding compound, wood flour filler	15	0.030	Rosin	10	0.015
Phenol formaldehyde, contains 20% cellulosic extender	3,840	0.120	Rubber, chlorinated	§	¥
Phenol formaldehyde, powdered	§	0.175	Rubber, crude, hard	50	0.025
Phenol furfural	10	0.025	Rubber, synthetic, hard, contains 33% sulfur	30	0.030
Phenolic resin	10	0.025	Shellac	10	0.020
Phthalic anhydride	15	0.015	Sodium resinate, dry size, grade X	160	0.045
Pitch, petroleum	25	0.045	Soap	60	0.045
Polycarbonate	25	0.025	Soy flour, from cyclone	100	0.060
Polyethylene wax, low molecular weight	35	0.020	Soy protein	100	0.040
Polyethylene, type D	80	0.025	Stearic acid, aluminum salt	15	0.015
Polypropylene, linear	30	0.020	Stearic acid, zinc salt	10	0.020
Polypropylene, molecular weight 0.6 million	400	0.055	Styrene modified polyester-glass fiber mixture (65-35)	50	0.045
Polypropylene, molecular weight 1.1 million	25	0.03	Styrene sulfonate, sodium	§	¥
Polypropylene, molecular weight 1.8 million	400	0.035	Styrene-maleic anhydride copolymer, nonsolvent process	20	0.030

TABLE 2.4   Explosion Characteristics of Dust Clouds (Continued)

Type of Dust	MIE (mJ)	MEC (oz/ft³)	Type of Dust	MIE (mJ)	MEC (oz/ft³)
Styrene-maleic anhydride copolymer, solvent process	50	0.045	Vinyl chloride-acrylonitrile copolymer (33-67), water emulsion product	15	0.035
Sucrose, chemically pure	100	0.045	Vinyl chloride-vinyl acetate copolymer molding compound, mineral filler	§	¥
Sugar, 20% flour, 30% starch	10	0.050	Violet 200 dye	60	0.035
Sugar, 30% egg white	45	0.060	Vitamin B₁, mononitrate	60	0.035
Sugar, powdered, commercial product	30	0.045	Vitamin C, ascorbic acid	60	0.070
Sugar, raw, light brown	40	0.035	Walnut shell, black	50	0.030
Sulfur	15	0.035	Wheat flour, commercial product	60	0.050
Terephthalic acid	20	0.050	Wheat starch, edible	25	0.045
Tetrafluoroethylene polymer (micronized)	§	¥	Wheat straw	50	0.055
Thorium	5	0.075	Wood flour	30	0.040
Thorium hydride	3	0.080	Wood flour, white pine	40	0.035
Titanium	25	0.045	Wood, birch bark, ground	60	0.020
Titanium hydride	60	0.080	Zinc, condensed	960	0.480
Uranium	45	0.060	Zirconium	15	0.060
Uranium hydride	5	0.060	Zirconium hydride	60	0.085
Urea formaldehyde molding compound, grade II, fine	80	0.085	β-Naphthalene-azo-dimethylaniline	50	0.020
Urea formaldehyde molding compound, granular	80	0.165	—	—	—
Urea, crystal, ground	§	¥	—	—	—

[a]The information in this table is from the Dorsett, Hartmann, Jacobson, and Nagy references (i.e., the Bureau of Mines reports).

the drum and ground were 0.5×10^{12} Ω and 80 pF, respectively, corresponding to a decay time constant of 40 sec. The charged shreddings inside the drum could have raised the potential of the drum to several kV or more. Assuming a potential of 3 kV, the electrostatic energy in this capacitance is 0.36 mJ, which is greater than the 0.24 mJ minimum ignition energy of a hexane/air atmosphere. It was believed that a spark between the charged drum and grounded shredder was the source of the ignition. It was recommended that the steel drum be grounded.

Case 4

Varnish solution in a large container was transferred to a smaller metal container where it was to be transported on a wooden dolly with rubber wheels. When a cover was placed over the smaller container, a fire erupted. Partially due to a layer of varnish on the bottom of the container, the resistance from the metal container to the ground was 10^{12} Ω. Assuming the potential of the charged drum was 6 kV and the container capacitance to ground was 120 pF, the electrostatic energy is about 2.2 mJ. The worker holding the lid was properly grounded. Therefore, if the cover was also conducting, a spark discharge could have been the source of the ignition between the grounded lid and charged drum. It was recommended that the metal container be grounded.

TABLE 2.5   Minimum Ignition Energy for Gases and Vapors[a] (at 1 atm ≈ 1 bar, 20 to 25°C)

Gas or Vapor	MIE (mJ)	Gas or Vapor	MIE (mJ)
1,3-Butadiene/air	0.13	Hydrogen/air	0.011
2,2-Dimethylbutane/air	0.25	i-Octane/air	1.35
2,2-Dimethylpropane/air	1.57	i-Pentane/air	0.21
2,3-Trimethylbutane/air	1.0	i-Propyl alcohol/air	0.65
2-Pentane/air	0.18	i-Propyl chloride/air	1.55
Acetaldehyde/air	0.376	i-Propyl mercaptan/air	0.53
Acetone/air	1.15	i-Propylamine/air	2.0
Acetylene/admixture in oxygen	0.0002	Isobutane/air	0.52
Acetylene/air	0.017	Iso-octane/air	1.35
Acrolein/air	0.137	Isopentane/air	0.70
Acrylonitrile/air	0.16	Isopropyl chloride/air	1.55
Allyl chloride/air	0.775	Isopropyl ether/air	1.14
Benzene/air	0.2	Isopropyl mercaptan/air	0.53
Butane/air	0.25	Isopropylamine/air	2.00
Carbon disulfide/air	0.009	Methane/admixture in nitrogen monoxide	8.7
Cyclohexane/air	0.22	Methane/admixture in oxygen	0.0027
Cyclopentadiene/air	0.67	Methane/air	0.28
Cyclopentane/air	0.238	Methanol/air	0.14
Cyclopropane/air	0.17	Methyl ethyl ketone/air	0.53
Diethyl ether/admixture in oxygen	0.0012	Methyl sulfide/air	0.48
Diethyl ether/air	0.19	Methylacetylene/air	0.11
Dihydropyran/air	0.365	Methylcyclohexane/air	0.27
Diisobutylene/air	0.96	n-Butane/air	0.38
Diisopropyl ether/air	1.14	n-Heptane/air	0.70
Dimethyl ether/air	0.29	n-Pentane/air	0.22
Ethane/admixture in oxygen	0.0019	n-Propyl chloride/air	1.08
Ethane/air	0.24	Pentane/air	0.49
Ethyl acetate/air	1.42	Propane/admixture in oxygen	0.0021
Ethylamine/air	2.4	Propane/air	0.25
Ethylene oxide/air	0.06	Propylene oxide/air	0.135
Ethylene/admixture in oxygen	0.0009	Propylene/air	0.28
Ethylene/air	0.07	Tetrahydrofuran/air	0.54
Ethyleneimine/air	0.48	Tetrahydropyran/air	0.22
Furan/air	0.225	Thiophene/air	0.39
Heptane/air	0.24	Triethylamine/air	0.75
Hexane/air	0.24	Triptane/air	1.00
Hydrogen sulfide/air	0.068	Vinyl acetate/air	0.7
Hydrogen/admixture in nitrogen monoxide	8.7	Vinylacetylene/air	0.082
Hydrogen/admixture in oxygen	0.0012	—	—

[a] The information in this table is from the Fenn and Hasse references. When different or multiple values were given, the smallest value was selected.

Case 5

While 120°C liquid diphenyl was being filled in a 100 m³ metallic cylindrical tank with thermal, exterior insulation, an explosion occurred. The dome of the tank was blown off and a large fire seen. The resistivity of the liquid was 10^8 Ω-m. Initially, the tank was cold and solidification of the liquid occurred along the inner surface of the tank during the filling operation. The solid version of diphenyl has a much greater resistivity thereby reducing the rate of discharge to the walls of the metallic tank. Although heating pipes were present in the tank, the tank would have to be at least one-half full for them to be effective. It is likely that a brush discharge occurred between the surface of the liquid and these pipes. Purging the tank with nitrogen or carbon dioxide might have been helpful.

Case 6

Organic flakes were poured from polyethylene bags into a stainless steel agitator vessel. After the 10 m³ vessel was about one-half filled, a burst of flame projected from the opening of the vessel. Through a low-pressure gas test, it was determined that a small leak of ethylene oxide passed inside the vessel via a previously used valve. Ethylene oxide has a very low minimum ignition energy. This explosion was believed to be from a brush discharge. It was recommended that all unused pipes be disconnected from the vessel. The vessel could also be purged with nitrogen, especially if idle for long periods of time.

Case 7

Sulfur, finely ground, was poured from paper bags into a filling pipe, 6 m in length, to an agitator vessel. The agitator vessel contained a water and methanol mixture. After the filling operation, an explosion occurred. It was determined that the exhaust system was not properly functioning. This explosion was believed to be from a brush discharge. The atmosphere could be purged with nitrogen, and the exhaust system should be properly maintained.

Case 8

Polluted toluene was accidentally pumped into a polyethylene drum for disposal. A flame shot from the drum and, unfortunately, the pump was not turned off causing a major fire and loss of property. The explosion was believed to be from a brush discharge. Plastic drums should not be used for storage of combustible liquids. Instead, metal drums that are properly grounded, such as sitting on steel grating, should be used. The worker holding the hose should also be grounded (e.g., wearing conductive shoes and standing on the steel grating).

Case 9

A pneumatically transported system used a grounded steel pipe, with a 5 cm diameter, to transport acrylic powder to a grounded bulk container. A few meters of the last portion of the pipe were replaced with polyethylene pipe. After a few hours of operation on a wet winter's day, an explosion occurred in the container. Although a 15 cm metal coupler used to connect two sections of the plastic portion of the pipe was not grounded, its capacitance between the coupler and ground was only about 12 pF. If the coupler's potential was 10 kV, then the potential energy associated with this capacitance is 0.6 mJ, which is low for powder/dust ignition. However, the outside of the plastic pipe was also wet from the winter's snow and rain. As a result, a number of locations on the outer surface of the plastic pipe were at ground potential. At a later investigation, long flashes were seen inside the plastic pipe and audible "cracks" heard during the transport operation. Therefore, the explosion was believed to be due to a propagating brush discharge. It was proposed that electrically insulating pipe not be used.

Case 10

A shock pressure resistant silo was partially pneumatically filled with a high resistivity (> 10^{12} Ω-m) organic powder when an explosion occurred. Although the silo was not damaged, the resultant fire spoiled the powder. The silo, 3 m in diameter with an 8 m³ volume, was constructed of stainless steel. The steel silo and all relevant conducting parts were grounded. No insulating or plastic parts were present. Cone discharges might be the source of the ignition.

Case 11

High resistivity ($\approx 10^{11}$ Ω-m) foam consisting of isobutylene and a polymer was emptied from an autoclave (pressure "cooker") into a steel container when spontaneous ignition occurred. The resistance from the autoclave and other relevant objects to ground was less than 100 MΩ. At the surface of the output stream, the magnitude of the electric field was 100 kV/m. Cone discharges might be the source of the ignition. Conducting grounding rods that are not too thin can be used to help discharge the material in the steel container. The container can also be purged using nitrogen.

A situation commonly encountered by the public is the filling of a gas can at a gas station. The low-conductivity fuel is tribocharged while passing through the connecting hose. While filling a gas can, the charge builds up inside the can. If the can is sufficiently conducting and in contact with a good ground (e.g., water saturated concrete), then the charge will "bleed" or decay to the ground. However, if the resistance to ground is large, then the charge will remain in the can for a period of time, and its potential relative to ground can be large. If a person (or object) approaches the can, then a discharge can occur between the can and person, possibly igniting the fuel vapor. For this reason, the gas can should not be filled while it is sitting on the top of a plastic truck bedliner, inside a carpeted or an insulated trunk, on a door step, or on other objects electrically insulated from the ground. The can should be filled while on the ground so that any charge in the gas can be, at least partially, dissipated while sitting on the ground.

Electrostatically ignited flash fires have occurred while filling the tank of a vehicle. Sometimes, especially during cold and dry days, individuals reenter their vehicle during fueling. While shifting on the seat or exiting the vehicle, they become tribocharged. A spark can then occur between them and the gas inlet. It is recommended that the individual touch a metal portion of the vehicle's surface away from the fuel inlet, thereby discharging themselves, before disconnecting the fuel nozzle. The situation is further aggravated when a plastic, poorly conducting gas tank or a plastic, poorly conducting inlet tube leading to the tank is used. Various methods have been proposed for reducing this hazard including the use of (i) tires with carbon black and other materials to increase the leakage resistance to ground, (ii) lower resistance filling hoses, (iii) grounded conductive parts near the inlet, and (iv) antistatic seats.

Because of the complexity and potential hazards associated with electrostatic discharges, a formal list of guidelines will not be provided. Even a simple guideline such as "ground all conductors" can in some cases result in an electrostatic hazard. With this said, in most electrostatic hazards, a spark discharge is believed to be the discharge type, and proper grounding of both metal objects and personnel is usually an important step in reducing the likelihood of this type of deleterious discharge. Grounding can dissipate excessive charge. Also, when two conducting objects are at the same potential, including 0 V, the electric field between them is zero (when there are no other field sources). For conductors, it has been recommended that the resistance to ground should be less than 100 MΩ. For very large conductors, such as tanker vessels, it has been recommended that the resistance to ground be less than 1 MΩ. Larger conductors have greater capacitance to ground. If the charge relaxation constant is equal to RC, then a smaller resistance to ground helps to compensate for a larger capacitance. Sometimes, the resistance to ground is intentionally increased to restrict fault current through any contacting personnel.

2.18   Measuring Charge

How can charge be measured? Is it necessary to contact directly a charged object to measure its charge? How can the charge transferred to a ground during a discharge be measured? [Secker; Taylor, '94; Cross; Takuma; Taylor, '01; Chubb; Hughes; Davidson; Keithley]

Charge, whether along the surface of a printed circuit board, on the sole of a shoe, or in a sugar silo, is a fundamental quantity that is of interest in many disciplines including electrostatic discharge. There are a number of methods of measuring charge, sufficient to fill a book. A few of the more common methods will be discussed in this section.

A simple, standard, and very useful method of determining the charge on an object (solid or liquid of any conductivity) is via a Faraday cup or pail. There are several versions of the Faraday pail. In the simplest version shown in Figure 2.53, a metal container of any convenient shape and size, isolated from ground through insulators, is connected to an instrument capable of accurately measuring voltage with little current drain through the instrument (such as an electrometer). An electrometer is a "highly refined dc multimeter." Rather than measuring the charge directly, a difficult task, a common technique is to measure the voltage across a known capacitance and then use $Q = CV$.

If a charged object is introduced into the container or metallic cup, this charge will induce or attract charge of the opposite sign on the inner surface of the cup as illustrated in Figure 2.54. If the cup was initially charge neutral, equal charge will exist on the outside of the cup (and the electrometer). It does not matter whether the charged object is insulative or conductive. According to Gauss' law, the total electric flux emanating from the object is a function of the total charge on the object. The object does not need to make contact with the cup for its charge to be measured. If the charged object contacts the inside of the cup, some of its charge will transfer to the metal cup and to the electrometer. The speed that this transfer occurs will be a function of the time constant of the system, which is a function of the electrical conductivity and permittivity of the charged object. The quantity of charge that is actually transferred to the cup is a function of the surface areas of the charged object and cup (i.e., the isolated

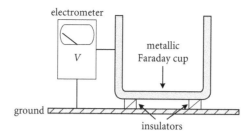

FIGURE 2.53 Faraday cup connected to an electrometer. The cup is electrically isolated from the ground through (very good) insulators.

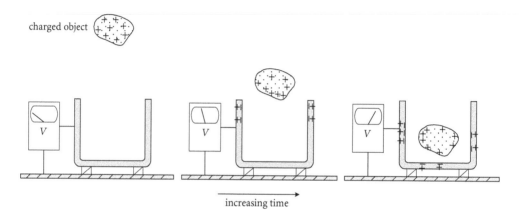

FIGURE 2.54 Introduction of a charged object into a Faraday cup. Notice the rise in the measured voltage.

capacitances of the object and cup). Nevertheless, even if contact to the cup is made, the electrometer will still measure the charge that was on the object.

Close examination of Figure 2.54 after the charge enters the Faraday cup will indicate that the total number of the negative charges on the inner surface of the cup is not quite equal to the total number of the positive charges on the charged object. To understand why this can occur, although probably not to the degree shown in this figure, the expected electric field distribution should be (at least, mentally) sketched. Although most of the electric fields from the charged object terminate on the inner surface of the cup, electric fields can leave the opening of the cup and terminate on the ground plane or other objects. These stray fields do not induce charge on the cup and, thus, the total charge measured by the electrometer will not be exactly equal to the charge on the object inside the cup. The illustrations in Figure 2.54 are also somewhat simplified in that negative charges are not shown induced on the ground plane from the positive charges on the outer surface of the cup. Also, no charge is shown being transferred to the electrometer. It is helpful to model this measurement system using capacitors.[23] Series and parallel resistances are present (e.g., ohmic resistance of the conductor leading to the electrometer and leakage resistance of the insulators), but they will be neglected in the model given in Figure 2.55 to simplify the analysis. The stray fields that do not terminate on the cup but instead terminate on the ground are represented by C_{og}. The capacitance C_{oc} is zero if the charged object is in contact with the cup. If other conductive bodies, such as a human hand, are nearby, stray fields could also terminate on them. The relationship between the charge, Q, on the object inserted in the cup and

---

[23]For these discussions, assume that the charged object inside the cup is conductive. This allows a capacitance to be easily defined between it and other metallic objects.

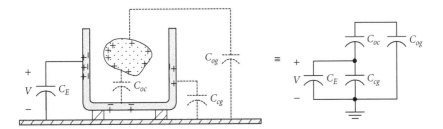

FIGURE 2.55 Parasitic capacitances between a charged conducting object, ground plane, and Faraday cup. The capacitance of the electrometer, C_E, is also shown.

the electrometer's voltage reading, V, is

$$Q = C_{eq}V = \left(C_E + C_{cg} + \frac{C_{oc}C_{og}}{C_{oc}+C_{og}} \right)V \qquad (2.129)$$

The charge Q is the total charge induced, not just the charge on the meter capacitance C_E. If the parasitic capacitances C_{cg}, C_{og}, and C_{oc} are small compared to the electrometer capacitance, C_E, then

$$Q \approx C_E V \quad \text{if } C_E \gg C_{cg}, C_E \gg C_{og}, C_E \gg C_{oc} \qquad (2.130)$$

In this case, the charge measured by the electrometer is essentially equal to the charge on the object inserted in the cup. When the electrometer capacitance is large, most of the positive charge shown on the outside of the cup would actually be on the upper plate of the electrometer capacitance. If the electrometer capacitance, which would include the capacitance of the cable leading from the cup to the meter, is not sufficiently large, an external capacitance can be added to increase its value. Since the stray capacitances are not easily predictable and can change with the test environment, it is usually preferable to "drown them out" with a known larger capacitance. However, as this capacitance increases, the voltage measured decreases for a given charge, Q. The lower limit for this voltage is determined by the capability of the meter and noise in the system. Commercial electrometers can measure charges from 10 μC to 1 fC (= 10^{-15} C).

The stray fields from the charged object to other objects, beside the cup, can be reduced by decreasing the size of the cup opening. However, this will reduce the size of the object that can be dropped or inserted inside the cup. The capacitance from the cup to the ground can be controlled (but not eliminated) by using an earthed or a grounded screen or shield as shown in Figure 2.56. An earthed screen is an electric shield that surrounds the Faraday cup but is conductively isolated from it. Since the capacitance between the two walls of this vessel, C_{cs}, is known and constant, it is easily compensated for in the

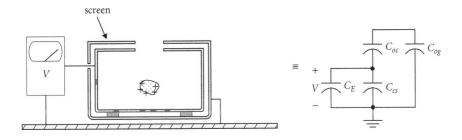

FIGURE 2.56 Shielded Faraday cup and its connection to an electrometer. The capacitive model of the system is also shown.

calculations. Because the screen is earthed or connected to ground, the potential difference between the screen and ground is zero: there is zero capacitance between the screen and ground. The relationship between the measured voltage and total induced charge is

$$Q = C_{eq}V = \left(C_E + C_{cs} + \frac{C_{oc}C_{og}}{C_{oc} + C_{og}} \right) V \qquad (2.131)$$

With this improved, double-wall, smaller opening version of the Faraday cup, C_{og} is smaller and C_{cs} is constant. By surrounding the cup with a screen, charges or fields external to the screen induce few charges on the outer surface of just the cup. Another advantage of the shielded cup is that charges not intended to be measured but still contacting the outer portion of the shield will be discharged to ground. With the unshielded cup, these charges will affect the voltage reading. The insulator separating the inner and outer walls should have a very high resistivity. The resistivity of this insulator, which is influenced by contaminants, will affect the rate of charge dissipation on the cup to ground. It will also affect the rate that the electrometer's voltage reading drops with time. After a measurement has been performed, the cup and electrometer capacitances should be discharged.

For electrostatic powder and liquid spraying operations, an average charge-to-mass ratio is frequently desired. After spraying into a Faraday cup, the total charge of the particles is measured. The total mass of the particles in the cup is then weighed, and the charge-to-mass ratio is obtained. Of course, this ratio does not provide an indication of the distribution of the charge on the particles. Some particles, for example, may have no charge. Furthermore, free ions (not attached to a powder or liquid particle) can also enter the cup. When these free ions have a higher mobility than the powder or liquid particles, a grounded screen or wire fence has been used to collect these free charged ions. Larger and heavier powder or liquid droplets should mostly pass through the screen if the mesh size is properly selected. One problem with this approach, however, is that eventually the powder or liquid will accumulate on the screen. If the powder or liquid is not very conductive, the accumulation on the screen can retain its charge for a period of time that is long compared to the time particles pass the screen (related to their velocity). The screen is essentially charged. The charge on the screen will generate a back ionization field that will affect the charges passing through the screen to the cup. One method of reducing the buildup on the mesh or wire fence is to pass water over the screen. If desired, the charge contained in this irrigation liquid can also be measured.

There are other methods of measuring the charge on an object. A noncontact method that is more sophisticated, but is useful for measuring the charge on smaller particles, is to first expose the charged particle to an electric field (or both an electric and a magnetic field if the charge is moving). The force on the small object, which is measured by its displacement or velocity, is obtained through Lorentz's force law

$$\vec{F} = q\vec{E} + q\vec{v} \times \vec{B} \qquad (2.132)$$

where $q, \vec{v}, \vec{E}$, and $\vec{B}$ are the charge, velocity, electric field, and magnetic flux density, respectively.

Sometimes a measure of the tribocharge generated between a conductor and an insulator can be determined by inserting an ammeter (e.g., electrometer) between the conductor and ground. In Figure 2.57, the belt is positively charged and the conducting roller is negatively charged. Assuming that the main path for the charge is through the ammeter, rather through other low-resistance paths or other objects, then a large percentage of the negative charge will pass through the ammeter. (Based on the capacitive coupling to other objects and the impedance of the ammeter, some of the negative charge will remain on the roller.) This method may also be used to measure the charge transferred to a bin or storage container, assuming the container is isolated from ground and the major path for the charge decay or flow is through the ammeter. To relate the current, I, to the surface charge density along the belt, first recall the general relationship between the convection current density, J_v, the volume charge density, ρ_V, and its velocity, U:

$$J_v = \rho_V U \quad \Rightarrow AJ_v = I = (A\rho_V)U \qquad (2.133)$$

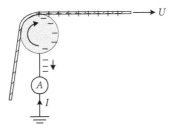

FIGURE 2.57 Measuring the current to ground to determine the surface charge on a moving belt.

where A is the cross-sectional area in which the current is passing. The quantity $A\rho_V$, which has units of $m^2(C/m^3) = C/m$, is the charge per unit length. In terms of an equivalent surface charge, $w\rho_s$ has units of $m(C/m^2) = C/m$, where w is the width of the belt. If the steady-state velocity of the belt is U, the relationship between the steady-state current, I, and surface charge density on the belt is

$$I = (w\rho_s)U \quad \Rightarrow \rho_s = \frac{I}{wU} \tag{2.134}$$

This result assumes that the current on the belt is mainly due to convection; in other words, the charges on the moving belt are mostly immobile. Instead of measuring the current directly with an ammeter, a voltmeter may also be used. The voltmeter is placed across a resistor R connected between the charged object and ground. The voltage reading, V, is then related to the measured current via $I = V/R$. If the current is small, this resistance may have to be large so the resultant voltage drop across it is sufficiently large to be measured by the voltmeter. (Of course, if R is too large, leakage resistances might not be negligible.) As with measuring safe leakage currents for electrical products, a capacitor can be used in parallel with this resistor to filter higher frequency noise signals. If the original charged object was grounded, the introduction of a resistor between the object and ground will affect the rate of charge decay on the object.

One of the oldest electrical instruments is the electroscope. Its operation is easily understood, and for this reason, it is used for demonstration purposes in science museums and elementary schools. There are many types of electroscopes, but probably the easiest to construct is the leaf electroscope. Two very thin sheets or leaves of a metal, such as gold foil, are connected to a metal rod.[24] A good insulator supports the metal rod and isolates it from the rest of the chassis. Because of the delicate nature of these thin sheets, which are often limp in appearance, higher quality electroscopes are enclosed in glass or other material so the sheets are not disturbed by air wind. Sometimes, as shown in Figure 2.58, the electroscope

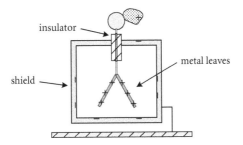

FIGURE 2.58 One example of an electroscope.

---

[24]Other less expensive but more durable materials can be used in place of the gold foil such as aluminum coated Mylar film.

is partially contained in a metallic shield or placed in the center of a metal ring. This shielding reduces the ability of external fields to charge inductively the electroscope and to affect the measurement. The object whose charge is to be measured is touched to a metal sphere, disk, Faraday cup, or other shape metallic object that is conductively connected to the rod. Again, the rod is connected to the thin metallic sheets or leaves. As a result, both of the sheets are then charged to the same sign as the charged object. Since like charges repel, these sheets spread outward. The angle of deflection of these sheets can be used to determine the charge transferred to the metal sheets. The isolated capacitance of the two thin sheets is very small, around 1 pF. Therefore, it will measure very small charges, around 0.01 pC. The capacitance of the leaves to ground will generally increase if a surrounding grounded shield is present. When a charged object is connected to an electroscope, the quantity of charge transferred to the electroscope is a function of the capacitance of the charged object relative to the total capacitance of the electroscope. The capacitance of the electroscope from the contact point (e.g., the sphere) to the leaves can be increased, for example, by using a larger sphere or disk. It is not necessary to touch the electroscope to the charged object to deflect the metal leaves. The sphere (for example) of the electroscope nearest to the charged object will be inductively charged to the opposite sign (but not the same magnitude) as the charged object. For charge to be conserved on the electroscope, the leaves must be charged to the same sign as the charged object. These mechanical electroscopes were common in the laboratory in the early-to-mid 1900's, but they are rarely used today for measurement of charge. Automatic recording of the charge is not easy (unless some optical technique is used).

The Faraday cup is a simple method of measuring the net charge on objects that are inserted, dropped, or even sprayed inside it. Unfortunately, there are situations where the Faraday cup cannot be used. For example, a moving conveyer belt that has been tribocharged obviously cannot be placed in a Faraday cup. In some cases, such as the spraying of large quantities of electrostatically-charged fluids or powders, the volume of material may be too great compared to the finite size of the cup. Since charge flow is current, the changing current can be monitored using classical transformer-based current probes. More specifically, the material can sometimes be passed directly through a ring, without any direct contact with the sensing ring. There is a mutual inductance, M, between the moving charge stream and ring, and the open-circuit voltage induced in the ring is given by $v = M\,di/dt$. In other situations, the distribution of charge along a surface is desired rather than the net charge, or the charge at a particular location is to be monitored.

A measurement of the electric field can indirectly determine the charge along a surface. Before the effect of a nearby ground is introduced, recall that the electric field from a uniform sheet of surface charge on a very large insulating slab and the electric field from a double layer of surface charge on an insulating slab, where the surface charges may have different magnitudes, are as given in Figure 2.59. The electric field inside the slabs is not shown. The directions shown for the electric fields outside the slab assume that $\rho_S > 0$ and $\rho_{S1} + \rho_{S2} > 0$. Now, if an electric field meter were placed near the slab with its

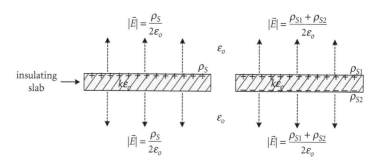

FIGURE 2.59 Electric field from charge on one and both sides of an insulating slab.

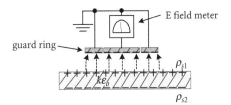

FIGURE 2.60 Measuring the net surface charge on the insulating slab via the electric field it generates.

sensor surface perpendicular to the field, the electric field measured would be

$$|\vec{E}| = \frac{\rho_{S1} + \rho_{S2}}{\varepsilon_o} \qquad (2.135)$$

Once the electric field is measured, the value for $\rho_{S1} + \rho_{S2}$ can immediately be determined. Notice that the expression for the measured field is twice the value of the actual field without the sensor probe. The field is doubled because the field generated from the induced charge on the probe itself is of the opposite sign. Unless other measurements are performed,[25] this electric field measurement cannot distinguish between surface charge on the two sides of the slab. If the surface charges were of the same magnitude but opposite sign, the field measured would be zero. The dielectric constant of the slab does not influence this flat-slab electric field measurement.

The distance from the field probe to the charged surface should be small in comparison to the surface area of the probe (and guard ring) so that the field is essentially uniform between the insulating slab and probe surface. If the distance between the probe and target surface is too large, the field to the probe will be nonuniform, and the previous expression relating the field to the charge would not be correct. Another factor that should be considered is the degree of flatness of the charged surface. For the field between the probe and surface to be approximately uniform, the distance between the probe and charged surface should be small compared to the radius of curvature of the charged surface. Furthermore, the surface area of the probe should be small compared to the area of the target or charged surface. What is not shown in Figure 2.60 is the effect of the meter on the charge distribution along the surface. In this idealized electrostatic situation, the charges are immobile along the insulating surface. Eventually, though, the charge concentration will be greater under the probe due to the attraction of the slab surface charge to the oppositely charged probe. The guard ring shown is helpful in producing a more uniform field at the sensor surface. Although the guard is shown connected to ground, it can also be connected to the same potential as the actual probe surface using a unity gain amplifier.

Other grounds complicate the analysis and will influence the value of the electric field measured from the surface charge along the slab. For example, imagine that the slab is in direct contact with a flat ground plane as shown in Figure 2.61. Although a field meter was previously shown, it can be replaced with an

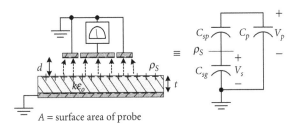

FIGURE 2.61 Measuring the surface charge on a slab in contact with the ground plane. Also shown is a capacitive model of the situation.

---

[25]The electric field from the surface charge on one side of the slab can be reduced by introducing a grounded object on that side.

electrometer since essentially a capacitor's voltage, V_p, is being measured. The surface area of the probe facing the charge is A, which does not include the area of the guard ring. Since the surface charge is along an equipotential, a capacitance can be defined between the surface charge and probe, C_{sp}, and between the surface charge and lower ground plane, C_{sg}. The capacitance C_p is the capacitance of the electric field meter and any connecting cable capacitance. The electric field in the slab is not shown. The relationship between the voltage at the surface of the slab and voltage measured by the probe (via the field meter) is

$$V_p = V_s \frac{C_{sp}}{C_{sp}+C_p} \quad \Rightarrow V_s = V_p \frac{C_{sp}+C_p}{C_{sp}} \tag{2.136}$$

using the standard expression for a capacitor voltage divider. The potential at the surface of the slab, V_s, will vary with C_{sp} due to changes in the position of the probe. The magnitude of the total charge along the surface (under the probe) is simply related to the surface charge:

$$Q = \rho_S A = C_{eq} V_s = \left(C_{sg} + \frac{C_{sp}C_p}{C_{sp}+C_p} \right) V_s \quad \Rightarrow \rho_S = \frac{V_s}{A}\left(C_{sg} + \frac{C_{sp}C_p}{C_{sp}+C_p} \right) \tag{2.137}$$

The relationship between the surface charge and measured probe voltage is then

$$\rho_S = \frac{V_p}{A}\frac{C_{sp}+C_p}{C_{sp}}\left(C_{sg} + \frac{C_{sp}C_p}{C_{sp}+C_p} \right) \tag{2.138}$$

The addition of a lower ground complicates the relationship considerably. The surface charge is a function of the capacitances C_{sg} and C_{sp}. The standard expression for the capacitance between two parallel plates will be used to estimate the capacitances C_{sg} and C_{sp}:

$$C_{sg} = \frac{k\varepsilon_o A}{t}, \quad C_{sp} = \frac{\varepsilon_o A}{d} \tag{2.139}$$

To use these capacitance expressions, the probe and lower ground plane should be close to the surface charge. If the meter or probe capacitance is large compared to the capacitance between the probe and surface, this expression simplifies to

$$\rho_S \approx \frac{V_p C_p}{A}\left(1+\frac{C_{sg}}{C_{sp}}\right) = \frac{V_p C_p}{A}\left(1+\frac{kd}{t}\right) \quad \text{if } C_p \gg C_{sp} \tag{2.140}$$

For many situations, the distance between the probe and surface is a few centimeters and C_{sp} is indeed small. Although the probe capacitance and probe area are known and the probe voltage is indicated via the field meter, the dielectric constant of the slab, thickness of the slab, and distance from the probe to the slab surface can vary with each measurement. Thus, two identical surface charges, one on a high plastic table with no nearby grounds and the other on an insulating mat sitting on a grounded metal table will provide two different results on the same field meter. If the slab thickness is very large, then

$$\rho_S \approx \frac{V_p C_p}{A} \quad \text{if } C_p \gg C_{sp}, t \gg kd \tag{2.141}$$

Of course, if the slab were very large, then C_{sg} would not be given by the simple parallel plate expression. Nevertheless, if the slab thickness is very large, then the field from the surface charge (directly below the probe) is mainly directed to the induced charges on the probe. Sometimes when the slab thickness is very small, the voltage or electric field measured by the meter might be very small:

$$\rho_S \approx \frac{V_p}{A}\frac{C_p C_{sg}}{C_{sp}} = \frac{V_p}{A}\frac{C_p\left(\frac{k\varepsilon_o A}{t}\right)}{\frac{\varepsilon_o A}{d}} = C_p V_p\left(\frac{kd}{At}\right) \quad \text{if } C_{sg} \gg C_{sp}, C_p \gg C_{sp} \tag{2.142}$$

$$V_p = \frac{\rho_S}{C_p}\frac{At}{kd} \quad \text{if } C_{sg} \gg C_{sp}, C_p \gg C_{sp} \tag{2.143}$$

A small slab thickness, t, and a large probe-to-surface spacing, d, will generate a small meter voltage, V_p. In other words, the electric field from the surface charge is mainly directed toward the lower ground plane. In this case, the electric field measured by a meter might be small or zero, but this does not imply that the surface charge is small or not present!

The resolution of these probes, or their ability to measure spatial variation in surface charge, is a function of the surface area, A, of the probe and the probe's distance from the charged surface. The literature indicates that the resolution is about 1.5 times the larger of two parameters: the probe diameter, $2\sqrt{A/\pi}$, or probe-to-sample spacing, d. Induced charge or induction field meters as just discussed have been designed to have resolutions near 50 μm. However, for these small resolutions, the distance between the probe and surface, d, is also small. Micropositioning methods must be used for gap distances less than about 300 μm. In this case, the probe (not the guard) is constructed of fine wire. For these small probe sizes, the output voltage is quite small since for a given surface charge, the voltage is proportional to the area of the probe:

$$V_p = \frac{\rho_S A}{C_p\left(1+\frac{kd}{t}\right)} \quad \text{if } C_p \gg C_{sp} \tag{2.144}$$

The classical tradeoff exists between spatial resolution and gain.

Measuring the volume charge for some general region is not an easy task since the test probe will nearly always have a significant effect on the measurement. Under some special conditions, the volume charge can be determined via an electric field measurement external to the volume. The measurement of this electric field can provide an indication of the likelihood of various electrostatic discharges. For uniformly distributed charges in spherical, very long cylindrical, or other classical, usually symmetrically shaped volumes, the analytical relationship between the charge and external field can be used. For example, the electric field outside a very long cylinder of charge is equal to

$$E_\rho = \frac{\rho_V a^2}{2\varepsilon_o \rho}\hat{a}_\rho \quad \text{if } \rho \ge a$$

where a is the radius of the uniform distribution of volume charge, ρ_V, and ρ is the distance from the center of the cylinder's axis. If an electric field probe is placed in a small opening, flush with the surface of a long cylindrical vessel containing uniformly distributed volume charge, then the volume charge density is equal to

$$\rho_V = \frac{2\varepsilon_o E_\rho}{a} \quad \text{if } \rho = a$$

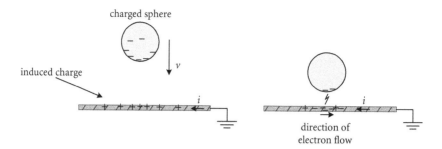

FIGURE 2.62 Current induced along a grounded plane during the approach of a charged sphere. Also shown is a discharge between the sphere and plane.

where *a* is the radius of the cylinder. The buildup of charge on the probe can be a problem. Measurement of the volume charge density has also been performed by using a probe inside the volume. The probe is surrounded by a sampling chamber or cage. Several of the surfaces of the cage should be along the expected equipotentials inside the volume to reduce some of the field distortion introduced by the cage. The cage can have significant effect on both the physical distribution of the volume charge, as well as the electric field in the vicinity of the cage. Sometimes, the effect of the cage and probe is studied by raising their potential in the volume charge and studying the change in the field reading. For liquids or other materials, moving at a known velocity, the volume charge can also sometimes be determined by measuring the convection current (if it is the dominant current mechanism).

When measuring the charge transferred during an electrostatic discharge, certain basic precautions must be taken. Imagine that a charged sphere (with a velocity of *v*) is brought near a metallic plane, which is grounded at some location along its width, and that a discharge occurs between the sphere and plane. In Figure 2.62, the sphere is assumed to be negatively charged. As shown in the figure, as the negatively charged sphere approaches the ground plane, positive (image) charges are induced on the ground plane. Negative charges traveling toward the actual ground connection, away from the sphere, result in a current, *i*, in the direction shown. When the sphere and ground plane are sufficiently close, an electrostatic discharge occurs. During the discharge, a *portion* of the overall negative charge on the sphere is transferred to the ground plane. These negative charges neutralize *some* of the positive charge induced along the ground plane. The current along the ground plane will increase during this discharge. Characteristics of this current, such as its rise time and peak value, are important in determination of the radiated and conducted noise content of the discharge, as well as its ability to damage other devices. By integrating this current at the ground connection, the total charge transferred through the ground connection can be determined:

$$i = \frac{dq}{dt} \quad \Rightarrow q = \int_0^t i\,dt \tag{2.145}$$

In Figure 2.63, a crude representation of the charge on the sphere and induced charge on the ground plane is shown immediately before and after the discharge. The magnitude of the induced charge on

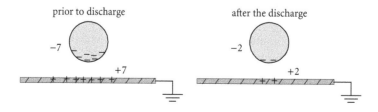

FIGURE 2.63 Rough sketch of the charge distributions before and after an electrical discharge.

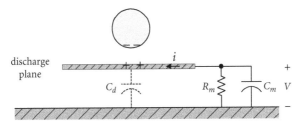

FIGURE 2.64 Model of the discharge plane and instrumentation.

the ground plane is shown exactly equal to the magnitude of the charge on the sphere. This implies that there are no other ground planes or objects nearby and all of the electric field lines are terminating on the ground plane. After the discharge, the total charge on the sphere has changed from –7 to –2 (the unit of the charge is not important for this discussion). Therefore, –5 charge transferred through the ground connection.

When a resistor (for current measurements) or capacitor (for charge measurements) and the measurement instrument is placed between the plane and ground connection, it influences the system. Also, the discharge plane shown has a resistance, (partial) inductance, and capacitance (to the major reference ground plane). The instrument and its connecting cable have resistance, capacitance, and inductance. All of the inductances, including the parasitic inductances of the current sensing resistor or charge sensing capacitor, are neglected to simplify the analysis. Assuming the discharge plane is electrically short, it can be modeled as a single capacitor. In Figure 2.64, the test instrument, connecting cable, and sensing resistor or capacitor are represented as R_m and C_m. The time constant for this low-pass system is given by

$$\tau = R_{eq}C_{eq} = R_m (C_d + C_m)$$

If the discharge current is desired, the voltage can be measured, using an oscilloscope, across a small current sensing resistor (R_m). So that the current response can be accurately measured, the cutoff frequency for this low-pass response should be much greater than the highest frequency of interest in the discharge current. (The highest frequency of interest is related to the change of the current with respect to time.) The cutoff frequency in Hz for this low-pass filter response is

$$f_c = \frac{1}{2\pi R_m (C_d + C_m)} \tag{2.146}$$

Assuming that the value for R_m is limited by the noise in the system and amplification of the instrumentation, small values for both C_d and C_m are desirable. Now, if the charge to the ground is to be measured directly, R_m should be very large. The resistance to the ground connection should be very large so that no charge is lost to the ground. The charge on both of the two upper plates of C_d and C_m is the total charge transferred during the discharge:

$$Q = (C_d + C_m) V = C_d V + C_m V = Q_d + Q_m$$

The charge measured by the test instrument is the charge across C_m, which is $Q_m = C_m V$. Therefore, when $C_m \gg C_d$, most of the charge is measured by the instrument, and $Q \approx Q_m$. Figure 2.65 and Figure 2.66 roughly show the distribution of charge between the two capacitances C_d and C_m before and after a discharge. Notice that the total charge on both upper plates of the capacitances is zero prior to the discharge. After the discharge, –5 units of charge are transferred to the upper plates of C_d and C_m. The two capacitors have been charged. Any resistance between the plates, including other leakage resistances, will eventually allow this excess charge to pass to the lower ground reference.

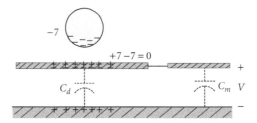

FIGURE 2.65 Charge distribution prior to the discharge.

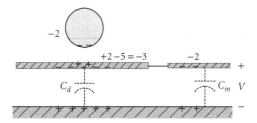

FIGURE 2.66 Charge distribution after the discharge.

The distribution of the −5 units of charge between the plates is a function of the size of these two capacitances. In this figure, $C_m < C_d$ so C_d carries more of the total charge. Finally, so that the discharge plane is nearly zero volts (i.e., about the same potential as the lower reference plane) during the transient current surge, the total impedance to ground,

$$Z = R_m \left\| \frac{1}{j\omega C_{eq}} \right. = \frac{R_m \dfrac{1}{j\omega (C_d + C_m)}}{R_m + \dfrac{1}{j\omega (C_d + C_m)}} = \frac{R_m}{1 + j\omega R_m (C_d + C_m)} \tag{2.147}$$

should be small. At low frequencies, well below the cutoff frequency, the impedance is $\approx R_m$. Again, a small value for R_m is desirable when performing a current measurement.

A discharge probe that has been used to measure the current directly and charge indirectly (via integration) is shown in Figure 2.67. It has a hemispherical shape. The discharge area, which is at the center of the inner conductor, is insulated from the surrounding outer conductor. The inner conductor is shown slightly higher so that the discharge will tend to pass mainly through this inner conductor. In one case, stainless steel was used as the hemisphere material, a current transformer was used to measure the current,

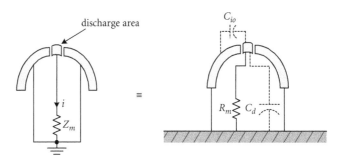

FIGURE 2.67 Hemispherical discharge probe used to measure the discharge current.

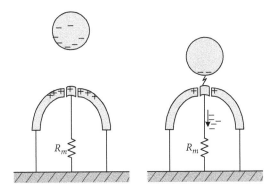

FIGURE 2.68   Charge distribution before and immediately after discharging to a hemispherical probe.

and ten 10 Ω, parallel carbon resistors (which is equivalent to 1 Ω and referred to as a resistor cage) were used to minimize the lead inductance of the resistor. The resistors and current transformer were electrically shielded.

The capacitance from the outer conductor to ground is zero since they are connected (ignoring any ohmic loss or inductance). There is, however, a very small capacitance from the inner conductor to ground: a voltage drop occurs across R_m when a current discharge passes through R_m. As previously discussed, a small discharge sensor capacitance, C_d, and small current sensing resistor, R_m, are desirable. There is also a mutual capacitance, C_{io}, between the inner conductor and surrounding outer conductor. Because of the close proximity of the surrounding outer conductor and its larger surface area, this mutual capacitance should be large *compared* to C_d, and $C_{eq} = C_d + C_{io} \approx C_{io}$. As the charged sphere approaches the probe, most of the discharge (hopefully) passes to the slightly protruding center conductor. The remaining charge on the sphere induces charge on the nearby center and outer conductor. If the surface area and protrusion of the inner conductor are small, most of the induced charge will exist on the outer electrode as shown in Figure 2.68.

2.19   Measuring the Electric Field

How can the electric field be measured? Why do some individuals state that the electric field has units of volts? [Taylor, '94; Secker; Cross; Vosteen; Taylor, '01; Hughes; Jonassen; Haus]

The electric field, like charge and voltage, is a fundamental quantity of interest in electromagnetic compatibility, electrostatics, and many other disciplines. In the section on charge measurement, the electric fields from a single charge sheet and double charge layer on an insulating slab were discussed. Furthermore, an entire table has been devoted to the electric field from various common charge distributions. As is probably expected, often the electric field and charge are determined using the same instrument. As with any measurement, the instrument itself will load down or affect its environment in some manner. With handheld instruments, the individual holding the meter can significantly affect the field and measurement.[26] The electric field measured is a function of whether the instrument is isolated, grounded, or handheld.

In the charge measurement discussion, an electrometer was used to determine the surface charge, ρ_s, as shown in Figure 2.69. The electric field can also be measured using this same setup. The variable A is the surface area of the probe connected to the electrometer. In terms of the given capacitances, the

---

[26]To reduce some of the field distortion generated by the proximity and size of the operator, long, insulating, low-dielectric constant extension rods are used. In some situations, if the measurement can be stored or recorded remotely, the instrument can be suspended using insulating string.

Electrostatic Discharge

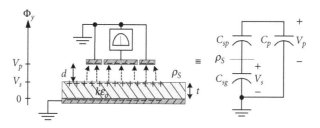

FIGURE 2.69 Electrometer to measure the charge and its electric field.

equipotential voltage along the surface of the slab is obtaining using voltage division for capacitors:

$$V_s = V_p \frac{C_{sp} + C_p}{C_{sp}} \tag{2.148}$$

The electric field between the probe and slab is, assuming a uniform electric field distribution under the probe,

$$|E_n| = |\nabla\Phi| = \frac{\Delta\Phi}{\Delta y} = \frac{|V_s - V_p|}{d} = \frac{\left|V_p \dfrac{C_{sp} + C_p}{C_{sp}} - V_p\right|}{d} = \frac{C_p}{C_{sp}} \frac{|V_p|}{d}$$

$$= \frac{C_p}{\dfrac{\varepsilon_o A}{d}} \frac{|V_p|}{d} = \frac{C_p}{\varepsilon_o A} |V_p| \tag{2.149}$$

This expression is not directly a function of the probe spacing, d, or slab thickness, t. The electric field can be determined from the measured voltage, C_p, and A.

The thickness of the slab and location of the ground will affect the measured electric field at the probe (through V_p). However, the previous probe setup can be used to determine the electric field from a wide distribution of charges, including charge on conducting boundaries. Of course, the probe itself may adversely affect the electric field and charge distribution near it. Close examination of Equation (2.149) indicates that the numerator is equal to the total charge measured by the probe, $C_p|V_p|$. The surface charge, ρ_S, along the conducting probe surface is equal to this charge divided by the surface area of the probe, A. At the conducting surface of the probe, the standard boundary condition applies, $|D_n| = \varepsilon_o |E_n| = \rho_S$, which is in complete agreement with Equation (2.149):

$$|E_n| = \frac{\rho_S}{\varepsilon_o} = \frac{\dfrac{Q_p}{A}}{\varepsilon_o} = \frac{C_p|V_p|}{\varepsilon_o A}$$

This result assumes that the surface charge is uniformly distributed along the probe's surface, so the electric field at the probe's surface is approximately uniform. If the field at the probe's surface is not uniform, which can easily occur when space or volume charge is near the probe, then error is introduced with the use of Equation (2.149). It is common to have field intensification at a probe. The size of the probe relative to the charged surface is a factor in intensification. It is important to reiterate that both the geometry of the field probe and geometry of the charged surface under test are important.

The previous method of measuring the electric field, although simple and low in cost, has the potential for error when in the presence of charged particles. Charged particles, such as airborne particulates and ions, when contacting the probe will charge the probe input capacitance, C_p. This will introduce an electric field deviation of

$$|\Delta E_n| = \frac{\Delta Q_p}{\varepsilon_o A} = \frac{C_p}{\varepsilon_o A}|\Delta V_p| \qquad (2.150)$$

where ΔQ_p is the increase or decrease in the probe charge due to the charged particles. A second problem with the simple induction probe is the drift in the instrument's reading. All amplifiers draw current, even quality electrometers. This current is labeled I_{in} in Figure 2.70. Over time, the change in voltage and final voltage across the capacitor C_p due to this input current are given by (assuming I_{in} is constant)

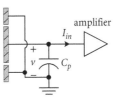

FIGURE 2.70 Loss of capacitor charge through the input of the amplifier.

$$\frac{dv}{dt} = \frac{1}{C_p}I_{in} \;\Rightarrow v(t) = \frac{I_{in}}{C_p}t + v(0) \qquad (2.151)$$

The percent change in the electric field measurement due to this drift is

$$E_{error\%} = 100\frac{|\Delta E_n|}{|E_n|} = 100\frac{\frac{C_p|\Delta V|}{\varepsilon_o A}}{|E_n|} = 100\frac{C_p}{\varepsilon_o A}\frac{|v(t)-v(0)|}{|E_n|} = 100\frac{C_p}{\varepsilon_o A}\frac{\frac{I_{in}}{C_p}t}{|E_n|}$$
$$= \frac{100 I_{in}}{\varepsilon_o A |E_n|}t \qquad (2.152)$$

If 5% error in the field measurement is acceptable over a 30 sec period of measurement, then the input bias current must be very small for a 40 kV/m electric field and 1 cm² probe surface:

$$I_{in} < \frac{\varepsilon_o A|E_n|E_{error\%}}{100t} = \frac{8.854\times10^{-12}\left(\frac{1}{10^4}\right)(40\times10^3)5}{100(30)} \approx 0.06\,\text{pA}$$

When using this type of field meter, the capacitor should be discharged before a measurement is taken (similar to capacitor integrator circuits). The capacitor can also be automatically discharged on a periodic basis. The meter should be zeroed in a zero-field region[27] or calibrated against a known field before use. This induction-type probe should not be used in those locations where charge from any source can contact and charge the probe's surface. Also, it should not be used for continuous monitoring unless the capacitor can be discharged frequently and periodically.

---

[27]For low-frequency electric fields, the human hand is a good shield; thus, a hand in front of the probe will usually provide a reasonable means of zeroing the field meter.

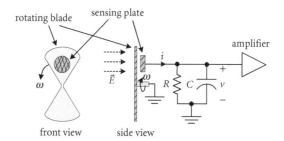

FIGURE 2.71 Field mill.

The field mill, also known as a shutter field mill, generating voltmeter, or rotating-vane field meter, is commonly used to measure the electric field. There are two types of these meters: shutter and cylindrical. For the shutter type of field mill meter, the field sensing plate(s) is placed behind a grounded rotating blade. As a result, the induced charge on the sensing plate is modulated by the rotating blade. The blade is periodically shielding the sensing plate from the electric field. A much simplified model of a field mill is shown in Figure 2.71. The expression for the output voltage, v, can be easily derived. Assume that the exposed area of the sensing plate to the electric field satisfies the relationship

$$a(t) = \frac{A}{2}[1 - \cos(2\omega t)] \tag{2.153}$$

where ω is the angular velocity of the blade and A is the area of the one circular sensing plate. (This area function has two maximums over the period $2\pi/\omega$.) At the conducting sensing plate, the relationship between the electric field normal to the plate, in the direction shown, and the surface charge on the plate is

$$\rho_S = -\varepsilon_o E_n$$

If the field is directed toward the plate, the surface charge is negative. If the field over the entire surface of the sensing plate is about the same or uniform, the total charge on the plate, Q, is

$$Q = a(t)\rho_S = -a(t)\varepsilon_o E_n$$

The current supplying this charge to the plate is[28]

$$i = -\frac{dQ}{dt} = \varepsilon_o \frac{d}{dt}[a(t)E_n] = \varepsilon_o \frac{d}{dt}\left\{\frac{A}{2}[1 - \cos(2\omega t)]E_n\right\} = A\varepsilon_o E_n \omega \sin(2\omega t)$$

$$= A\varepsilon_o E_n \omega \cos\left(2\omega t - \frac{\pi}{2}\right) \tag{2.154}$$

The amplitude of the current is directly proportional to the blade rotation frequency. Using standard phasor analysis, the sinusoidal steady-state voltage across the input of the amplifier is readily determined

---

[28]To understand the sign of $i = -dQ/dt$, imagine that the area is increasing linearly such as At so that $i = -dQ/dt = -d(-At\varepsilon_o E_n)/dt = A\varepsilon_o E_n > 0$. In this case, the total negative charge on the plate is increasing linearly with time. Negatively charged electrons must supply this charge to the plate. Current is in the opposite direction of electron flow.

(the frequency of the source is 2ω):

$$V_s = ZI_s = \left(R \left\| \frac{1}{j2\omega C}\right.\right)\left(A\varepsilon_o E_n \omega \angle -\frac{\pi}{2}\right) = \frac{R}{1+j\omega 2RC}(-jA\varepsilon_o E_n \omega)$$

$$= \frac{-jRA\varepsilon_o E_n \omega}{1+j\omega 2RC}\left(\frac{1-j\omega 2RC}{1-j\omega 2RC}\right) = \frac{-2R^2 CA\varepsilon_o E_n \omega^2 - jRA\varepsilon_o E_n \omega}{1+\omega^2 4R^2 C^2}$$

$$= \frac{\sqrt{\left(-2R^2 CA\varepsilon_o E_n \omega^2\right)^2 + \left(-RA\varepsilon_o E_n \omega\right)^2}}{1+\omega^2 4R^2 C^2}\angle \tan^{-1}\left(\frac{-E_n}{-2RC\omega E_n}\right)$$

$$= \frac{AR\varepsilon_o |E_n| \omega}{\sqrt{1+\omega^2 4R^2 C^2}}\angle \tan^{-1}\left(\frac{-E_n}{-2RC\omega E_n}\right)$$

The angle was not further simplified so that the proper quadrant of the angle could be determined, based on the sign of E_n. In this polar form, the time-domain result is written "by inspection" (the phrase dreaded by most students):

$$v(t) = \frac{AR\varepsilon_o |E_n| \omega}{\sqrt{1+\omega^2 4R^2 C^2}}\cos\left[2\omega t + \tan^{-1}\left(\frac{-E_n}{-2RC\omega E_n}\right)\right]$$

$$= \frac{AR\varepsilon_o |E_n| \omega}{\sqrt{1+\left(\dfrac{\omega}{\dfrac{1}{2RC}}\right)^2}}\cos\left[2\omega t + \tan^{-1}\left(\frac{-E_n}{-2RC\omega E_n}\right)\right] \tag{2.155}$$

The sign or direction of the electric field can be sensed. For low blade speeds, the peak amplitude of the voltage is a function of the blade speed:

$$|v(t)| \approx AR\varepsilon_o |E_n| \omega \quad \text{if } \omega \ll \frac{1}{2RC} \tag{2.156}$$

For high blade speeds, however, the peak amplitude of the voltage is not a function of the blade speed or the resistance:

$$|v(t)| \approx \frac{A\varepsilon_o |E_n|}{2C} \quad \text{if } \omega \gg \frac{1}{2RC} \tag{2.157}$$

At higher frequencies, the capacitive reactance is much smaller than the resistance parallel to it. The sensitivity or peak output voltage is a function of the capacitance. The smaller the capacitance, the greater the output voltage. However, the speed of the blades must increase as C decreases to satisfy $\omega \gg 1/(2RC)$. If the motor's speed were slightly unstable or not accurately known, there would be an advantage in operating at higher speeds. Because the voltage is time varying, the signal to the amplifier can be ac coupled, and thus drift is not an issue. Also, since the output voltage is based on time-varying charge buildup on the sensing plate, some unwanted slow-changing charge buildup on the sensor is not a major issue: this meter can be used where some space or ion charge is present. Air can also be used to help purge the sensing plate of dust and other particulates. Calibrating the instrument on a regular basis in

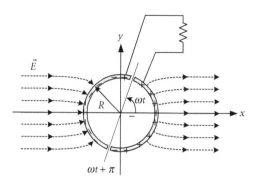

FIGURE 2.72   Cylindrical field mill.

a field-free region is a good idea when in the presence of highly insulating charged particles. There are obvious disadvantages with this field meter: the electrical noise, weight, size, reliability, and cost of the motor and its circuitry.

A cylindrical field mill involves two rotating half-cylinders, both conducting. The applied electric field is perpendicular to the axis of the half-cylinders as shown in Figure 2.72.[29] The input circuitry, such as the previously given parallel *RC* circuit in shunt with an amplifier, is placed across or shunt with the two halves of the cylindrical electrodes. As will be seen, the total charge induced on each half of the cylinder will vary periodically with its position, resulting in a current through the connecting circuitry. Again, the physical rotation of the cylinders modulates the incident electric field. The analysis of the cylindrical field mill is slightly more complicated than the previous shutter type. As seen in Figure 2.72, the conducting cylinder does distort the field around it. Far from the cylinder, however, the electric field is uniform. For an electric field in the *x* direction, on the outer surfaces, negative surface charges are induced where $x < 0$ and positive surface charges where $x > 0$. The charge concentration is the greatest at $(x = \pm R, y = 0)$ where the electric field is the strongest. The cylinder is rotating at a frequency of ω rad/sec around the *z* axis. Fortunately, analytical expressions for the potential and hence electric field around a perfectly conducting, infinitely long cylinder with its axis perpendicular to an applied uniform electric field, $E_o \hat{a}_x$, are available:

$$\Phi = -E_o R \left(\frac{\rho}{R} - \frac{R}{\rho} \right) \cos \phi \qquad (2.158)$$

$$\vec{E} = -\nabla \Phi = -\frac{\partial \Phi}{\partial \rho} \hat{a}_\rho - \frac{1}{\rho} \frac{\partial \Phi}{\partial \phi} \hat{a}_\phi - \frac{\partial \Phi}{\partial z} \hat{a}_z$$

$$= E_o \left[1 + \left(\frac{R}{\rho} \right)^2 \right] \cos \phi \hat{a}_\rho - E_o \left[1 - \left(\frac{R}{\rho} \right)^2 \right] \sin \phi \hat{a}_\phi \qquad (2.159)$$

Free space surrounds the cylinder. This result, given in the cylindrical coordinate system, clearly shows that the field is a maximum at $\phi = 0$ and $\phi = \pi$ since the $-\hat{a}_\phi$ term is zero. (The variable ϕ is the standard angle from the *x* axis, and ρ is the radial distance from the origin.) The surface charge anywhere along the cylinder is obtained using the standard boundary condition for a conductor in free space:

$$\rho_S = \varepsilon_o E_n$$

---

[29]More carefully stated, the applied field in this figure is perpendicular to the *yz* plane.

where E_n is the outward normal component of the electric field to the cylinder's surface. In this cylindrical coordinate system, the $\hat{a}_\rho$ component is entirely normal to the surface so

$$\rho_S = \varepsilon_o E_o \left[1 + \left(\frac{R}{\rho}\right)^2\right] \cos\phi \Bigg|_{\rho=R} = 2\varepsilon_o E_o \cos\phi \qquad (2.160)$$

The sign of the surface charge agrees with that shown in Figure 2.72. The total charge on one-half of the cylinder is obtained by integrating over the half-cylinder's surface:

$$Q = \iint \rho_S \, ds = \int\limits_0^L \int\limits_{\omega t}^{\omega t+\pi} (2\varepsilon_o E_o \cos\phi)\rho \, d\phi \, dz \Big|_{\rho=R} \qquad (2.161)$$

$$= 2\varepsilon_o E_o LR \left[\sin(\omega t + \pi) - \sin(\omega t)\right] = -4\varepsilon_o E_o LR \sin(\omega t)$$

where L, the length of the half-cylinders, is much greater than R, and end effects are neglected. When $\omega t = 0$, the total surface charge on the cylinder half is zero: equal number of field lines are terminating and beginning on the cylinder's surface. For an initially charge-neutral cylinder, the charge on the other half of the cylinder must be equal in magnitude but opposite in sign to this result. The current into the circuitry is

$$i = -\frac{dQ}{dt} = 4\varepsilon_o E_o LR\omega \cos(\omega t) \qquad (2.162)$$

The remaining analysis is identical to the shutter field mill. The electrical connections to the sensors are more complicated than the shutter version because of their rotation.

Another instrument for measuring the electric field is the vibrating field, vibrating capacitor, or reed meter shown in Figure 2.73. An aperture is placed in front of the sensing plate, and either the sensing plate or aperture is mechanically oscillated. When the sensor is periodically swept parallel to the aperture's opening, the analysis is similar to the shutter field mill. When the sensor is vibrated normally to the aperture and field, the analysis is different. The electric field passing through to the sensor plate is "attenuated" or diffused as its passes through the guard aperture to the sensing plate. This affects the strength and distribution of the surface charge on the sensor plate. The electric field is no longer uniform at the sensor plate. As the distance between the sensor and aperture plate increases, the capacitance between them decreases. This capacitance is a function of the spacing and width of the sensor plate. If the width of the plate is small or comparable to the width of the aperture opening, the capacitance will be small. The open-circuited voltage from the sensor plate to the ground is given by $V = Q/C$. Since the sensor plate is oscillating with

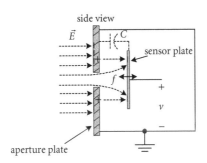

FIGURE 2.73 Vibrating field meter.

a frequency of f, the voltage will also be oscillating. The total charge, Q, on the sensor plate will be a function of the electric field strength and the distance between the two plates. According to one reference, the vibrating field meter can be built smaller in size than the field mill (mm vs. cm sensor diameters). If the electric field to be measured is from a surface charge near a ground plane, then the capacitance between the sensor plate and this surface charge is also varying with time.

Free-body field meters are available to measure time-varying electric fields via current between two conducting halves of a body. They are sometimes used to estimate the current through an individual in the presence of 60 Hz electric fields in commercial and residential settings where the electric field strength can vary over a wide range. The two conducting electrodes can be, for example, rectangular or hemispherical in shape. The charge induced on one-half of the body is

$$Q = A\varepsilon_o E_o \cos(\omega t) \tag{2.163}$$

where A is a constant proportional to the surface area of the conducting body. The constant A will be determined for a sphere. The expressions for the potential and electric field around a perfectly conducting sphere of radius R in an applied electric field $\vec{E} = E_o \hat{a}_z$ are

$$\Phi = -E_o R \left[\frac{r}{R} - \left(\frac{R}{r} \right)^2 \right] \cos\theta \tag{2.164}$$

$$\vec{E} = -\nabla\Phi = -\frac{\partial\Phi}{\partial r}\hat{a}_r - \frac{1}{r}\frac{\partial\Phi}{\partial\theta}\hat{a}_\theta - \frac{1}{r\sin\theta}\frac{\partial\Phi}{\partial\phi}\hat{a}_\phi$$

$$= E_o \left[1 + 2\left(\frac{R}{r} \right)^3 \right] \cos\theta\,\hat{a}_r - E_o \left[1 - \left(\frac{R}{r} \right)^3 \right] \sin\theta\,\hat{a}_\theta \tag{2.165}$$

Free space surrounds the sphere. The surface charge is obtained as before using the electric field normal to the sphere's surface:

$$\rho_S = \varepsilon_o E_o \left[1 + 2\left(\frac{R}{r} \right)^3 \right] \cos\theta \Bigg|_{r=R} = 3\varepsilon_o E_o \cos\theta \tag{2.166}$$

When the electric field is in the direction shown in Figure 2.74, the surface charge on the lower hemisphere is negative while it is positive on the upper hemisphere. If the electric field is varying sinusoidally

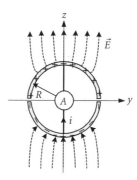

FIGURE 2.74   Spherical free-body field meter.

as $\vec{E} = E_o \cos(\omega t) \hat{a}_z$, the total charge on the lower hemisphere is[30]

$$Q = \iint \rho_s ds = \int_0^{2\pi} \int_{\frac{\pi}{2}}^{\pi} [3\varepsilon_o E_o \cos(\omega t) \cos\theta] r d\theta r \sin\theta d\phi \Big|_{r=R} = -3\varepsilon_o \pi R^2 E_o \cos(\omega t) \qquad (2.167)$$

The polarity of the total charge on the lower hemisphere changes when the direction of the applied field changes. Assuming both hemispheres were initially charge neutral, the total charge on the upper hemisphere is $-Q$. The current between the two hemispheres is

$$i = -\frac{dQ}{dt} = -3\varepsilon_o \pi R^2 E_o \omega \sin(\omega t) \qquad (2.168)$$

The current and sensitivity increase with the square of the radius. This current, of course, will generate a voltage drop across the ammeter, and the voltage across the two halves of the sphere will not be zero — a source of error in using the given expression.

There are other methods of measuring the electric field including the use of electrical-optical effects. One extremely simple and novel method is the corona meter. For the corona meter, a sharp probe is placed in the electric field. When the electric field is sufficiently large near the probe tip, corona occurs. This probe can be connected in series with a large resistor (e.g., 10 GΩ) that is connected to a neon tube (e.g., with a breakdown voltage of 90 V) in parallel with a capacitor (e.g., 0.01 μF). The current, due to the corona, charges the capacitor and eventually turns on the neon tube. The capacitor then discharges through the tube, the capacitor is again recharged over time, and the cycle repeated. The flashing rate is a rough measure of the electric field. Many years ago, a gas tube connected directly to a sharp electrode, was used as a radio frequency energy detector. The capacitance to ground was supplied by the user in contact with the device.

Surveying the commercial sites, it is apparent that there are many types of hand-held electric field, voltage, and static charge meters. They are useful, but they have limitations. Some of these meters have insulating rods or long handles so that the operator has less influence on the actual field. If possible, the operator should be grounded so that any excess charge on the operator will be dissipated. Excess charge can also be dissipated if the instrument is grounded and the operator is in direct contact with the instrument. However, a grounded meter may generate more field distortion than a floating meter. Many of the simpler expressions for field meters assume the field at the sensor is uniform.

Electric field meters are usually calibrated in a uniform electric field. Electric field has units of V/m. Unfortunately, because some static charge or field meters have units of volts, certain people incorrectly believe that the measured electric field also has units of volts. This is absolutely not correct. Some of these meters will measure the voltage in free space under certain restricted conditions.

2.20 Measuring Voltage

How can the voltage be measured along both conducting and insulating surfaces containing surface charge? [Cross; Jonassen; Secker; Keithley; Schwab]

It would seem that a discussion of how to measure voltage would be much too elementary for a more advanced book in electrical engineering. Even weaker students can use a multimeter or an oscilloscope to measure most voltages at a given point relative to ground. With electrostatics, though, high voltages and small quantities of charges are common. Furthermore, the voltage along an insulating surface or in free space at given distance from a ground is sometimes desired. In these situations, a more in-depth understanding of the measurement of voltage is required.

---

[30]The radius of the sphere, R, should be electrically small compared to the wavelength of ω if the sphere's surface is to be considered equipotential.

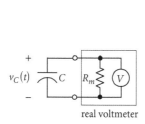
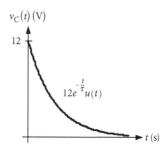

real voltmeter

FIGURE 2.75   Depletion of the charge on a capacitor through the multimeter.

Generally, measurements are either contact or noncontact. Direct contact measurements will first be discussed. It is important to remember that all instruments have a finite input impedance. The effect of a common multimeter on a measurement is sometimes introduced in basic circuit laboratories. Imagine that a high-quality 0.01 µF capacitor is charged to 12 V using a dc supply, and the supply is removed. Then, a common multimeter is connected across the capacitor to measure its voltage as shown in Figure 2.75. The voltage across the meter initially indicates a voltage of 12 V, but soon the meter voltage begins to drop in value. Although the shunt resistance of the capacitor representing the dielectric loss is partially responsible for this decay in voltage, for a good capacitor, the finite resistance of the multimeter is mostly responsible. If the multimeter has a resistance of 100 MΩ, which on the surface seems high, the classical exponential drop in the voltage is given by

$$v_C(t) = 12e^{-\frac{t}{\tau}} \quad t \geq 0 \quad \text{where } \tau = R_m C \approx (100 \times 10^6)(0.01 \times 10^{-6}) = 1\,\text{sec}$$

The time constant of the response is 1 sec. After just 5 sec, the initial 12 V is about 0.08 V. The initial charge and the charge at 5 sec are

$$Q(t=0) = CV = (0.01 \times 10^{-6})12 = 0.12\,\mu\text{C}, \quad Q(t=5) \approx (0.01 \times 10^{-6})0.08 = 0.8\,\text{nC}$$

The negative charge on one plate of the capacitor passes through the multimeter to neutralize the positive charge on the other plate of the capacitor. The drop in charge is dramatic. In many electrostatic discharge-related situations, the total stored charge is not significant. Depletion of such large quantities of charge is often unacceptable since it is a dramatic perturbation. Ordinary multimeters do not have the necessary high megaohm input resistance required for measurements of this type. Electrometers, though, have input resistances from 10^{14} Ω (100 TΩ) to 10^{16} Ω (10,000 TΩ) with input capacitances of 10 to 20 pF and less. For a 100 TΩ input resistance, the time constant is much greater:

$$\tau = R_m C \approx (100 \times 10^{12})(0.01 \times 10^{-6}) = 1\,\text{Ms} \approx 12\,\text{days}$$

For such large input resistances, the capacitor dielectric losses, contaminants, and connecting cable losses are important.

The main difficulty with using an electrometer for electrostatic voltage measurements is that they are mainly designed for low-level measurements. Some electrometers can measure voltages in the 100 to 200 V range. Voltages in the kV range, however, are common in electrostatics. High-voltage probes and voltmeters typically have input resistances of 500 MΩ to 1 to 3 GΩ, much less than a good electrometer. The first impulse might be to use a resistor divider, as shown in Figure 2.76, to step down the voltage to the safe limit of the electrometer. For two series resistors R_1 and R_2 with a source

FIGURE 2.76   Resistor divider to step down a high voltage.

voltage of V_s across them, the voltage across R_2 is

$$V_2 = V_s \frac{R_2}{R_1 + R_2} \tag{2.169}$$

Precision high-voltage divider networks are available with 10 to 25 kV ratings. If R_1 = 99 MΩ and R_2 = 1 MΩ, the voltage division step-down ratio is 100:1. If R_1 = 199.8 MΩ and R_2 = 200 kΩ, the voltage division step-down ratio is 1000:1. Dividers can step down a kV voltage to a much lower value. The voltage rating of the two resistors should be carefully noted and, especially, the voltage rating of R_1 since most of the voltage drop usually occurs across this resistor. Again, so that the resistor divider does not excessively load down the circuit under test, its net resistance, $R_1 + R_2$, ignoring the electrometers resistance, should be large. The previous MΩ resistor divider values are usually much too low. If the voltage to be measured is 25 kV and if a current drain of 1 μA is acceptable, then $R_1 + R_2 = 25\,\text{kV}/(1\,\mu\text{A}) = 25\,\text{G}\Omega$. Note that even a constant current of 1 μA over a period of 5 sec corresponds to a loss of 5 μC:

$$i = \frac{dQ}{dt} \quad \Rightarrow Q = \int_0^5 (1 \times 10^{-6})\,dt = 5\,\mu\text{C}$$

This can be a significant quantity of charge in many electrostatic situations. The loss of this charge can be acceptable, though, when measuring the output of a high-voltage power supply.

It is also possible to step down a high voltage by using a capacitor divider as illustrated in Figure 2.77. The step-down voltage using capacitor voltage division is

$$V_2 = V_s \frac{C_1}{C_1 + C_2} \tag{2.170}$$

where C_2 can be the input capacitance of the electrometer or other instrument measuring V_2 (and supplemented by a shunt capacitance if required).[31] As an example, if C_1 = 25 pF and C_2 = 0.1 μF, then the voltage step-down ratio is about 4000:1. Unlike a resistor voltage divider where most of the voltage drop is across the larger resistance, with capacitor voltage dividers, most of the voltage drop is across the smaller capacitance. The voltage rating of both capacitors should be carefully selected. The main advantage of using a capacitor voltage divider is that the dc current is very low, assuming the leakage resistances of the capacitors are sufficiently high. In addition, the power dissipated is less than the resistor divider, again assuming the leakage resistances are high.

FIGURE 2.77 Capacitor divider to step down a high voltage.

---

[31] The capacitors C_1 and C_2 can also correspond to the capacitances between conducting objects such as spheres or cylinders with rounded end caps.

The field meters previously discussed (and other voltmeters) can sometimes be used to determine the voltage without directly contacting the circuit. The electric field and voltage are related via

$$V = -\int_a^b \vec{E} \cdot d\vec{L}$$

If the electric field is uniform with a magnitude of E_o, then the voltage drop over some distance d in the direction of the field is

$$\Delta V = E_o d \tag{2.171}$$

Therefore, if the field is uniform between the measurement location and a conducting charged object, then the voltage of the conducting object might be obtained from the voltage reading on the meter. It is important to state that the meter was (probably) calibrated in a particular field distribution (e.g., a uniform field). Unless the field between the meter and charged object under test is similar to the calibration field, the voltage reading or field reading (with the spacing d) is probably not the voltage of the charged object. When the object is not conducting, its potential is not usually clearly defined unless it happens to be along an equipotential surface. Another method of using a field meter to determine the voltage of a nearby conducting object is to adjust the potential of its guard and sensor electrodes until the field measured is small. When the meter's guard and sensor potential is equal to the target object, the electric field between them should be small.

When a conducting sphere at a high voltage is moved toward a grounded conducting sphere, a spark can occur between the two spheres. The diameter of the spheres and their spacing will determine this minimum flashover or sparking potential (of the high-voltage sphere relative to the grounded sphere). Of course, the typical factors that affect the air breakdown such as the surrounding pressure (i.e., air density) and contaminants on the surfaces of the spheres will affect these results. In Figure 2.78, the relationship between the spacing between two identical spheres and this minimum sparking potential of the high-voltage sphere is shown for 2, 5, and 10 cm diameter spheres. Typically, the results are provided in table form as a function of specific sphere diameters and spacing. These smooth curves were obtained from this data. An unknown high voltage could be measured by connecting the voltage across the two spheres and then reducing their spacing until a spark occurs. However, when the unknown voltage is connected across the spheres, based on the capacitance of the spheres relative to the capacitance of the

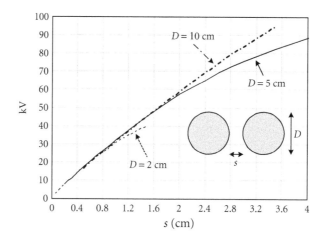

FIGURE 2.78 Flashover voltage for identical spheres as a function of their spacing and diameter.

circuit of the unknown voltage, the spheres could excessively load down the circuit. This method of measuring a high voltage is used as a standard for calibrating high-voltage instruments. The curves in Figure 2.78 assume a temperature of 20° C at a pressure 1.013 bar. The voltage given is the peak value and can be applied to ac voltages, dc voltages of either polarity, and full negative "standard" impulses (even though the results do vary somewhat with the signal polarity and shape). The surfaces of the spheres should be smooth and far from nearby objects. So that the electric field is not too nonuniform between the spheres, the spacing between the spheres should not be too large (e.g., $s < D$). Otherwise, other breakdown mechanisms can occur such as corona. Other conductor shapes or profiles may also be used that may provide a more uniform electric field between the conducting electrodes.

There are other methods of measuring a high voltage, including utilizing the force of the electric field on a conducting object, measuring the intentional distortion produced by a grounded probe in a space region, and using weak radioactive sources to generate ions. For time-varying voltages including impulses, the resistor and capacitor dividers can still be used. Transformers can also be used to step down high voltages. The parasitic resistances, inductances, and capacitances, however, should be considered. For very high frequencies or quickly changing signals, any connecting cables that are not electrically short should be impedance matched. Finally, when using even a simple resistor voltage divider, the variation of the elements with respect to temperature and their overall stability should be considered.

2.21 Measuring Bulk and Surface Resistivity

Discuss the standard methods of measuring the bulk and surface resistivity. [Jonassen; Valdes; Smits; Cross]

Although charge can dissipate through many different paths, often for high-resistivity materials, it will discharge along the surface of the material. For this reason, many materials are described by a surface resistivity, ρ_s in Ω, instead of or in addition to a bulk resistivity, ρ in Ω-m. Because surface resistivity and its measurement are frequently misunderstood, the simpler concept of bulk resistivity will first be discussed. The following discussions concerning resistance and resistivity assume that the material is ohmic in nature. Because many insulative materials are not ohmic, sometimes the measurements to determine the resistivity are performed at specified voltages. Also, the frequency of the test voltage or current is assumed sufficiently low so that the skin effect is not a factor. However, this does not imply that the bulk and surface resistivities cannot be defined or measured at higher frequencies. When actually performing a surface resistivity measurement, factors such as the humidity can have a dramatic effect on the resistivity of insulators. The pressure of the probes on the sample can also cause deviations in the measurement.

Recall that the resistance of a rectangular-shaped object when the current is uniformly distributed throughout it is defined as

$$R = \frac{l_{th}}{\sigma A} = \frac{\rho l_{th}}{A} \tag{2.172}$$

As shown in Figure 2.79, l_{th} is the length, A is the cross-sectional area, σ is the conductivity, and $\rho = 1/\sigma$ is the resistivity of the object.[32] To use this expression, the current density, $\vec{J}$, should be constant or uniform over the object. The resistance of a material is a function of the dimensions of the object and resistivity of the material. The resistivity is independent of the cross-sectional dimensions of the object. Resistivity has units of Ω-m:

$$\rho = \frac{RA}{l_{th}} \quad \left(\frac{\Omega\text{-m}^2}{\text{m}} = \Omega\text{-m} \right) \tag{2.173}$$

---

[32]Unfortunately, the same variable is commonly used for resistivity, charge density, and the radial cylindrical coordinate variable.

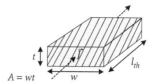

FIGURE 2.79   Uniform current distribution over an object of length l_{th} and cross-sectional area wt.

The bulk, volume, or body resistivity is also given by Equation (2.173). The bulk resistance can be measured by placing the material under test between two electrodes, labeled the central and counter electrodes in Figure 2.80. A known voltage, V, is applied across these two electrodes and resultant current, I, through the electrodes measured. The ratio of this voltage to current is the bulk resistance, R:

$$R = \frac{V}{I} = \frac{\rho t}{A}$$
(2.174)

where t, the thickness of the sample, is the path length between the electrodes, and A is the cross-sectional area of the central electrode. The current is passing through the bulk of the material. The screen or guard electrode actually surrounds the central electrode, which can be cylindrical in cross section. The guard electrode is at ground potential and is designed to reduce the fringing fields between the central and counter electrode. In Figure 2.80, a small fringing in the current density is shown near the outer surface of the central electrode near the guard electrode. Between the central and counter electrode the current density is mostly uniform. Between the guard and counter electrode, however, the current density is not uniform due to the fringing near the guard electrode's outer surface.

Another term seen in the literature is the sheet resistance. It is sometimes used when describing thin-film resistors, ESD coatings, and protective coatings for glass. The sheet resistance, R_s, is defined as the ratio of the resistivity to the thickness and has the units of $\Omega/\square$, which will be explained shortly:

$$R_s = \frac{1}{\sigma t} = \frac{\rho}{t}$$
(2.175)

In terms of the resistance,

$$R = \frac{\rho l_{th}}{wt} = R_s \frac{l_{th}}{w}$$
(2.176)

where w is the width and l_{th} is the length of the sample. Unfortunately, the variable R_s is also frequently used for the surface resistance to be described next. To confuse matters further, this definition is sometimes

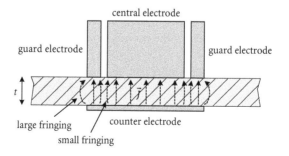

FIGURE 2.80   Guard electrodes to reduce the effects of fringing.

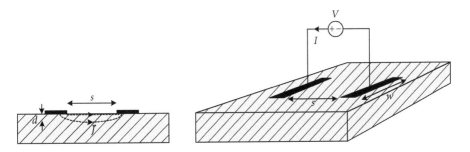

FIGURE 2.81   Resistance measurement along the surface of an object using two parallel linear electrodes.

used incorrectly to determine the resistance of a thin sheet of material by using the bulk resistivity of the material. The bulk properties of the material are not necessarily appropriate along its surface, especially when surface contaminates are present, the material is very thin, or the material is poorly conducting.

The resistance of a material can also be measured along its surface, but it is then referred to as the surface resistance. The resistance along the surface of an object could be measured by applying a voltage, V, between two parallel plates in contact with the surface of the object, and the resultant current, I, between the plates measured. The surface resistance is then

$$R_{sm} = \frac{V}{I} \tag{2.177}$$

The measurement of this resistance is shown in Figure 2.81. To relate this measured resistance to a quantity referred to as the surface or sheet resistivity, the field and current distributions between the electrodes are required. The magnitude of the electric field between the two parallel plates and in the material near the surface, ignoring fringing, is just $|\vec{E}| = V/s$, where s is the spacing between the parallel electrodes.[33] The magnitude of the current density, ignoring fringing, is $|\vec{J}| = I/(wd)$, where w is the width of the electrodes and d is the approximate depth of the current into the material. Obviously, the current between the plates will not be entirely just along the surface, and the current will diffuse some distance into the material. Although this diffusion will not abruptly stop (unless a zero conductivity region is encountered), the depth of significant amount of this diffusion is given as d. To "convert" a volume current density to a surface current density, the volume current is multiplied by the depth of the current, d: $\vec{K} = d\vec{J}$. The volume conduction current and electric field are related via Ohm's law, $\vec{J} = \sigma\vec{E} = \vec{E}/\rho$. Therefore,

$$\rho = \frac{|\vec{E}|}{|\vec{J}|} = d\frac{|\vec{E}|}{|\vec{K}|} \tag{2.178}$$

The surface resistivity *could* be defined as $\rho_s = \rho/d$ or the ratio of the bulk resistivity to the depth of current diffusion (or thickness of the sample under test). However, it is somewhat misleading to relate the surface and bulk resistivity in this simple and naïve manner. An alternative definition for the surface resistivity from electromagnetic field theory is the ratio of the electric field along the surface to the current along the surface:

$$\rho_s = \frac{|\vec{E}|}{|\vec{K}|} = \frac{\dfrac{V}{s}}{\dfrac{I}{w}} = R_{sm}\frac{w}{s} \tag{2.179}$$

---

[33]Since the tangential component of the electric field is continuous along an interface, the tangential electric field in the material right below the interface has this same value.

As is easily seen by examining the units in this expression, the surface resistivity has units of just Ω. Thus, given the surface resistivity and the surface dimensions w and s, the surface resistance can be determined. The measured resistance along the surface is not necessarily equal to the surface resistivity. But, when the spacing, s, between the electrodes is set equal to the width, w, of the electrodes, the resistance measured over this surface square, $\square$, of material is

$$\rho_s = R_{sm}\frac{s}{s} = R_{sm} \tag{2.180}$$

This surface resistivity still just has units of Ω. This surface resistance is frequently given as $\Omega/\square$ or Ω/sq just to indicate the width of the linear parallel electrodes are equal to the distance between them for the measurement. The surface resistivity may also be given the units of $\Omega/\square$ to distinguish it from any other resistance measurement. It would be perfectly correct (and some consider preferable) not to indicate the $\square$ "unit," but it would be absolutely *not* correct to state the resistivity was in Ω/m^2. Notice that as long as the distance between the electrodes is equal to their width, the resistance is the same, independent of the size of the square. Thus, if the thicknesses and surface resistivities of two objects are the same, the resistance of a 1 m by 1 m object is the same as a 1 cm by 1 cm object. The surface resistance or surface resistivity is a function of the thickness of the material. For example, typical values for surface resistivity for copper are 0.39 Ω for 1 mil thickness, 0.10 Ω for 2 mil thickness, and 0.06 Ω for 3 mil thickness. As expected, as the thickness increases, the surface resistance decreases. This variation of the resistance with thickness is due to the diffusion of the current into the sample, and the current's interaction with the backside surface of the sample.

The electric fields between the two linear electrodes previously discussed are not quite perfectly uniform between the electrodes. Fringing fields exist near the ends of the electrodes. (This fringing is different from the current diffusion into the material.) When linear electrodes are used to measure the surface resistivity, correction factors are available. To avoid *some* of the uncertainty of these fringing fields, the test electrodes can be two concentric, cylindrical-shaped electrodes. The ends of these concentric cylinders are placed on the surface of the material to be measured as shown in Figure 2.82. The resistance is measured between the two cylinders by applying a precisely known voltage (e.g., 100 V) across the electrodes and the current between the electrodes measured. The basic expression for the resistivity is easily obtained. Using Laplace's equation in the cylindrical coordinate system, the electric field between two infinitely long concentric electrodes is (where the variable r is used to denote radial distance from the center instead of ρ to avoid confusion)

$$\vec{E} = \frac{V}{r\ln\left(\dfrac{a}{b}\right)}\hat{a}_r \quad b < r < a$$

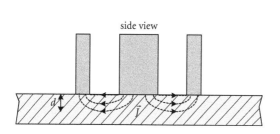

side view

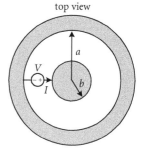

top view

FIGURE 2.82   Resistance measurement along the surface of an object using concentric cylindrical electrodes.

The electric fields at the end of these cylindrical electrodes when placed on top of the material to be measured are approximated using this infinitely long result. Obviously, the actual electric field will spread out and not be entirely in the radial direction. The current density in the material at the ends of the electrodes is also approximated as being mainly in the radial direction, even though the current will diffuse into the sample as shown. If the current from the inner to the outer electrode is I, the current density at any r location from the center (between the electrodes) is

$$\vec{J} = \frac{I}{area}\hat{a}_r = \frac{I}{\int_0^d \int_0^{2\pi} r \, d\phi \, dz}\hat{a}_r = \frac{I}{2\pi rd}\hat{a}_r \qquad (2.181)$$

where d is the diffusion depth in the sample where the current is significant. If the current is considered a surface current, then it has no depth. The surface current is then

$$\vec{K} = \vec{J}d = \frac{I}{2\pi r}\hat{a}_r \qquad (2.182)$$

Again, the actual current will diffuse a nonzero distance into the material and not be a surface current. Since the conduction current and electric field are related via Ohm's law, $\vec{J} = \sigma\vec{E} = \vec{E}/\rho$ and the current and electric field are in the same direction, the surface resistivity is equal to

$$\rho_s = \frac{|\vec{E}|}{|\vec{K}|} = \frac{V}{r\ln\left(\frac{a}{b}\right)}\frac{2\pi r}{I} = \frac{V}{I}\frac{2\pi}{\ln\left(\frac{a}{b}\right)} = R_{sm}\frac{2\pi}{\ln\left(\frac{a}{b}\right)} \qquad (2.183)$$

where R_{sm} is the measured resistance. For one standard cylinder, $a = 1.125"$, $b = 0.6"$, and the relationship between the resistivity and measured resistance is simple:

$$\rho_s = R_{sm}\frac{2\pi}{\ln\left(\frac{1.125}{0.6}\right)} \approx 10R_{sm} \qquad (2.184)$$

When the contact resistance between the material to be measured and test electrodes is important, the four-probe (point or terminal) method should be used to measure the resistance and resistivity. The theory that relates the voltage measured to the current injected for the four-probe method, as applied to measuring the earth's resistivity, was introduced by Frank Wenner in 1915. This analysis assumed that the earth was infinitely deep. It was developed in the 1950's for less restrictive geometries by Valdes and Smits. Briefly, in the four-probe method of measuring resistance and resistivity, two of the probes are used to inject current into the sample while the other two probes are used to measure the resultant voltage drop across the surface of the sample. The resistance is a function of the ratio of this measured voltage to this injected current. Rectangular and circular samples are shown in Figure 2.83. For a

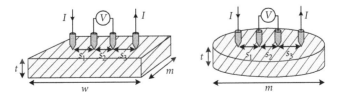

FIGURE 2.83 Four-probe method of measuring resistance.

seminfinite volume and flat surface with four probes in a straight line (i.e., collinear), the measured resistance, R_{sm}, is related to the material's resistivity, ρ, via the expression

$$\rho = \frac{2\pi R_{sm}}{\dfrac{1}{s_1} - \dfrac{1}{s_1 + s_2} + \dfrac{1}{s_3} - \dfrac{1}{s_2 + s_3}} \quad \text{if } w, m, t \to \infty \qquad (2.185)$$

(This expression can be obtained from the general resistivity equation from Wenner.) The cross-sectional dimensions of the probes should be small compared to the distances between them. Obviously, the sample cannot be seminfinite. However, if the nearest side or the bottom surface is at least five times the largest spacing between the probes, the previous expression should provide very good results without the need for a correction factor. When the probe spacings are equal, this expression reduces to

$$\rho = 2\pi s R_{sm} \quad \text{if } w, m, t \to \infty, \ s = s_1 = s_2 = s_3 \qquad (2.186)$$

Again, if the distance from any probe to any side, including the lower surface, is greater than about $5s$, this expression will provide very good results. For many thin samples, the location of the bottom side, which determines the thickness, is not far away compared to the probe spacing. When one or more boundaries are not far away, edge-effect correction factors should be applied. These correction factors are based on a uniform sample and are determined using the method of images. For thin samples where $t/s \leq 0.5$, the surface and bulk resistivity are, with the correction factor,

$$\rho_s = 2\pi s R_{sm} \left(\frac{1}{2s \ln 2} \right) \approx 4.53 R_{sm} \quad \text{if } w, m \to \infty, s \geq 2t \qquad (2.187)$$

$$\rho = 2\pi s R_{sm} \left(\frac{t}{2s \ln 2} \right) \approx 4.53\, t R_{sm} \quad \text{if } w, m \to \infty, s \geq 2t \qquad (2.188)$$

Again, besides the lower surface, the sides should be $5s$ from the nearest probe. These expressions are not a function of the probe spacing. The surface or sheet resistivity, ρ_s, has units of Ω and the bulk resistivity, ρ, has units of Ω-m. The surface resistivity expression given is also used for the idealistic infinitely thin sample. For thicker samples, the correction factor F plotted in Figure 2.84 can be used:

$$\rho \approx 4.53\, t R_{sm} F \quad \text{if } w, m \to \infty \qquad (2.189)$$

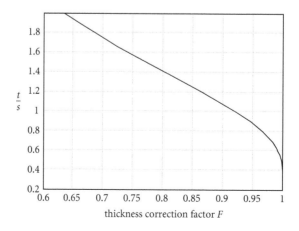

FIGURE 2.84 Four-point probe correction factor F for a finite thickness sample.

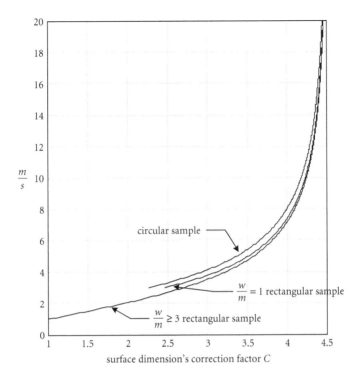

FIGURE 2.85 Four-point probe correction factor C for a finite length and width rectangular sample and finite diameter circular sample.

As seen in this plot, for $t/s \leq 0.5$, the correction factor F is about one. For infinite thickness but finite values for the surface dimensions w and m, the expression for the surface resistivity can be improved by the use of the correction factor C:

$$\rho_s \approx R_{sm} C \qquad (2.190)$$

where C is plotted in Figure 2.85. The probes are evenly spaced and centered or symmetrically placed on the sample surface. For large ratios of m/s (and $w \geq m$), the correction factor for both the rectangular and circular samples approaches the value previously given, 4.53. Both of the factors F and C can be applied to correct for both finite thickness and finite surface dimensions:

$$\rho \approx t R_{sm} CF \implies R_{sm} = \frac{\rho}{tCF} \qquad (2.191)$$

The measured resistance along the surface is a function of the bulk resistivity, thickness of the sample, and thickness and surface dimension correction factors.

For materials such as delicate antistatic films, direct contact with the material to perform a surface measurement may not be possible. Sometimes, charge can be introduced (e.g., sprayed via a high-voltage electrode) or induced on the material, and the charge decay measured via its electric field. The decay rate is a measure of the conductivity of the material. This method can also be used for moving materials if the velocity is known. (See the discussion on convection charge flow.)

2.22   Maximum Body Voltage and Typical Capacitances

The maximum voltage that a human body can be charged to is given in one reference as 25 kV.[34] Determine a possible theoretical basis for this number. What are typical values for capacitance to ground of various objects? [Jonassen; Glor; Lüttgens; Strojny; Cross; ANSI/IEEE; Kessler; Gandhi, '82]

It is instructive to first model the human body as an isolated spherical conductor. To relate the 25 kV to this spherical object, the capacitance of an isolated spherical conductor is required. The capacitance between two spherical concentric electrodes of radii a and b is given by

$$C = \frac{Q}{V} = \frac{4\pi\varepsilon}{\dfrac{1}{a} - \dfrac{1}{b}}$$

If the radius of the outer electrode, b, is much greater than the inner conductor, a, the capacitance reduces to that for an "isolated" sphere:

$$C \approx 4\pi\varepsilon a \quad \text{if } b \gg a$$

(Actually, the shape of the *distant* outer electrode is irrelevant.) Since a typical capacitance of the human body is about 150 pF, the radius of the spherical body in air corresponding to this value is

$$a = \frac{C}{4\pi\varepsilon_o} = \frac{150 \times 10^{-12}}{4\pi\,(8.854 \times 10^{-12})} \approx 1.3\text{ m}$$

This corresponds to a surface area of $4\pi a^2 \approx 23\text{ m}^2 \approx 250\text{ ft}^2$. Even for corpulent individuals, this surface area value and radius are too large. Continuing, though, the maximum charge that can be placed on an isolated human body is determined by the maximum electric field. The electric field at the surface of a charged spherical object of radius a in air is

$$E_r = \frac{Q}{4\pi\varepsilon_o a^2} = \frac{CV}{4\pi\varepsilon_o a^2} = \frac{4\pi\varepsilon_o aV}{4\pi\varepsilon_o a^2} = \frac{V}{a} \tag{2.192}$$

If the value 3 MV/m is used for the breakdown field level of air, an approximation for the maximum voltage is

$$\frac{V_{max}}{a} < 3 \times 10^6 \quad \Rightarrow V_{max} < 3 \times 10^6 a \tag{2.193}$$

When $a = 1.3$ m is used, the maximum voltage is 3.9 MV. This value is clearly not correct since most sources indicate that the maximum voltage is around 25 kV. The radius in this equation is the radius of curvature at the location of the body where the air breaks down. The curvature at the tip of a finger (or knuckle) is much less than 1.3 m and more typically around 1 cm or less. For a radius of 1 cm, the maximum voltage is around 30 kV. A maximum voltage of 25 kV corresponds to a radius of curvature of about 0.8 cm.

The actual surface charge distribution on a body is a function of the distribution of the electric field. Assuming free space around a conducting body, the electrostatic surface charge at any point along the body is determined from

$$D_n = \varepsilon_o E_n = \rho_s$$

---

[34]Some sources indicate a maximum voltage ranging from 38 to 50 kV.

where E_n is the electric field normal to the body and ρ_S is the surface charge density. If the electric field is strongest near the tip of the finger, then the charge has the greatest density at the finger tip. The total charge on the finger tip, assuming the field strength is approximately constant over the tip, is the surface area of the tip multiplied by the surface charge density. When tribocharging occurs between the shoes and floor and no other conducting objects are nearby, the field is likely to be the strongest between the bottom side of the feet and floor. Probably, in this case, most of the total charge on the body is on the bottom side of the feet.

Instead of modeling the human body as an isolated capacitance far from the ground, imagine that the body is modeled as cylinder of length l_{th} and radius a. The distance between the bottom of the cylinder and large, flat ground plane is h. The cylinder's axis is perpendicular to the ground plane. If the individual is isolated via the thin soles of their shoes from the ground plane, then it is reasonable to use the following approximation for the body's capacitance:

$$C \approx \frac{2\pi\varepsilon_o l_{th}}{\ln\left(\frac{l_{th}}{a\sqrt{3}}\right)} \quad \text{if } l_{th} \gg h$$

For a "typical" 5'8" (≈ 1.7 m) male, a capacitance of 150 pF corresponds to a body radius of

$$a = \frac{l_{th}}{\sqrt{3}}e^{-\frac{2\pi\varepsilon_o l_{th}}{C}} \approx \frac{1.7}{\sqrt{3}}e^{-\frac{2\pi(8.854\times10^{-12})(1.7)}{150\times10^{-12}}} \approx 0.5\,\text{m}$$

Although still rather large, this cylinder model is more reasonable than the isolated sphere model.

Probably, the 25 kV limit is based on experimental measurements and a parallel RC model. Imagine that a current source is responsible for the charging of the human body. The source of the current is the tribocharging between the shoes of the individual and the floor covering. The actual process is quite complicated since the charge transferred per step is not constant and is highly dependent on the material of the shoes and floor, as well as the humidity. However, as a first-order model, assume that a constant charge of the same polarity is transferred per step equal to Δq and the number of steps per second is n. Therefore, the current into the body is

$$I_s = \frac{dq}{dt} \approx n\Delta q \quad \left(\frac{n \text{ steps}}{\text{s}}\frac{\Delta q}{\text{step}}\frac{C}{} = n\Delta q\frac{C}{\text{s}}\right) \tag{2.194}$$

If the capacitance of the body to the ground is C and resistance between the body and ground is R, then the voltage of the body from Equation (2.10) is

$$v_C(t) = I_s R\left(1 - e^{-\frac{t}{\tau}}\right)u(t) = n\Delta qR\left(1 - e^{-\frac{t}{\tau}}\right)u(t) \quad \text{where } \tau = RC \tag{2.195}$$

$$v_C(t) \approx n\Delta qR \quad \text{if } t > 3\tau \tag{2.196}$$

The resistance shown in Figure 2.86 mostly represents the leakage through and over the shoes to the ground. This resistance limits the upper voltage across the body since it "bleeds off" the charge on the capacitor.

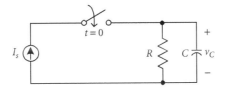

FIGURE 2.86 Simple model of the tribocharging of a walking individual. The RC circuit represents the individual.

TABLE 2.6   Maximum Body Voltage as a Function of the Body Resistance and Charge Transferred per Step

Δq	Maximum Body Voltage = $n\Delta qR$ ($n = 2$)						
R	100 MΩ	1 GΩ	10 GΩ	100 GΩ	1 TΩ	10 TΩ	100 TΩ
0.1 nC	0.02 V	0.2 V	2 V	20 V	200 V	2 kV	20 kV
1 nC	0.2 V	2 V	20 V	200 V	2 kV	20 kV	—
10 nC	2 V	20 V	200 V	2 kV	20 kV	—	—
100 nC	20 V	200 V	2 kV	20 kV	—	—	—

There is considerable variation in most of the parameters in Equation (2.195). The capacitance of the human body is frequently taken between 100 to 400 pF. The resistance from the body to the ground is a function of many factors including the shoe material (e.g., synthetic or leather) and floor composition (e.g., rug or conductive-wax coated tile floor). This decay resistance, which limits the upper voltage on the body, has been given values of less than 100 MΩ for "conductive" shoes to 1 TΩ for "ordinary" shoes. Another source states that for antistatic footwear on noninsulating floors the resistance ranges from 50 kΩ to 50 MΩ. The resistance through the shoes is generally much greater than the resistance of the body. For ESD events, the resistance of the body varies from 350 Ω to 2 kΩ, but 150 Ω is often used. The step rate, n, is a function of the speed of the person but 2 steps per sec is a reasonable walking rate.[35] The last parameter, the charge collected per step is also quite variable. One reference indicated an average value of 50 nC per step. Another reference provided the average charge collected for quartz and Teflon cylinders rolling down 12" of various types of carpet (for 12 and 14% relative humidity). The average charge buildup was around 1 to 2 nC over 12".

In Table 2.6, the maximum value of the body voltage is tabulated for a matrix of values for Δq and R when $n = 2$. Since the maximum voltage on the body is generally around 20 to 30 kV or less, voltages much greater than this range are not shown. Although the maximum voltage is not a function of the capacitance of the body relative to ground, the capacitance determines the rate of the charge and voltage increase. The larger the RC product, the greater the buildup (and discharge) time.

The actual voltage response of a person walking across a floor is not a smooth rising exponential but a rising exponential with a "noisy" signal riding on it. As a leg rises during the walk, the capacitance, C, to ground of this leg and, hence, the capacitance of the person, decreases (since the capacitance generally decreases as the spacing increases). For a given charge, Q, on the bottom of the sole, the voltage must increase if the capacitance decreases ($Q = CV$). Then, as the leg drops, the capacitance increases and voltage decreases. (When a charged printed circuit board is lifted off a table, assuming the ground reference is the floor, the voltage of the board increases since its capacitance to ground decreases.)

The quantity of charge that can be generated by a human is a function of the activity and environmental conditions. For example, walking across a vinyl floor can generate a potential of 12 kV when the relative humidity is 20% while only 250 V when the relative humidity is 80%. Other activities such as picking up a bubble bag, walking across a carpet, or spraying freeze spray on a circuit board can produce voltages from 10 to 30 kV in 20% relative humidity. (The conductivity of the air and the surface conductivity of objects increases with increased humidity. This increase in humidity often decreases the maximum voltage.) Any charge in the material is more likely to be dissipated in a high-humidity environment.

The capacitances given in Table 2.7 are only rough values since capacitance is defined (and measured) between two conducting objects, and the proximity and shape of the ground will affect these results. For the human body, factors such as shoes, clothing, and sex seem to have little impact on the capacitance. The "conductive" environment has the greatest impact. Conductive objects and other

---

[35]To determine whether this number was reasonable, the author left his den chair and walked around the living room several times. Over a period of one minute, 83 steps were recorded, corresponding to $n \approx 1.4$.

TABLE 2.7   Approximate Capacitances for Typical Items of Interest in ESD (Also Listed Are the Corresponding Energies Stored in This Capacitance at 1 kV and 10 kV)

Item	Approximate Capacitance (pF)	Stored Energy @ 1 kV (mJ)	Stored Energy @ 10 kV (mJ)
10 cm flange	10	0.005	0.5
2,000 gal tank truck	1,000	0.5	50
3.6 m diameter tank with insulative lining	100,000	50	5,000
Automobile with insulating tires	500	0.25	25
Drum (200 liter)	100–300	0.05–0.15	5–15
Fork-lift truck	1,000	0.5	50
Funnel	10–100	0.005–0.05	0.5–5
Medium size containers (250–500 liter)	50–300	0.025–0.15	2.5–15
Metal drum on insulated support	200–400	0.1–0.2	10–20
Metal flange on plastic or glass tube	10–30	0.005–0.015	0.5–1.5
Person on insulated platform	60–105	0.03–0.053	3–5.3
Person sitting and close to grounded objects	110–130	0.055–0.065	5.5–6.5
Person sitting and far from conductive objects	80–90	0.04–0.045	4–4.5
Person standing and close to grounded objects	140–165	0.07–0.083	7–8.3
Person standing and far from conductive objects	100–110	0.05–0.055	5–5.5
Person with insulating shoes	100–400	0.05–0.2	5–20
Shovel	20	0.01	1
Single screw	1	0.0005	0.05
Small container (bucket to 50 liter drum)	10–100	0.005–0.05	0.5–5

persons within 0.5 m seem to have the greatest influence, which is reasonable considering the relative dimensions of a person. The expression $(1/2)CV^2$ was used to determine the stored electrical energy results in Table 2.7.

2.23   *RLC* Discharge Model

Use SPICE to provide a simulation of the ESD current using the model shown in Figure 2.87. [Boxleitner]

It is well known that real ESD events have a nonzero rise time and sometimes oscillation similar to switch arcing. For this reason, a simple *RC* model is often inadequate since its response cannot be oscillatory. Furthermore, to simulate accurately a real ESD event, the inductance of not only the torso but also the finger and arm should be considered. Therefore, a single-discharge model involving only a series *RLC* circuit will not always adequately model a human ESD event. To take into account multiple discharges, the more elaborate model given in Figure 2.87 is considered the minimum level of acceptable complexity. (Of course, other simpler and more complex models are also used.) In addition to the shape, size, and condition of the human body, some of the component values are also a function of the proximity, size, and shape of nearby conducting objects. Obviously, once the current or voltage of the actual event is known or measured, the parameters can be adjusted to fit the response. It is not a simple matter,

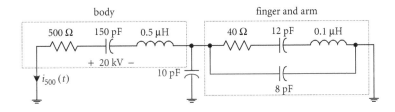

FIGURE 2.87   Electrostatic discharge model for a human body charged to 20 kV.

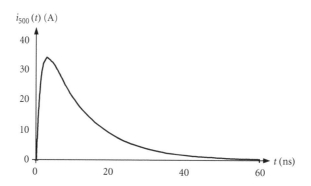

FIGURE 2.88   SPICE simulation results for the circuit in Figure 2.87.

however, to predict accurately the response of a human ESD event in a nonlaboratory or unfamiliar setting.

The SPICE results shown in Figure 2.88 indicate that the rise time for the current through the 500 Ω resistor is about 1.5 ns, pulse width is about 12 ns, and maximum amplitude is about 34 A. The waveforms used in many test standards have rise times less than 1 ns, total widths of 60 ns, and peak currents up to 30 A. The source of the initial energy is the 20 kV across the 150 pF capacitor.

Since no oscillation is present, the curve can be represented by a double-exponential function as shown in Mathcad 2.3. The upper break frequency of the spectrum for this double-exponential function is equal to β:

$$\beta = k\alpha = 11\,(82 \times 10^6) \approx 900 \times 10^6 \text{ rad/s} \approx 140 \text{ MHz}$$

$C := 47 \qquad \alpha := 82 \cdot 10^6 \qquad k := 11$

$t := 0, 1 \cdot 10^{-10} .. 1 \cdot 10^{-7}$

$x(t, k) := C \cdot \left(e^{-\alpha \cdot t} - e^{-\alpha \cdot k \cdot t} \right)$

ESD Current Model

$A \quad \underline{x(t, k)}$

$h := 1 \cdot 10^{-11}$

$g := 1 \cdot 10^{-8}$

t
s

$A(k) := -\dfrac{\ln(k)}{k-1} \qquad t_m(k) := \dfrac{-A(k)}{\alpha} \qquad d(t, k) := \dfrac{t}{t_m(k)} \qquad x_r(t, k) := \dfrac{e^{A(k) \cdot d(t, k)} - e^{k \cdot A(k) \cdot d(t, k)}}{e^{A(k)} - e^{k \cdot A(k)}}$

$pw_1(k) := \text{root}\!\left(x_r(g, k) - 0.5, g\right) \qquad\qquad pw_2(k) := \text{root}\!\left(x_r(h, k) - 0.5, h\right)$

$pw(k) := pw_1(k) - pw_2(k) \qquad\qquad \text{rise}(k) := \text{root}\!\left(x_r(h, k) - 0.9, h\right) - \text{root}\!\left(x_r(h, k) - 0.1, h\right)$

$pw(k) = 1.194 \times 10^{-8} \qquad \text{rise}(k) = 1.529 \times 10^{-9} \qquad x\!\left(t_m(k), k\right) = 33.618$

MATHCAD 2.3   Electrostatic discharge current modeled using a double-exponential function.

At least 95% of the energy is contained up to this break frequency. An electrostatic discharge is a source of both radiated and conducted emissions.

2.24   ESD Rules-of-Thumb and Guidelines

Provide a possible, simple theoretical basis for each of the following ESD-related rules-of-thumb and guidelines:

(a) for holes and slots, the longest dimension of the opening should be less than 2 cm
(b) shields should be at least 0.025 mm thick
(c) a voltage of about 200 V per in of wire is generated
(d) a wire possesses an impedance of about 15 Ω per cm
(e) exposed, ungrounded metal objects and nearby electronics/wires should be separated by at least 1 cm
(f) exposed, grounded metal objects and nearby electronics/wires should be separated by at least 1 mm
    [Boxleitner; Bendjamin; Paul, '92(b)]

All of these statements are based on typical time characteristics of an ESD event. Experiments have shown that for many electrostatic discharges, the rise time ranges from 200 ps to 70 ns, pulse width ranges from 0.5 to 10 ns, total duration ranges from 100 ns to 2 μs, and peak current ranges from 1 to 200 A. For human-to-metal discharges, the rise times are usually greater than 3 ns and less than 25 ns. The actual response, including whether oscillation occurs, is dependent on the characteristics of the ESD source, approach speed, and discharge path. An uncharacteristically brief (possible) theoretical basis for each of the given rules-of-thumb and guidelines follows:

(1) Holes and seams should be as small as possible to reduce field coupling and field energy transfer. Generally, the largest opening should be small compared to the wavelength of the highest frequency of interest. The highest frequency of interest, assuming the ESD signal can be represented using a double-exponential function, is

$$\text{``highest frequency of interest''} \approx \frac{1}{2\pi\tau_r}\text{ Hz}$$

where τ_r is the rise time of the signal. The range of rise times (and durations and peak amplitudes) for an ESD event is extremely broad. A statistical approach is sometimes the most reasonable. To simplify matters, a typical rise time of 1 ns will be initially assumed. This corresponds to a highest frequency of interest of about 160 MHz and a free-space wavelength of about 1.9 m. If the opening dimension was smaller than one-half of this wavelength, the dimension would be less than about 94 cm. If the opening dimension were electrically short or less than one-tenth of the wavelength, the dimension opening would be less than about 19 cm. It appears that the 2 cm guideline must be based on a more conservative, smaller typical rise time. If a rise time of 200 ps is assumed, the highest frequency of interest and corresponding free-space wavelength are about 800 MHz and 38 cm, respectively. One-half and one-tenth of this wavelength corresponds to about 19 cm and 3.8 cm, respectively. Thus, the 2 cm guideline seems reasonable. The actual highest frequency of interest might be higher than $1/(2\pi\tau_r)$ and in the GHz range. Also, some references indicate that the largest dimension should be less than $\lambda/20$ or $\lambda/50$.

(2) For plane wave shielding, the shielding thickness should be several skin depths thick. The skin depth for a good conductor is

$$\delta = \frac{1}{\sqrt{\pi f \mu \sigma}}$$

where μ is the permeability and σ is the conductivity of the shield. The skin depths for copper, a nonmagnetic material, at 160 MHz and 800 MHz are

$$\delta = \frac{1}{\sqrt{\pi(160\times10^6)(4\pi\times10^{-7})(5.8\times10^7)}} \approx 5\,\mu m = 0.005\,mm \quad @\,160\,MHz$$

$$\delta = \frac{1}{\sqrt{\pi(800\times10^6)(4\pi\times10^{-7})(5.8\times10^7)}} \approx 2\,\mu m = 0.002\,mm \quad @\,800\,MHz$$

If the shield is constructed of aluminum instead of copper, the skin depths are larger since its conductivity relative to copper is $\sigma_r = 0.61$. The skin depth for aluminum at 160 MHz is about 0.007 mm. If the aluminum shield were 3δ thick at 160 MHz, it would be about 0.02 mm thick. This thickness is very close to the rule-of-thumb given.

(3) For a lightning pulse, an estimate for the peak value of the voltage across the length of a grounding strap for a peak current amplitude of I_m is

$$v_{max} \approx L\frac{di}{dt} \approx L\frac{I_m}{\tau_r}$$

where the self partial inductance of the round solid conductor is given by

$$L = 2\times10^{-7}l_{th}\left[\ln\left(\frac{2l_{th}}{r_w}\right)-1\right]$$

The length and radius of the conductor are l_{th} and r_w, respectively. The inductance is not a linear function of the length of the conductor, but it is a weak function of the radius. For #20 AWG copper with a length of $1" \approx 0.0254$ m, $L \approx 19$ nH. For a length of $2" \approx 0.051$ m, $L \approx 46$ nH. For #28 AWG copper with a length of $1" \approx 0.0254$ m, $L \approx 24$ nH. For a length of $2" \approx 0.051$ m, $L \approx 55$ nH. Typically, the inductance of a piece of wire is between 15 nH/in and 30 nH/in. Using this range of inductances, the maximum voltage range per inch across the wire for two specific current amplitudes and rise times are

$$75 < v_{max} < 150 \text{ V/inch} \quad \text{if } I_m = 1,\ \tau_r = 200 \text{ ps}$$

$$300 < v_{max} < 600 \text{ V/inch} \quad \text{if } I_m = 20,\ \tau_r = 1 \text{ ns}$$

The 200 V/inch rule-of-thumb appears reasonable (for wires that are electrically short at the highest frequency of interest). If this wire corresponds to a "pigtail" connection between the outer shield of a cable and ground, significant voltage "bounce" or drop can occur during an electrostatic discharge.

(4) The magnitude of the impedance of an isolated wire that is electrically short is

$$|Z| = \sqrt{r_{AC}^2 + (2\pi f L)^2}$$

The highest frequency of interest for an ESD event is such that the ac resistance, r_{AC}, is usually insignificant compared to the inductive reactance, $2\pi f L$. For example, at 160 MHz, the ac resistance and inductance for a #20 AWG copper wire 1 cm in length are

$$r_{AC} \approx \frac{l_{th}}{2r_w}\sqrt{\frac{\mu f}{\pi\sigma}} \approx 0.013\,\Omega,\ \ L = 2\times10^{-7}l_{th}\left[\ln\left(\frac{2l_{th}}{r_w}\right)-1\right] \approx 6\text{ nH}$$

At 160 MHz, the inductive reactance is about 6 Ω, which is much greater than 0.013 Ω. At 800 MHz, the inductive reactance is about 30 Ω. Thus, a 15 Ω/cm rule-of-thumb is reasonable for ESD events.

(5) An ungrounded metal object that is exposed to an ESD can rise to a large potential. For example, a floating reset button, when touched by a charged person, can rise to 10 kV or more. If the potential of the metal object rises to 20 to 30 kV, the upper voltage range for a person, this signal can arc or spark across to other objects such as wires or electronic devices. The uniform, electric field breakdown strength in air is about 3 MV/m,

$$|E| \approx \frac{\Delta V}{d} \approx 3\,\text{MV/m}$$

where d is the distance between and ΔV the potential across the two metallic objects (this field strength is different for nonuniform fields). A potential difference of 20 kV corresponds to a separation distance of about 0.7 cm, and 30 kV to a distance of 1 cm. The given guideline is therefore reasonable.

(6) The separation distance between exposed, grounded metal objects and nearby electronics/wires can generally be less than the previous 1 cm rule-of-thumb. When a good-conducting grounded chassis is touched by a charged person, for example, its potential should not rise very high. The given guideline may refer to a 12' long cable shield or grounding strap connected from the actual ground to the grounded metal object. Using the previous 200 V per in rule-of-thumb, 12' corresponds to a voltage drop of 2.4 kV across the connecting strap. For field strength of 3 MV/m, the separation distance corresponds to 0.8 mm. This value is close to the given guideline.

2.25 Raindrop Bursts, P-Static, and Corona Noise

How can droplets of rain affect the noise level? [Nanevicz, '64; Nanevicz, '82; Gunn; Hara; Watt; Spencer]

Precipitation static or p-static is noise that is partially due to the direct impact of charged particles onto an antenna. Typical charged particles include rain droplets, snowflakes, ice fog, dust, or engine exhaust. (Refer to Table 2.8 for some average values.) As a charged particle approaches another object, the electric field can become sufficiently intense to cause a discharge between them. The object can be conducting and grounded, conducting and floating, or even insulating. This discharge is a source of strong electrical noise usually of concern from the low frequencies (LF) to the very high frequencies (VHF).

Electrical precipitation noise is also generated when charged particles hit nearby objects, such as a supporting tower, a mast, or an antenna. Furthermore, if the object is not or cannot be grounded (e.g., flying aircraft) to dissipate the charge, noise can also be generated some time after impact. This noise is often referred to as corona noise. The charge level on these objects can build to such high levels that breakdown can occur. In addition to ungrounded metal objects, this type of breakdown can originate

TABLE 2.8   Average Charge per Droplet [Gunn]

Year of Documentation	Quiet Rain	Shower Rain	Electrical Storm Rain	Quiet Snowfall	Squall Snowfall
1921	+0.08 pC −0.18 pC	+0.58 pC −1.8 pC	+2.7 pC −2.0 pC	+0.03 pC −0.02 pC	+1.9 pC −1.6 pC
1932	—	+2.1 pC −2.2 pC	+2.3 pC −2.4 pC	—	—
1938	+0.73 pC −1.0 pC	+0.43 pC −0.77 pC	—	—	+3.5 pC −1.9 pC
1949	—	—	+5.0 pC −6.3 pC	+0.22 pC −0.33 pC	—

from insulating objects (e.g., windshield or radome), metal objects with an insulative coating (e.g., "rubber ducky" antenna), and isolated conducting objects (e.g., metal clamp connecting two fiberglass mast sections). On insulating objects, the charge buildup can be a source of a surface breakdown to adjacent objects. When conducting objects are not grounded or not bonded together, the potential difference between them can be sufficiently large to cause breakdown.

Several techniques have been used to reduce the level of corona noise and p-static. To reduce the corona noise, the conductivity of insulating objects and coatings can sometimes be increased (e.g., via carbon embedded filling) to allow the charge to be dissipated more quickly. Wick dischargers, which are graphite-coated strings with bar tips, can be effective in redistributing larger discharges into smaller level discharges.[36] Stronger high-resistance rods with perpendicular, sharp bare tips near their end, can be even more effective. When placed at locations where the field strength is expected to be intense, these rods can slowly dissipate the charge. Although discharges still occur, the rod's high resistance reduces the rate of the discharge (similar to increasing the time constant of an *RC* circuit). To reduce the level of p-static, semiinsulating coatings or shrouds can be placed around conducting objects, such as a vertical antenna, to reduce the level and likelihood of the initial discharge to conducting objects. It is interesting to note that sometimes the noise level will temporarily drop during lightning discharges possibly because the lightning, streamers from the lightning, or induced currents are neutralizing or dissipating some of the charge.

2.26   Locating Weaknesses with a "Zapper"

How can an ESD gun, or a "zapper," be used to help determine intermittent failures in a system? [Boxleitner]

A common method of diagnosing an intermittent failure is through the use of an ESD generator.[37] These guns are also used to check the susceptibility of a product or device to ESD. If the intermittent failure cannot be repeated and the source of the problem not determined by turning nearby devices on and off, the ESD gun can be used to force a failure. The region of interest can be divided into various sectors or cells, and the ESD gun discharged to these cells. The discharge should begin at a low level (e.g., 5 kV) at each of the cells. If the failure cannot be produced, a progressively larger and larger voltage should be used until a failure is produced. Once the ESD-susceptible cell is located, the region near the cell can be examined and EMC "fixes" implemented. For example, if the cell contains input/output cables and a fan vent, then chokes and waveguide shielding may be possible fixes. The ESD generator will almost always eventually produce a failure, but this failure may not be the actual intermittent failure. Also, there is the possibility of permanently damaging or weakening the system under test when using this test-to-failure procedure.

2.27   Surround, Ground, and Impound

One guideline seen in ESD control programs is to surround, ground, and impound. Discuss the rational for this guideline. [Anderson, '94; Boxleitner; Jonassen]

Although there are many sources of electrostatic discharge, humans are believed to be the major source. Humans can easily be charged to the kV level, yet a person cannot usually perceive less than 2 to 3 kV — a level capable of damaging many sensitive electronic components. Unfortunately, components sensitive to 100 V or less are becoming more commonplace, especially with the rapid decrease in the dimensions of components. It is not necessary for the unaided human eye to see a spark for damage to occur.

An ESD event has both radiated and conducted components. The direct conductive component is probably the most obvious and of major concern in most situations. However, the conductive component

---

[36]This assumes that any stress on the wick does not break up and isolate the conductive coating into segments.

[37]A "poor man's" ESD generator is the tip of stainless steel tweezers held by a hand in a latex glove!

can easily capacitively or inductively couple to other components or conductors. The insidious aspect of these lower level discharges is that the damage may not be immediately apparent. The damage may be latent.

The surround, ground, and impound guideline states that for ESD-sensitive objects:

(1) "the object should be surrounded by conductive shielding materials when not in a static-safe area
(2) all humans and objects that may come in contact with the sensitive components should be grounded
(3) all static-generating materials in the static-safe area should be removed or impounded"

It is a good general strategy to follow this guideline. By completely enclosing the sensitive object with conductive materials when not in a protected area, the object is partially shielded from electric fields, plane waves, and some direct conductive discharges. (Refer to the later discussion on shielded bags.) By following the second part of the guideline, excessive charge is dissipated on personnel and objects. Charge can build up through various means such as tribocharging between the floor and shoes. When a potential difference exists between two conductors, sparking between them can occur. Grounding can be accomplished through various means such as through grounding straps. Conductive floor coatings also help in reducing the resistance to ground. Finally, by following the last part of the guideline, additional sources of electric field and tribocharging can be eliminated. Although all materials can be charged, insulating materials will tend to retain their charge for a long period of time.

It is common for plants, research labs, and even smaller university laboratory facilities to have an ESD program. The program usually includes educating and training personnel (and visitors) on ESD, posting of ESD warning signs, checking to ensure that the ESD tools are functioning on a regular basis, monitoring of electric field levels, and reviewing manufacturing and shipping guidelines. Often, antistatic bags, wrist straps, special floor coatings, dissipative work surfaces, and ionizers are part of the program.

2.28 Wrist and Ankle Straps

Wrist straps are used to discharge operators. The straps are sometimes connected to the nearest large metallic machine or object that is connected to ground via its power cord. How effective would the wrist strap be if the power cord was disconnected? [Koyler; Jonassen; Sclater; Smith, '99; Kallman; Gaertner]

Ungrounded personnel are frequently cited as the origin of 40% of ESD-related damage. Most individuals are aware of the common methods of tribocharging such as walking across a carpet or rubbing a balloon against a sweater. There are many unexpected methods of tribocharging, and a few are listed in Table 2.9.

Although electrostatic discharge is a complex subject, earth bonding or grounding can reduce the number of discharges considerably. Wrist straps are an inexpensive method of dissipating charge and reducing discharges from the body. Wrist straps are essentially a grounding strap for humans. Although a human can be grounded or partially discharged by touching a large metallic body, which is not sensitive to ESD, a grounding strap is more reliable and continuous. Wrist straps should (normally) be used by anyone touching ESD-sensitive components. The metal contact electrode of the wrist strap should be in direct contact with the skin and not be over clothing.

TABLE 2.9 Unexpected Sources of Tribocharging and ESD

Opening a vinyl package	Shifting on a padded chair
Picking up a plastic bag	Pulling off a piece of plastic tape
Spraying aerosol freeze spray	Jingling coins inside a pocket
Pulling apart two Styrofoam cups	Pulling a product from packaging peanuts
Pulling off a hat	Handling of tweezers with latex gloves

One of the first questions that should be asked when grounding an individual is, "What should the resistance to ground be?" Assume the resistance to ground for a body is about 100 MΩ and capacitance of the body to ground is about 200 pF. The time constant of this simple *RC* model of the human body is

$$\tau = RC = (100 \times 10^{6})(200 \times 10^{-12}) = 20 \text{ msec}$$

After several time constants, a charge on a body is mostly dissipated. The response time of an individual is usually in the 100's msec range. From this simple calculation, the < 100 MΩ grounding resistance limit used in industry appears reasonable.

Although there is a nonzero resistance between an individual and a "ground," the actual resistance is quite variable. A wrist strap, if properly maintained *can* provide a more reliable and consistent resistance to ground.[38] Although a lower resistance is usually desirable to dissipate quickly charge buildup on the body, a lower limit is usually in the 250 kΩ to 1 MΩ range. A resistor is usually inserted between the strap and ground connection to limit the current through the body in case of accidental contact with an energized or a hot voltage conductor. This additional resistor is also used with static-dissipative mats. This ground connection through a resistor is referred to as a "soft" ground. When a person contacts the ground other than through the soft ground of the wrist strap (e.g., leaning a bare arm on a metal grounded chassis), the resistance to ground can be much lower than that of the soft-ground resistance value.

If the wrist strap is connected to ground through a large metallic machine, disconnecting the power cord will disconnect the strap wearer from the ground. However, if the machine is large compared to the operator, most of the body charge will transfer to the machine. The actual final charges on the individual and machine are a function of the capacitances of the two objects and their initial charges. (See the energy and capacitance discussion.)

When personnel are not stationary, a wrist strap tether is usually impractical. For mobile workers, heel straps, foot grounders, or conductive footwear can be used in conjunction with static-dissipative floors. Static-dissipative shoes are preferred over rubber or plastic (insulating) shoes, but are much more expensive than heel and foot straps. Heel and foot straps pass under the sole of the shoe(s) and the conductive electrode contacts the foot or ankle. Soft grounds are again used. Wrist straps should be used with foot straps when working at a lab bench since the feet are commonly lifted from the floor.

The connection and wiring resistance for these straps should be periodically checked. This includes checking the tightness of the snaps, electrical continuity of the connections and conductors, and cleanliness of the electrodes.

2.29 Floor Coatings

How dangerous is a concrete or tile floor from an ESD perspective? When the humidity is high, is the potential for ESD greater? [Kessler]

Painted concrete, varnished wood, vinyl tile, and of course, many carpets are not typically very conductive. Topical antistats can be applied to carpets, seats, clothing, and floors to increase (at least temporarily) their conductivity. There are two major ESD requirements for floors: (i) the floor should not increase the tribocharging of objects traveling across it; (ii) the floor should appropriately dissipate charge on objects in contact with the floor. Concrete and tile floors are considered an ESD hazard since they do not provide the necessary consistent path to dissipate charge. The conductivity of concrete will vary with the age of the concrete and local water table.

Antistatic and static-dissipative floor finishes in conjunction with foot straps are a low-cost method of static limiting. The rate of charge transfer can be partially controlled via the conductivity (and surface

---

[38]One reference experimentally determined that the resistance to ground for the human body should be less than 50 MΩ in order to keep the body from charging to voltages greater than 100 V.

resistivity) of the finish. However, it is usually not desirable to have floors with a high conductivity. They can be expensive and potentially dangerous. Furthermore, if the floor is too conductive, sparks can occur between charged objects and the floor. These sparks must be avoided, especially in an explosive environment.

During humid conditions, film production on the floor can occur, which reduces the surface resistivity of the floor. Wax buildup and cleanliness also affect the floor's performance. According to product literature, with proper care, an antistatic floor finish can keep the voltage level of workers to 200 V at 50% humidity.

Since tribocharging is involved between the floor and moving objects, the surface resistivity is not the only measure of the effectiveness of the floor in limiting charge buildup. Tribocharging is a complicated and not always predictable phenomena that is very sensitive to contaminates.

2.30 Pink, Black, and Shielded Bags

Discuss the advantages and disadvantages of using pink, black, and shielded bags for protecting electronic components and devices from ESD. [Koyler; Greason, '87; Maissel; Static]

Special bags (and other containers such as tote boxes, bins, and tubes) are used when assembling and shipping sensitive electronic components and devices to protect them from accidental electrostatic discharge. ESD can occur from sources outside the bag, inside the bag, and from the bag itself. It is therefore difficult to classify the effectiveness of a bag by using just one parameter or test.

Probably the most common method of classifying a bag is through its surface resistivity measured in Ω (or $\Omega/\square = \Omega/\text{sq}$ as discussed previously). However, the discharge or decay of charge on a bag can occur through other paths besides along its surface (e.g., through its bulk and along conductive internal layers). Also, the surface resistivity was probably determined using only one test voltage. Unfortunately, the materials used for many bags are not ohmic. This implies that the resistivity will be a function of the applied voltage and electric field, as well as the charge placed on the material's surface. The electric field and quantity of charge generated via ionization of the medium between a charged object and bag will vary based on many factors including the shape of the object. Obviously, these factors limit the usefulness of using just one value for the surface resistivity in classifying a bag's effectiveness. Another factor complicating the use of a dc surface resistance measurement is that ESD is not a dc process. It is dynamic with a wide range of frequency components. A high-frequency surface resistance can be substantially different from a low-frequency surface resistance.

Another method of characterizing a bag is to introduce charge on the bag and then determine, via the bag's voltage, the time required for most of the charge to dissipate to a nearby ground. (The charge on an object connected to a known voltage can be determined through its capacitance to ground from $Q = CV$.) By comparing discharge times, the discharge effectiveness of different bags can be compared. The obvious advantage of this time-based approach is that it is more illuminating to know the time required for charge to dissipate from the bag than the resistivity of the surface of the bag. Also, charge dissipation can include both surface and volume mechanisms. There are, however, several disadvantages of this approach. The rate of charge decay from the surface of an object is a function of the (i) electrical properties of the bag, (ii) shape and size of the bag, (iii) shape, size, and location of nearby grounds, and (iv) shape, size, location, and electrical properties of other objects. (See the static-dissipative mat discussion.) Although the rates of discharge can be compared in a known standardized setting, they will not necessarily be the same as the discharge in the bag's real environment. Also, as with the surface resistivity, for nonohmic bag materials, the rate of discharge will be a function of the test voltage used to charge the bag.

Table 2.10 classifies materials based on their surface and bulk resistivity. It is important to state that these range definitions are not universal. Various associations and vendors may have different range definitions. Furthermore, the surface resistivity is a measure of the ability of charge to move on and through the material. However, a material with a low resistivity (a moderate conductivity) does not necessarily have a low tendency to tribocharge. For insulators, or nonconductors, the surface resistivity is often taken as greater than 10^{12} Ω.

TABLE 2.10 Material Classification Based on Surface and Bulk Resistivity

Classification	Surface Resistivity	Bulk Resistivity
Shielding	0.01–1 Ω	< 10 Ω-m
Conductive	0–10^5 Ω	< 10^2 Ω-m
Static dissipative	10^5–10^9 Ω	10^2–10^9 Ω-m
Antistatic	10^9–10^{14} Ω	
Insulative	> 10^{14} Ω	

The charge dissipation time can be roughly related to the surface resistivity if the decay rate or time constant of the discharge is ohmic and equal to RC. The resistance, R, is mainly determined by the "net" resistance of the bag (and not by the path to the ground). The capacitance, C, is between the bag and ground. If a flat bag of cross-sectional dimensions w by l_{th} of constant resistivity ρ_s is parallel with, and a distance d from, a flat ground plane, this time constant is

$$\tau = RC \approx \left(\frac{\rho_s l_{th}}{w} \right) \left(\frac{\varepsilon_o l_{th} w}{d} \right) = \frac{\varepsilon_o \rho_s l_{th}^2}{d} \qquad (2.197)$$

According to this simple model, as the length of the bag increases, the charge decay time increases. As the distance d increases, however, the charge decay time decreases. The electric field between the bag and ground plane decreases as their separation distance increases. As a result, as this distance increases, the attraction of the charge on the bag to the ground plane decreases. Thus, the charge on the bag can more easily dissipate through the bag to the grounding conductor (at one end of the bag). Negligible charge dissipates through the leakage resistance in the air between the bag and ground plane. Again, this model is a dramatic simplification. For example, the resistance "seen" by each charge to the ground connection will vary based on the distance and position of the ground connection(s). Although an ideal grounding conductor or strap is shown in Figure 2.89, frequently the contact resistance between the bag and ground connection(s) will vary with bag dimensions, contact pressure, surface contaminates, and surface roughness. The connection to ground could also be a human finger touching the bag. The resistance and capacitance (and inductance) to ground is distributed similar to a lossy transmission line. For a flat 15 by 15 cm bag surface, a distance of 1 mm from a flat ground, with a resistivity of 10^5 Ω, the time constant is about 20 μs. For a resistivity of 10^9 Ω, which is the upper resistivity range for static-dissipative materials, $\tau \approx 0.2$ sec.[39] Using this simple model, as the size of the bag increases, the time of discharge increases. After 3–5τ, most of the charge is considered dissipated for an exponential rate of decay. The actual time required to discharge a bag, though, is a function of the acceptable quantity of charge on the bag.

The effectiveness of an ESD-protective bag is usually determined by its ability to protect its contents against external direct conductive discharges, external (often electric) fields, and radiated and conductive

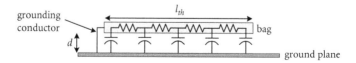

FIGURE 2.89 Modeling an ESD bag as a distributed RC circuit.

---

[39]One reference indicated that for resistivities less than 10^{12} Ω, surface charge should decay within 2 sec or less. As seen from this discussion, the shape and size of the object and its environment affect this decay time.

emissions due to the tribocharging of the bag with itself and with internal components. There are several categories of bags designed to protect against these ESD threats: "pink," "black," and shielded.

A "pink" or "pink poly" bag consists of material with a resistivity ranging from 10^9 to 10^{11} Ω. As is seen in Table 2.10, pink bags are in the antistatic category and for this reason are referred to as antistatic bags. The plastic polyethylene used to manufacture pink bags is actually clear. These bags are frequently colored to distinguish them from standard insulative bags. A chemical antistatic agent is applied to the polyethylene to lower its resistance. When sensitive electronic components are placed in an ordinary plastic bag (e.g., bubble bag), the rubbing of the bag with itself (e.g., when opening and closing the bag) and with internal and external objects tends to build up charge on both the bag and internal components. When components slide around inside the bag, the components and bag can become charged. This charge can build to such a critical level that arcing or air breakdown occurs, which can damage some components. The antistatic material in the pink bag reduces the tribocharging. Also, it allows the charge on the bag to distribute more easily over the bag, and it allows for the charge to more easily be dissipated when connected to a ground (or other objects of lower potential). The antistatic agent for antistatic bags can be topically treated or treated on the polyethylene layers. Some bags can have antistatic properties on both sides of the bag.

Pink or antistatic bags do not protect internal components from external conductive or radiated disturbances. A strong external electric (or magnetic) field or conductive discharge will penetrate the bag. Although the bag will have some effect on the external electric field, since its dielectric constant is not equal to its free-space surroundings, the field will still penetrate into the bag. Pink bags are used to transport ESD-sensitive components into areas where the electric field environment is controlled and where there is no danger of a direct conductive discharge to the bag. They are also used to transport ESD-*insensitive* components that are in close proximity to ESD-sensitive components. (Since the pink bags are antistatic, the bags themselves should not generate much charge or be a source of ESD.) These bags have limited reuse and storage lifetime, and they are sensitive to environmental factors such as humidity. To summarize, pink bags are for static dissipation not shielding.

Because of the unpredictable nature of tribocharging, the surface resistivity is not the only measure of the tendency of a material to dissipate charge or generate tribocharge. (Tribocharging is very sensitive to contaminates.) A material that is antistatic is one that tends not to produce significant tribogenerated charge with itself or other objects. Although 10^9 Ω would not be considered a low resistivity, a grounded object constructed of a material with a low resistivity will tend to dissipate very quickly any charge generated or placed on it. Also, a material with a low resistivity will have a greater electric field (and plane wave) shielding ability.

Another category of bag is the dissipative bag. They have resistivities from 10^4 to 10^{11} Ω. For the lower resistivities, the charge of the bag can be dissipated more quickly to ground, relative to the pink bag. They have some of the advantages and disadvantages associated with the "black" bag.

The resistivity of "black" bag material ranges from 10^3 to 10^4 Ω. Black, or Faraday-cage, bags are constructed of carbon-loaded polyethylene, a conductive material. Black bags are unfortunately opaque and black and, thus, their contents cannot be determined without removal from the bag. Charge placed on grounded black bags will dissipate quickly to ground. Also, the black bag will provide some shielding of external *electric* fields (and plane waves) depending on the thickness and conductivity of the bag. The charge on the conductive bag can rearrange itself to shield the contents of the bag from external fields.

Although tribocharging between the black bag and internal and external objects (or with the bag itself) is not usually an issue, the high conductivity of the bag allows for arcing to it. A charged object approaching the bag can induce charge on the bag. The electric field from this charge distribution and the ability of the free charge on the bag to respond quickly can produce a sparkover to the bag. Since the internal contents can be in direct contact with the metallic bag, even an external discharge can pass directly to the contents. This bag is sometimes used when there is no danger of a charged object approaching or touching the bag.

The last category is the shielded, metallized shielded, or static shielded bag. A shielded bag will often consist of two layers. The first outer layer is of high conductivity: resistivity of 10^4 Ω or less. The second inner layer is much less conductive, with resistivities ranging from 10^6 to $10^{10}\ \Omega$. A shielded bag has the advantages of the pink and black bags. There are two versions of the shielded bag: metal-in and metal-out.[40] The metal-in or buried metal version consists of a conductive material (e.g., aluminum, nickel, or copper) buried between polyester and polyethylene. The metal-out version has the metal on the outside of the bag sometimes with an abrasion coating. The metal-out version is generally better in dissipating charge on the exterior of the bag. The metal-out version is also easier to see through. The metal-in version is less likely to produce a spark to outside objects since its outer surface is less conductive.

For metal-out bags, on the interior layer of polyethylene, an antistatic agent is applied on its inner surface to reduce tribocharging. The outer layer is often a metallized polyester with an antistatic coating. The inner layer helps reduce charge transfer from the conductive outer layer to the bag's contents. These bags are semitransparent and thus allow determination of their contents without their removal from the bag. Not surprisingly, shielded bags are (currently) more costly than pink or black bags. Versions of these bags also exist that contain a moisture vapor barrier. This feature helps reduce the effect of moisture absorption in both the short term and long term (e.g., storage). When an ESD-sensitive electronic component or device is carried into an uncontrolled area, the shielded bag should be used.

Other factors that should be considered when selecting an electrostatic bag are the bag's reusability, cost, recycleability, strength, resistivity variation with humidity, transparency, size, and style. An informal test used to determine the reusability of a bag is to measure the resistance of the bag between several different contact points after it has been crumpled several times.

2.31   Static-Dissipative Work Surfaces

A static-dissipative mat of thickness d, conductivity σ, and relative permittivity ε_r is placed in direct contact with a grounded metallic plane (as shown in Figure 2.90). If a charge of density ρ_S is on the surface of the mat, describe the time variation of this charge. [Crowley, '90]

Static-dissipative work surfaces are designed to remove or dissipate charge to ground. For example, the placement and removal of a Styrofoam cup (an insulating material) or sweater on a surface can place charge on that surface. In ESD-sensitive environments, this charge should be discharged to ground. The derivation of the dissipation relationship for a thin mat in contact with a ground plane is straightforward.

The initial volume charge density in the mat is assumed zero; therefore, for ohmic mats the volume charge density remains zero. The current density in the mat below the surface charge is a function of the conductivity and electric field in the mat (assuming the mat is thin so that the field is uniform):

$$\vec{J} = \sigma\vec{E} = \sigma E_y \hat{a}_y$$

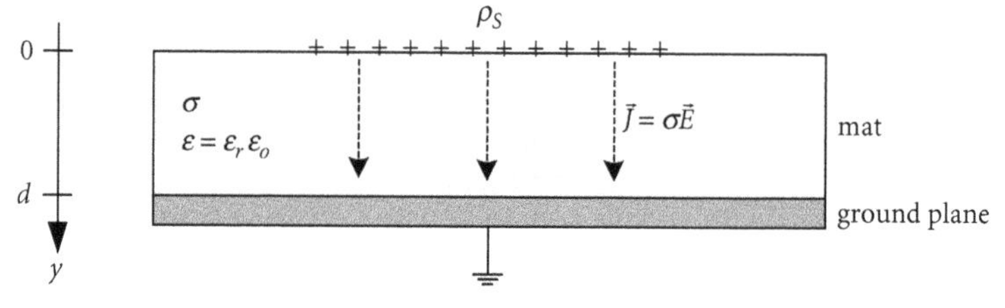

FIGURE 2.90   Surface charge residing on a static-dissipative mat in direct contact with a large ground plane.

---

[40]Probably after the publication of this book, other shielded bags will become available that contain even more layers.

By using this equation, it is assumed that the mat is ohmic in nature. The electric field in the mat is obtained through the electromagnetic boundary condition

$$D_{n2} - D_{n1} = \rho_S$$

where D_{n2} is the flux density in the mat and D_{n1} is the flux density above the mat. The charge on the mat-air interface is a surface charge. Although there is an electric field both above ($y < 0$) the mat in the air and in ($0 < y < d$) the mat, the field is much more intense in the mat. The ground plane at $y = d$ is close to the surface charge while the ground above the surface charge (not shown) is assumed comparatively far away. Therefore,

$$D_{n2} = \varepsilon_r \varepsilon_o E_{n2} = \varepsilon_r \varepsilon_o E_y \approx \rho_S \quad \text{if } D_{n2} \gg D_{n1}$$

and

$$J_y = \sigma E_y = \sigma \frac{\rho_S}{\varepsilon_r \varepsilon_o}$$

The boundary condition involving the conservation of charge

$$J_{n2} - J_{n1} = -\frac{d\rho_S}{dt}$$

may also be applied realizing that the conductivity of the air above the material is zero and hence $J_{n1} = 0$:

$$\sigma \frac{\rho_S}{\varepsilon_r \varepsilon_o} = -\frac{d\rho_S}{dt}$$

The solution to this simple differential equation is the standard decaying exponential:

$$\int_{\rho_S(0)}^{\rho_S(t)} \frac{d\rho_S}{\rho_S} = -\int_0^t \frac{\sigma}{\varepsilon_r \varepsilon_o} dt \quad \Rightarrow \ln\left[\frac{\rho_S(t)}{\rho_S(0)}\right] = -\frac{\sigma}{\varepsilon_r \varepsilon_o} t$$

$$\rho_S(t) = \rho_S(0) e^{-\frac{\sigma}{\varepsilon_r \varepsilon_o} t} = \rho_S(0) e^{-\frac{t}{\tau_m}} \quad \text{where } \tau_m = \frac{\varepsilon_r \varepsilon_o}{\sigma} \tag{2.198}$$

The variable τ_m is referred to as the time constant for the mat, and $\rho_S(0)$ is the initial charge density along the surface of the mat. Equation (2.198) is not a function of the dimensions of the mat since the electric field and current density are in the same direction.

The relative permittivity of most mats is in the range of 3 to 10. The conductivity, however, can vary from at least 10^{-12} 1/Ω-m to 100 1/Ω-m (or more). If the response time of a human is 0.5 sec, the time constant should be several times less than 0.5 sec. For a mat of relative permittivity equal to 5, the range of conductivities required to dissipate the charge by three time constants is about

$$3\tau = 3\left(\frac{5\varepsilon_o}{\sigma}\right) < 0.5 \quad \Rightarrow \sigma > 3 \times 10^{-10} \text{ 1/}\Omega\text{-m}$$

If the mat's conductivity is smaller than this value, the time required to dissipate the charge on the surface through the mat may be too large. However, if the conductivity of the mat is too large, the mat can be

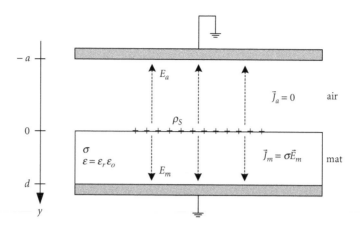

FIGURE 2.91 Ground plane both above and below a charged static-dissipative mat.

rapidly inductively charged. Recall that a charged object above a conducting plane induces charge of the opposite sign on the plane. The electric field between the charged object and conducting plane may be sufficiently large to cause an electrical discharge. Therefore, a (homogeneous) mat should not be a good conductor such as stainless steel, copper, and some carbon-embedded plastics. Although even real insulators can be inductively charged, the time required for the charge on the insulating surface to "gather" below the object is usually large compared to the response time of objects approaching the mat. The conductivity of the mat should be large enough to dissipate quickly any charge on its surface but not too large as to cause an inductive-related discharge.

When a ground plane exists both above and below the mat, as shown in Figure 2.91, then the surface charge decays at a different rate. The distance $d + a$ is assumed small compared to the other dimensions so that end effects can be neglected. For this more general case, the algebra is slightly more involved. As before, the charge conservation equation at the air-mat interface is used:

$$J_{nm} - J_{na} = -\frac{d\rho_S}{dt}$$

Again, in the air above the mat the current density is zero and $J_{na} = 0$. However, now the upper grounded electrode at $y = -a$ will have some effect on the electric field distribution. Assuming the initial volume charge densities in both the mat and air are again zero, the volume charge in both of these regions will always remain zero (assuming the mat is ohmic in nature). Therefore, Laplace's equation can be used to determine the general potential functions in both regions:

$$\nabla^2 \Phi_m = \frac{d^2\Phi_m}{dy^2} = 0 \quad \Rightarrow \Phi_m(y,t) = C_1 y + C_2$$

$$\nabla^2 \Phi_a = \frac{d^2\Phi_a}{dy^2} = 0 \quad \Rightarrow \Phi_a(y,t) = C_3 y + C_4$$

Since there are four constants of integration, four boundary conditions are required. Two of the boundary conditions are simple since the potential is zero at the two electrodes:

$$\Phi_m(d,t) = 0, \quad \Phi_a(-a,t) = 0$$

Even with a single layer of surface charge present, the potential cannot jump or change suddenly over zero distance. In other words, the potentials must be equal near either side of the interface

$$\Phi_m(0,t) = \Phi_a(0,t)$$

When the surface charge is not varying with time, the boundary condition relating the normal flux densities to the surface charge is used:

$$D_{nm} - D_{na} = \rho_S \quad @\ y=0 \tag{2.199}$$

Since the surface charge is varying with time, the charge conservation equation is used with this flux density boundary condition:

$$J_{nm} - J_{na} = -\frac{d\rho_S}{dt} = -\frac{d}{dt}(D_{nm} - D_{na}) = -\frac{d}{dt}(\varepsilon_r \varepsilon_o E_{ym} - \varepsilon_o E_{ya}) \quad @\ y=0$$

As previously stated, because the conductivity of the air above the mat is zero, there is no conduction current in the upper region and $J_{na} = 0$. Using $\vec{E} = -\nabla\Phi$ and $\vec{J} = \sigma\vec{E}$, the final boundary condition can be written entirely in terms of the potentials:

$$J_{ym} = \sigma E_{ym} = -\varepsilon_r \varepsilon_o \frac{dE_{ym}}{dt} + \varepsilon_o \frac{dE_{ya}}{dt} \quad @\ y=0$$

$$-\sigma \frac{d\Phi_m}{dy} = \varepsilon_r \varepsilon_o \frac{d}{dt}\left(\frac{d\Phi_m}{dy}\right) - \varepsilon_o \frac{d}{dt}\left(\frac{d\Phi_a}{dy}\right) \quad @\ y=0 \tag{2.200}$$

The first three boundary conditions will be applied first:

$$C_1 d + C_2 = 0, \ C_3(-a) + C_4 = 0, \ C_2 = C_4$$

resulting in

$$C_1 = -\frac{aC_3}{d}, \ C_4 = aC_3, \ C_2 = aC_3 \tag{2.201}$$

Application of the final boundary condition (2.200) is more complicated so a few steps will be shown. Substituting the expressions for the potentials, realizing that the constants of integration are only constant with respect to the variable y, and using (2.201)

$$-\sigma C_1 = \varepsilon_r \varepsilon_o \frac{dC_1}{dt} - \varepsilon_o \frac{dC_3}{dt} \quad @\ y=0$$

$$\frac{a\sigma C_3}{d} = -\frac{a\varepsilon_r \varepsilon_o}{d}\frac{dC_3}{dt} - \varepsilon_o \frac{dC_3}{dt} \quad @\ y=0$$

$$\frac{dC_3}{dt} = -\frac{a\sigma C_3}{d\varepsilon_o\left(\frac{a\varepsilon_r}{d}+1\right)} \quad @\ y=0$$

The solution of this differential equation is simply obtained:

$$\int_{C_3(0)}^{C_3(t)} \frac{dC_3}{C_3} = -\int_0^t \frac{a\sigma dt}{d\varepsilon_o\left(\frac{a\varepsilon_r}{d}+1\right)}$$

$$\ln\left[\frac{C_3(t)}{C_3(0)}\right] = -\frac{a\sigma t}{d\varepsilon_o\left(\frac{a\varepsilon_r}{d}+1\right)} \quad \Rightarrow C_3(t) = C_3(0)\, e^{-\frac{a\sigma}{d\varepsilon_o\left(\frac{a\varepsilon_r}{d}+1\right)}t}$$

To determine the new constant of time integration, $C_3(0)$, and Equations (2.199) and (2.201) are utilized:

$$\varepsilon_r\varepsilon_o E_{ym} - \varepsilon_o E_{ya} = \rho_S(0) \quad @\ y=0, t=0$$

$$-\varepsilon_r\varepsilon_o \frac{d\Phi_m}{dy} + \varepsilon_o \frac{d\Phi_a}{dy} = \rho_S(0) \quad @\ y=0, t=0$$

$$\frac{a\varepsilon_r\varepsilon_o}{d}C_3(0) + \varepsilon_o C_3(0) = \rho_S(0) \quad \Rightarrow C_3(0) = \frac{\rho_S(0)}{\varepsilon_o\left(1+\frac{a\varepsilon_r}{d}\right)}$$

To determine the surface charge density variation with the time, the difference in the flux densities at the interface is required:

$$\rho_S(t) = \varepsilon_r\varepsilon_o E_{ym} - \varepsilon_o E_{ya} = -\varepsilon_r\varepsilon_o \frac{d\Phi_m}{dy} + \varepsilon_o \frac{d\Phi_a}{dy} = -\varepsilon_r\varepsilon_o C_1(t) + \varepsilon_o C_3(t)$$

$$= \frac{a\varepsilon_r\varepsilon_o}{d}C_3(t) + \varepsilon_o C_3(t) = C_3(t)\,\varepsilon_o\left(1+\frac{a\varepsilon_r}{d}\right) \qquad (2.202)$$

$$= C_3(0)\,e^{-\frac{t}{\tau_c}}\varepsilon_o\left(1+\frac{a\varepsilon_r}{d}\right) = \rho_S(0)\,e^{-\frac{t}{\tau_c}}$$

where the time constant for the surface charge decay is

$$\tau_c = \frac{d\varepsilon_o\left(\frac{a\varepsilon_r}{d}+1\right)}{a\sigma} = \frac{\varepsilon_r\varepsilon_o}{\sigma}\left(1+\frac{d}{a\varepsilon_r}\right) = \tau_m\left(1+\frac{d}{a\varepsilon_r}\right) \qquad (2.203)$$

The variable τ_m is the time constant of the decay of the surface charge on the mat if the upper grounded electrode was far away ($a\varepsilon_r \gg d$). (If volume charge were present in the ohmic mat, its decay rate would always be equal to τ_m.) The additional electrode has increased the time constant and, hence, slowed the rate of decay of the surface charge. Therefore, the decay of *surface* charge on an ohmic material is a function of the electrical properties of the material and the shape, size, location, and electrical properties of nearby objects. Electric fields from other sources and electrodes can have an influence on the surface charge.

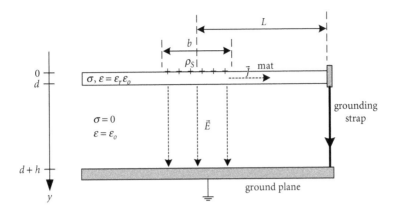

FIGURE 2.92 Static-dissipative mat located a distance of h from the ground plane. The mat is connected to the ground plane at one end through a grounding strap.

To simplify the previous analysis, the static-dissipative mat was assumed homogeneous. Commercial mats for both table and floor applications are available with one material or multiple layers of different materials. A single sheet of vinyl or rubber is mechanically simple and durable. A two-layer mat would allow the top layer to be chemically resistant, for example, while allowing the lower layer to act as a foam cushion. A three-layer mat could include a conductive inner layer. Hard mats constructed of fiberglass are also available.

As shown in Figure 2.92, a static-dissipative mat of thickness d, conductivity σ, and relative permittivity ε_r is placed at a distance h above a grounded metallic plane. The mat is connected to this grounded plane at one edge. If a charge of density ρ_S is on the surface of the mat, determine the time constant that describes the variation of the charge with respect to time. [Crowley, '90]

In many situations, a static-dissipative mat is not in direct contact with a ground. Instead, it is connected to a ground via a grounding strap at one end. In this analysis, it is simpler to work with an equivalent RC model rather than using the charge conservation equation. Since the region below the mat is not conductive, the charge must travel along the mat to the grounding strap to be dissipated. Obviously, since R's will be used to model the various parts of the system, this derivation assumes the materials are ohmic in nature. In general, insulators are not ohmic in nature, and the derived equations should be used with caution for these materials.

The equivalent RC model for this system, shown in Figure 2.93, is obtained by assuming that the surface charge is distributed over some region of length b and width w located at an average distance of

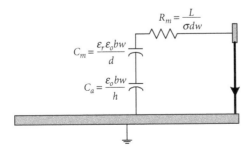

FIGURE 2.93 Circuit model for the situation in Figure 2.92.

L from the grounding strap. The time constant for this system is

$$\tau_c = R_m C_{eq} = R_m \left(\frac{C_m C_a}{C_m + C_a} \right) = \frac{L}{\sigma dw} \left(\frac{\dfrac{\varepsilon_r \varepsilon_o bw}{d} \dfrac{\varepsilon_o bw}{h}}{\dfrac{\varepsilon_r \varepsilon_o bw}{d} + \dfrac{\varepsilon_o bw}{h}} \right) = \frac{\varepsilon_r \varepsilon_o L}{\sigma d} \left(\frac{b}{\varepsilon_r h + d} \right) \tag{2.204}$$

Unlike (2.198), which is only a function of the electrical properties of the mat, this time constant is also a function of the dimensions of the mat. This is true since the current density, which is parallel to the mat, and the main component of the electric field, which is perpendicular to the mat, are not in the same direction. If the mat is thin compared to its distance to the ground plane, this time constant reduces to

$$\tau_c \approx \frac{\varepsilon_r \varepsilon_o L}{\sigma d} \left(\frac{b}{\varepsilon_r h} \right) = \frac{\varepsilon_o}{\sigma} \left(\frac{Lb}{hd} \right) \quad \text{if } d \ll \varepsilon_r h \tag{2.205}$$

This result is independent of the dielectric constant of the mat. The capacitance of the air region between the mat and ground is much smaller than the capacitance of the mat. Since these capacitors are in series, the net capacitance is dominated by the mat-to-ground capacitance.

As the distance between the surface charge and ground strap, L, increases, the time required to dissipate the charge increases. As the distance between the mat and ground plane, h, increases, the time required to dissipate the charge decreases. The attraction of the charge to the ground plane decreases as h increases, and the electric field between the mat and ground decreases as h increases. The location of the ground plane relative to the mat is quite important. The system time constant can easily be several orders of magnitude greater than the material time constant:

$$\frac{\tau_c}{\tau_m} = \frac{\dfrac{\varepsilon_r \varepsilon_o L}{\sigma d} \left(\dfrac{b}{\varepsilon_r h + d} \right)}{\dfrac{\varepsilon_r \varepsilon_o}{\sigma}} = \frac{L}{d} \left(\frac{b}{\varepsilon_r h + d} \right) \approx \frac{L}{d} \left(\frac{b}{\varepsilon_r h} \right) \quad \text{if } d \ll \varepsilon_r h \tag{2.206}$$

If the charge is very near the ground strap, the time required to dissipate the charge can be much less than τ_m, while if the charge is very far from the ground strap, the time required can be much greater than τ_m. For example, assuming that $b = 10d$ and $\varepsilon_r = 5$, then

$$\frac{\tau_c}{\tau_m} = \frac{L}{d} \left(\frac{10d}{5h} \right) = \frac{2L}{h} = \begin{cases} 0.2 & \text{if } d \ll \varepsilon_r h, L = h/10 \\ 20 & \text{if } d \ll \varepsilon_r h, L = 10h \end{cases}$$

Edge grounding is sometimes avoided because of this variation in the discharge time.

For safety reasons, a current-limiting resistor is placed between the mat and ground plane. (Personnel that accidentally contact a high voltage should not be put at risk by coming in contact with a good electrical ground.) The resistance value is selected so that the maximum current passing between the workstation and ground plane is less than about 0.5 mA rms. For the ground plane mat shown in Figure 2.90, the resistor is placed between the mat and ground. For the end-grounded configuration shown in Figure 2.92, the resistor is placed in series with the grounding strap. In either case, the added resistance often has little effect on the discharging rate of the mat (assuming the mat is not too conductive).

2.32   Sugar Charge Decay

Sugar is tribocharged during mixing and transportation. In a partially filled container with grounded walls, eventually the charge in the sugar will dissipate to the walls. Does any charge flow to the sugar-air interface? Use the simple configuration shown in Figure 2.94 to determine the variation with respect to time of the volume charge in the sugar, surface charge at the interface, and electric field in the air. [Jones, '89]

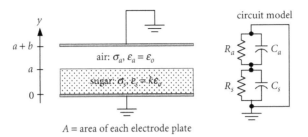

FIGURE 2.94 Simple model of charged sugar in a grounded container.

Tribocharging of materials can result in an electrostatic-based hazard. As will be seen in this analysis, both the dimensions and electrical time constant of the material affect the electric fields and the rate of the surface charge decay. A one-dimensional simple model in the Cartesian coordinate system is used so that relatively simple but yet insightful results can be obtained. In a real system, cylindrical or other shaped containers may be used with possibly the sides, bottom, and top of the container grounded.

Before Poisson's equation and boundary conditions are used to solve for the charge and electric field distributions, a circuit's model will be presented. It is important to state that the use of both resistances in the model and the exponential relationship for the charge decay imply that the sugar is ohmic; that is, it can be described using $\vec{J} = \sigma \vec{E}$. Insulating materials are frequently not adequately described by this expression. Without this assumption, however, problems of this type are very difficult to solve, at least analytically. The resistances and capacitances shown in Figure 2.94 are easily obtained using the standard expressions for parallel plate geometry (and assuming a and b are small compared to the other dimensions so that edge effects can be ignored):

$$R_s = \frac{a}{\sigma_s A}, \ C_s = \frac{k\varepsilon_o A}{a}, \ R_a = \frac{b}{\sigma_a A} = \infty, \ C_a = \frac{\varepsilon_o A}{b} \tag{2.207}$$

The conductivity of the air above the sugar is zero. For a single RC circuit, the time constant is $\tau = RC$. Therefore, the time constants of the sugar and air portions of the circuit are

$$\tau_s = R_s C_s = \frac{k\varepsilon_o}{\sigma_s}, \ \tau_a = R_a C_a = \infty \tag{2.208}$$

However, the time constant of the entire circuit is

$$\tau_c = \left(R_s \| R_a\right)\left(C_s + C_a\right) = \left(\frac{a}{\sigma_s A}\right)\left(\frac{k\varepsilon_o A}{a} + \frac{\varepsilon_o A}{b}\right)$$

$$= \left(\frac{k\varepsilon_o}{\sigma_s}\right)\left(1 + \frac{a}{bk}\right) = \tau_s \left(1 + \frac{a}{bk}\right) \tag{2.209}$$

After examining the circuit in Figure 2.94, it is clear that all four components are in parallel. Notice that the overall circuit time constant is not merely the sum of the two individual time constants. It is greater than the time constant of the sugar portion of the circuit. However, if b, the spacing between the upper ground electrode and sugar-air interface is very large compared to a, then $\tau_c \approx \tau_s$. The charge in the sugar will mostly be attracted to the lower grounded electrode when the upper grounded electrode is far away. For a fast discharge rate to the ground plane, the time constant of the sugar should be small. The time constant of the sugar is small for large sugar conductivities, as expected.

The volume charge density inside the sugar will be assumed initially uniformly distributed with a constant value of $\rho(0)$. Since the sugar is also assumed ohmic, or at least approximately, the charge decay everywhere in the sugar will be exponential in nature:

$$\rho(t) = \rho(0)e^{-\frac{\sigma_s}{k\varepsilon_o}t} = \rho(0)e^{-\frac{t}{\tau_s}} \tag{2.210}$$

As previously discussed, external fields, boundaries, and the initial shape of the volume charge do not affect the distribution or rate of decay of the *volume* charge inside the (ohmic) sugar! To determine the expressions for the potential and electric field in both the sugar and air, Poisson's equation can be solved for each of the regions. For this simple geometry neglecting end effects, the potential is only varying in the y direction. Furthermore, there is no volume charge in the air region. The general solution for Poisson's equation is solved in the standard way:

$$\nabla^2\Phi_s = \frac{d^2\Phi_s}{dy^2} = -\frac{\rho(t)}{k\varepsilon_o} = -\frac{\rho(0)e^{-\frac{t}{\tau_s}}}{k\varepsilon_o} \Rightarrow \Phi_s(y,t) = -\frac{\rho(0)e^{-\frac{t}{\tau_s}}}{2k\varepsilon_o}y^2 + C_1 y + C_2$$

$$\nabla^2\Phi_a = \frac{d^2\Phi_a}{dy^2} = 0 \Rightarrow \Phi_a(y,t) = C_3 y + C_4$$

Since there are four constants of integration, four boundary conditions are required. Two of the boundary conditions are simple since the potential is zero for all time at the two electrodes:

$$\Phi_s(0,t) = 0, \ \Phi_a(a+b,t) = 0 \tag{2.211}$$

Even with surface charge present (but not a double layer of surface charge), the potential cannot jump or change suddenly over zero distance, so the potentials near the interface must be equal for all time:

$$\Phi_s(a,t) = \Phi_a(a,t) \tag{2.212}$$

When the surface charge is not varying with time, the boundary condition relating the normal flux densities to the surface charge is used:

$$D_{na} - D_{ns} = \rho_S \ @ y = a$$

Because the surface charge is varying with time, the boundary condition based on the charge conservation equation is used with this flux density boundary condition:

$$J_{na} - J_{ns} = -\frac{d\rho_S}{dt} = -\frac{d}{dt}(D_{na} - D_{ns}) = -\frac{d}{dt}(\varepsilon_o E_{ya} - k\varepsilon_o E_{ys}) \ @ y = a \tag{2.213}$$

Since the conductivity of the air above the sugar is zero, there is no conduction current in the upper region and $J_{na} = 0$. Using $\vec{E} = -\nabla\Phi$ and $\vec{J} = \sigma\vec{E}$, Equation (2.213) can be written entirely in terms of the potentials:

$$J_{ys} = \sigma E_{ys} = \varepsilon_o\frac{dE_{ya}}{dt} - k\varepsilon_o\frac{dE_{ys}}{dt} \ @ y = a$$

$$-\sigma_s\frac{d\Phi_s}{dy} = -\varepsilon_o\frac{d}{dt}\left(\frac{d\Phi_a}{dy}\right) + k\varepsilon_o\frac{d}{dt}\left(\frac{d\Phi_s}{dy}\right) \ @ y = a \tag{2.214}$$

The first three boundary conditions, Equations (2.211) and (2.212), will be applied first

$$-\frac{\rho(0)e^{-\frac{t}{\tau_s}}}{2k\varepsilon_o}(0)^2 + C_1(0) + C_2 = 0, \ C_3(a+b) + C_4 = 0$$

$$-\frac{\rho(0)e^{-\frac{t}{\tau_s}}}{2k\varepsilon_o}a^2 + C_1a + C_2 = C_3a + C_4$$

resulting in

$$C_2 = 0, \ C_4 = -(a+b)C_3, \ C_1 = -\frac{C_3b}{a} + \frac{a\rho(0)e^{-\frac{t}{\tau_s}}}{2k\varepsilon_o} \tag{2.215}$$

Applying the final boundary condition, (2.214), requires more work:

$$-\sigma_s\frac{d}{dy}\left[-\frac{\rho(0)e^{-\frac{t}{\tau_s}}}{2k\varepsilon_o}y^2 + C_1y + C_2\right] = -\varepsilon_o\frac{d}{dt}\left[\frac{d}{dy}(C_3y + C_4)\right]$$

$$+k\varepsilon_o\frac{d}{dt}\left\{\frac{d}{dy}\left[-\frac{\rho(0)e^{-\frac{t}{\tau_s}}}{2k\varepsilon_o}y^2 + C_1y + C_2\right]\right\} \ @ \ y = a$$

The "constants" of integration are constant with respect to y not t:

$$\sigma_s\frac{\rho(0)e^{-\frac{t}{\tau_s}}}{k\varepsilon_o}y - \sigma_sC_1 = -\varepsilon_o\frac{dC_3}{dt} + k\varepsilon_o\frac{d}{dt}\left[-\frac{\rho(0)e^{-\frac{t}{\tau_s}}}{k\varepsilon_o}y + C_1\right] \ @ \ y = a$$

Substituting the expression for C_1 given in Equation (2.215), which is given as a function of C_3 and the charge density, and simplifying,

$$\frac{dC_3}{dt} = -\frac{\sigma_sC_3}{\varepsilon_o\left(\dfrac{a}{b}+k\right)}$$

The time constant relation $\tau_s = k\varepsilon_o/\sigma_s$ was used to reduce to this form. The solution to this differential expression is easily obtained:

$$\int_{C_3(0)}^{C_3(t)}\frac{dC_3}{C_3} = -\int_0^t\frac{\sigma_s\,dt}{\varepsilon_o\left(\dfrac{a}{b}+k\right)}$$

$$\ln\left[\frac{C_3(t)}{C_3(0)}\right] = -\frac{\sigma_st}{\varepsilon_o\left(\dfrac{a}{b}+k\right)} \Rightarrow C_3(t) = C_3(0)e^{-\frac{\sigma_s}{\varepsilon_o\left(\frac{a}{b}+k\right)}t} = C_3(0)e^{-\frac{t}{\tau_c}}$$

Therefore, all of the constants of integration are a function of the entire circuit's time constant. To determine the initial value for C_3 at $t = 0$, $C_3(0)$, the initial surface charge along the sugar-air interface must be specified. Assuming this initial surface charge is equal to zero and employing the standard, normal, flux density boundary condition,

$$\varepsilon_o E_{ya} - k\varepsilon_o E_{ys} = 0 \quad @ \; y = a, \rho_S(0) = 0, t = 0$$

$$-\varepsilon_o \frac{d\Phi_a}{dy} + k\varepsilon_o \frac{d\Phi_s}{dy} = 0 \quad @ \; y = a, \rho_S(0) = 0, t = 0$$

$$-\varepsilon_o C_3(0) e^{-(0)} + k\varepsilon_o \left[-\frac{\rho(0)}{k\varepsilon_o} a - \frac{C_3(0) e^{-(0)} b}{a} + \frac{a\rho(0) e^{-(0)}}{2k\varepsilon_o} \right] = 0 \;\; \Rightarrow C_3(0) = -\frac{\rho(0) a}{2\varepsilon_o \left(1 + \dfrac{bk}{a} \right)}$$

With all of the constants of integration obtained, the expressions for the electric field in both the sugar and air and the expression for the surface charge can be determined:

$$\vec{E}_s(t) = -\frac{d\Phi_s(y,t)}{dy} \hat{a}_y = \left[\frac{\rho(0) e^{-\frac{t}{\tau_s}}}{k\varepsilon_o} y - C_1 \right] \hat{a}_y$$

$$= \left[\frac{\rho(0) e^{-\frac{t}{\tau_s}}}{k\varepsilon_o} y - \frac{b}{a} \frac{\rho(0) a e^{-\frac{t}{\tau_c}}}{2\varepsilon_o \left(1 + \dfrac{bk}{a} \right)} - \frac{a\rho(0) e^{-\frac{t}{\tau_s}}}{2k\varepsilon_o} \right] \hat{a}_y \tag{2.216}$$

$$= \left[\frac{\rho(0) e^{-\frac{t}{\tau_s}}}{k\varepsilon_o} \left(y - \frac{a}{2} \right) - \frac{b\rho(0) e^{-\frac{t}{\tau_c}}}{2\varepsilon_o \left(1 + \dfrac{bk}{a} \right)} \right] \hat{a}_y$$

$$\vec{E}_a(t) = -\frac{d\Phi_a(y,t)}{dy} \hat{a}_y = -C_3(0) e^{-\frac{t}{\tau_c}} \hat{a}_y = \frac{a\rho(0) e^{-\frac{t}{\tau_c}}}{2\varepsilon_o \left(1 + \dfrac{bk}{a} \right)} \hat{a}_y \tag{2.217}$$

$$\rho_S(t) = D_{ya}(t) - D_{ys}(t) = \varepsilon_o E_{ya}(t) - k\varepsilon_o E_{ys}(t) \quad @ \; y = a$$

$$= \varepsilon_o \frac{a\rho(0) e^{-\frac{t}{\tau_c}}}{2\varepsilon_o \left(1 + \dfrac{bk}{a} \right)} - k\varepsilon_o \left[\frac{a\rho(0) e^{-\frac{t}{\tau_s}}}{2k\varepsilon_o} - \frac{b\rho(0) e^{-\frac{t}{\tau_c}}}{2\varepsilon_o \left(1 + \dfrac{bk}{a} \right)} \right] \quad @ \; y = a \tag{2.218}$$

$$= \frac{a\rho(0)}{2} \left(e^{-\frac{t}{\tau_c}} - e^{-\frac{t}{\tau_s}} \right) \quad @ \; y = a$$

It is clear from Equation (2.217) that the electric field in the air is not a function of y, and it decays exponentially with a time constant of τ_c not τ_s. The surface charge distribution, on the other hand, is a double exponential and a function of both the overall circuit time constant and sugar's time constant. After several τ_s and if $\tau_c \gg \tau_s$, the exponential term corresponding to the sugar time constant is small compared to the exponential term corresponding to the overall circuit time constant:

$$\rho_S(t) \approx \frac{a\rho(0)}{2} e^{-\frac{t}{\tau_c}} \quad \text{if } \tau_c \gg \tau_s, t > 3\tau_s \text{ to } 5\tau_s \qquad (2.219)$$

Under these conditions, the surface charge decays exponentially with a time constant corresponding to the overall circuit. As with the static-dissipative mats shown in Figure 2.91 and Figure 2.92, the dissipation of surface charge is a function of the environment, including the location of the electrodes. For the surface charge to dissipate quickly, the overall circuit time constant should be small. The sugar's conductivity should be large and the distance to the nearby discharging electrode in direct contact with the sugar should be small.

As the sugar tank is initially being filled, the decay rate of the sugar on the surface is essentially determined by the time constant τ_s since $b \gg a$:

$$\tau_c = \tau_s\left(1 + \frac{a}{bk}\right) \approx \tau_s \quad \text{if } bk \gg a \qquad (2.220)$$

However, as the sugar level approaches the upper, grounded tank surface, $b \ll a$, and the circuit time constant becomes very large:

$$\tau_c \approx \tau_s \frac{a}{bk} \quad \text{if } bk \ll a \qquad (2.221)$$

As the sugar approaches the upper surface or the fill fraction increases, the rate that the charge decays from the surface will decrease: the surface charge is strongly attracted to the upper grounded electrode.

It is clear from this analysis that the electric fields and surface charge are a function of both an intrinsic and extrinsic time constant. The intrinsic time constant is the standard time constant of the sugar, which is only a function of the permittivity and conductivity of the sugar. The extrinsic time constant, however, is a function of dimensions and electrical properties. The effect of any dust cloud above the sugar/air interface has not been modeled. Furthermore, if an electrical discharge does occur, other processes such as partial discharges and space-charge limiting will also influence the surface charge decay and electric fields.

2.33   Capacitance Measurement for Multiple Conductors

In most situations, there are more than two conductors. How is capacitance between multiple objects defined and measured? [Greason, '87; Ramo]

The basic definition for capacitance between two objects is given in most circuits and electromagnetic texts as

$$C = \frac{Q}{V} \qquad (2.222)$$

where Q is the total charge on either conductor and V is the potential difference between the two conductors.[41] The capacitance between two conductors can be easily measured using a capacitance meter

---

[41]This capacitance is assumed a positive quantity. If the ratio of Q to V is negative, negate the result.

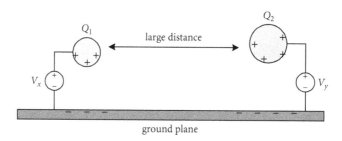

FIGURE 2.95   Charging of two conducting objects using two voltage supplies.

or bridge for values down to a few pF. When there are three or more conducting objects, the system is modeled using self and mutual capacitances.

Rather than introducing this subject by presenting the general relationships for the mutual and self capacitances for n conducting objects, the two-conductor problem with a ground plane will first be studied. After this scenario is understood, the n-object case is obtained by induction. Imagine, as shown in Figure 2.95, that two conducting objects (with a common ground plane) are charged by two voltage supplies V_x and V_y. The objects are far from each other so that the electric field lines between them are negligible. The charge that is transferred to each object is a function of the capacitance of the respective object with respect to the ground plane:

$$Q_1 = C_x V_x \quad \text{and} \quad Q_2 = C_y V_y \qquad (2.223)$$

(As long as the two voltage sources are connected to the two conductors, the conductor's potentials are fixed, independent of their position along the ground plane.) In this discussion, this situation is referred to as a two-body problem even though, including the ground plane, there are three conductors. Some other references, by including the ground plane or reference, label this a three-body problem. If both voltage supplies are now disconnected from the conductors, the charge on the conductors will remain at Q_1 and Q_2 (at least for some period of time based on the leakage resistances to the ground). However, with the supplies removed, the two conductors are now floating: the potential of floating conductors is not fixed and can change based on their environment. Imagine, as shown in Figure 2.96, that the two charged conductors are now brought sufficiently close together so that the electric field coupling between them is no longer negligible. The voltage of each conductor relative to the ground plane is a linear function of the two charges:

$$V_1 = s_{11}Q_1 + s_{12}Q_2$$
$$V_2 = s_{21}Q_1 + s_{22}Q_2 \qquad (2.224)$$

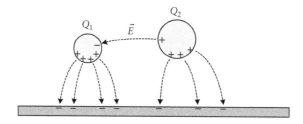

FIGURE 2.96   Two charged floating conductors. Note the electric field coupling between them.

The *coefficients of potential* s_{11}, s_{12}, s_{21}, and s_{22} are a function of the distance between the two objects, position of the ground plane, and shape of all of the objects (including the ground plane). The charge on the two objects can be easily solved for as function of the voltages:

$$Q_1 = c_{11}V_1 + c_{12}V_2$$
$$Q_2 = c_{21}V_1 + c_{22}V_2$$

(2.225)

where c_{11}, c_{12}, c_{21}, and c_{22} are the *coefficients of capacitance*, which can be written in terms of the coefficients of potential. The coefficients of capacitance are not generally equal to the standard self and mutual capacitance terms seen most elsewhere, including many other locations in this book.

These coefficients of capacitance are normally determined experimentally (since only the simplest shapes can be determined analytically). To measure these capacitances, first, all objects are connected to the ground reference with the exception of one. Second, a known voltage is applied to the one ungrounded object. Third, the charge transferred to each of the objects from the ground reference, including the ungrounded object, is measured, using an electrometer. (It is not necessary to measure all of these charges at the same time.) This process of applying a test voltage and measuring the charge transferred to each of the objects is repeated for each of the objects (except the reference conductor). For example, referring to Figure 2.97, imagine that object 1 is connected to a one volt source while the charge transferred, by induction, to the grounded object 2 and the charge transferred to object 1, by the application of the 1 V source, are measured:

$$Q_1 = c_{11}(1) + c_{12}(0) \Rightarrow c_{11} = Q_1$$
$$Q_2 = c_{21}(1) + c_{22}(0) \Rightarrow c_{21} = Q_2$$

(2.226)

The variable V_2 is zero since object 2 is grounded. From these two charge measurements, two of the four coefficients of capacitance can be determined. The coefficients c_{11} and c_{22} are referred to as the *coefficients of self capacitance* of object 1 and object 2, respectively. The remaining coefficients are referred to as the *coefficients of induction* or *coefficients of mutual capacitance* (but *not* just mutual capacitance). Also, note that the coefficients of self capacitance are always positive while the coefficients of induction are always negative or zero. The charge induced on object 2 is negative when a positive voltage is applied to object 1. If a negative voltage were applied to object 1, the induced charge on object 2 would be positive, and the coefficient of induction would still be negative. The remaining coefficients c_{12} and c_{22} can be determined by applying a known voltage (in this case 1 V) to object 2 and grounding object 1:

$$Q_1 = c_{11}(0) + c_{12}(1) \Rightarrow c_{12} = Q_1$$
$$Q_2 = c_{21}(0) + c_{22}(1) \Rightarrow c_{22} = Q_2$$

(2.227)

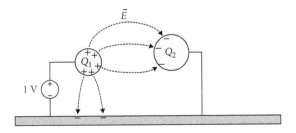

FIGURE 2.97 Determination of two of the coefficients of capacitance.

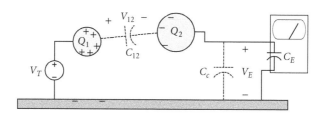

FIGURE 2.98 Parasitic capacitances during the measurement of Q_2.

This measured Q_2, for example, is not necessarily equal to the previously measured Q_2. However, from Green's reciprocation theorem, $c_{12} = c_{21}$. On the professor's blackboard, the value of the applied test voltage is not important. In practice, the range of the test (precision) voltage supply, the range of the charge-measuring electrometer, and electrical discharges around and between the objects determine the range of acceptable test voltages.

Because the capacitance between objects can often be very small, parasitic capacitances, including the capacitance of the electrometer, connecting cable, and any other nearby objects, must be considered. Also, the objects should be discharged or grounded prior to the tests. To appreciate the effect of the capacitance of the cable and electrometer, imagine that c_{21} is to be measured. As shown in Figure 2.98, the electrometer is connected between object 2 and ground so that Q_2 can be measured. The voltage source applied to conductor 1 is V_T. In Figure 2.98, C_c is the cable or connecting line capacitance and C_E is the electrometer's capacitance (the relationship between c_{12} and C_{12} will be discussed momentarily). The charge Q_2 is equal to the sum of the charge on the cable capacitance and electrometer capacitance:

$$Q_2 = Q_c + Q_E$$

Since the voltage across the cable and electrometer is the same,

$$V_E = \frac{Q_c}{C_c} = \frac{Q_E}{C_E}$$

it immediately follows that

$$Q_2 = \frac{C_c}{C_E}Q_E + Q_E = Q_E\left(1 + \frac{C_c}{C_E}\right) \approx Q_E \quad \text{if } C_E \gg C_c \tag{2.228}$$

If the cable capacitance is small compared to the electrometer's capacitance, the measured charge is approximately Q_2. For object 2 to be nearly at ground potential, $V_E \approx 0$. Using KVL,

$$-V_T + V_{12} + V_E = 0 \quad \therefore V_T \approx V_{12} \quad \text{if } V_E \approx 0$$

In other words, nearly all of the test voltage drops across the capacitance between the two conductors. Using a capacitor voltage divider,

$$V_{12} = V_T \frac{C_c + C_E}{C_c + C_E + C_{12}} \approx V_T \quad \text{if } C_c + C_E \gg C_{12}$$

Therefore, if $C_E \gg C_c$ and $C_E \gg C_{12}$ then the charge measured by the electrometer can be used to determine the coefficient c_{12}:

$$Q_2 \approx Q_E \approx c_{21}(V_T) + c_{22}(0) \quad \Rightarrow c_{21} \approx \frac{Q_E}{V_T} \quad \text{if } C_E \gg C_c, C_E \gg C_{12} \tag{2.229}$$

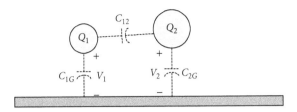

FIGURE 2.99 Circuit capacitances between all three conductors.

To measure the coefficient of self capacitance c_{11}, a floating or battery operated electrometer is inserted between the voltage source V_T and object 1. The electrometer charge is simply related to the coefficient of self capacitance using similar arguments:

$$Q_1 \approx Q_E \approx c_{11}(V_T) + c_{12}(0) \quad \Rightarrow c_{11} \approx \frac{Q_E}{V_T} \quad \text{if } C_E \gg C_c, C_E \gg C_{1G} \tag{2.230}$$

where C_{1G} is the circuit capacitance from object 1 to ground.

In theory, these coefficients of self capacitance and induction are straightforward to measure. Most engineers, though, are not familiar with these coefficients. Fortunately, the relationships between the standard "circuit capacitances" used in many books and the coefficients of self capacitance and induction are obtained with nominal effort. Referring to Figure 2.99, the charge on object 1 is a function of object 1's capacitance to ground (and its voltage to ground) and the capacitance from object 1 to object 2 (and the voltage between these two objects):

$$Q_1 = C_{1G}V_1 + C_{12}(V_1 - V_2) = (C_{1G} + C_{12})V_1 - C_{12}V_2$$

Comparing this expression to Equation (2.225), the relationships between the circuit and coefficients of capacitance are obtained:

$$c_{11} = C_{1G} + C_{12} \quad \text{and} \quad c_{12} = -C_{12}$$

$$C_{1G} = c_{11} + c_{12}$$

The charge on object 2 is

$$Q_2 = C_{2G}V_2 + C_{12}(V_2 - V_1) = -C_{12}V_1 + (C_{2G} + C_{12})V_2$$

and by comparison to (2.225)

$$c_{21} = -C_{12} \quad \text{and} \quad c_{22} = C_{2G} + C_{12}$$

$$C_{2G} = c_{22} + c_{21}$$

By reciprocity, $c_{12} = c_{21}$ and $C_{12} = C_{21}$. To summarize, for two conductors and one reference plane,

$$c_{11} = C_{1G} + C_{12}, \; c_{22} = C_{2G} + C_{12}, \; c_{12} = -C_{12} = -C_{21} \tag{2.231}$$

$$C_{1G} = c_{11} + c_{12}, \; C_{2G} = c_{22} + c_{12}, \; C_{12} = -c_{12} = -c_{21} \tag{2.232}$$

For quick reference, for three objects above a ground plane, the relationships between these variables are

$$c_{11} = C_{1G} + C_{12} + C_{13}, \quad c_{22} = C_{2G} + C_{12} + C_{23}, \quad c_{33} = C_{3G} + C_{13} + C_{23}$$

$$c_{12} = -C_{12} = -C_{21}, \quad c_{13} = -C_{13} = -C_{31}, \quad c_{23} = -C_{23} = -C_{32} \tag{2.233}$$

$$C_{1G} = c_{11} + c_{12} + c_{13}, \quad C_{2G} = c_{22} + c_{12} + c_{23}, \quad C_{3G} = c_{33} + c_{13} + c_{23}$$

$$C_{12} = -c_{12} = -c_{21}, \quad C_{13} = -c_{13} = -c_{31}, \quad C_{23} = -c_{23} = -c_{32} \tag{2.234}$$

$$Q_1 = c_{11}V_1 + c_{12}V_2 + c_{13}V_3$$

$$Q_2 = c_{21}V_1 + c_{22}V_2 + c_{23}V_3 \quad \text{where } c_{ij} = c_{ji} \text{ when } i \neq j \tag{2.235}$$

$$Q_3 = c_{31}V_1 + c_{32}V_2 + c_{33}V_3$$

where the voltages are with respect to the reference conductor.

As previously stated, by reciprocity, $c_{ij} = c_{ji}$ when $i \neq j$. However, there are other special cases where the coefficients of capacitance are simply related. Although these examples involve only two conductors and a ground, the concepts can be extended to three or more conductors. Probably the most obvious case is when the two conducting objects are identical (including their position relative to the ground).[42] In this case, $c_{11} = c_{22}$ and

$$Q_1 = c_{11}V_1 + c_{12}V_2$$

$$Q_2 = c_{12}V_1 + c_{11}V_2$$

The charge on the two identical objects is not necessarily the same unless $V_1 = V_2$. Another simple and obvious case is when the two objects are far from each. The two conductors are then considered isolated, and there is little electrical coupling between them. In this case, $C_{12} \approx 0$, and

$$c_{11} = C_{1G} + C_{12} \approx C_{1G}, \quad c_{22} = C_{2G} + C_{12} \approx C_{2G}, \quad c_{12} = -C_{12} \approx 0 \tag{2.236}$$

Another special case is when the distance to ground for either object is large and one object is much larger than the other. Then, the coefficient of mutual capacitance between the two objects is simply related to the coefficient of self capacitance. For example, if object 1 is large compared to object 2, then object 1's surface area and capacitance are also comparatively large. (The self capacitance of an object is a strong function of its surface area. Generally, two objects with the same surface area have about the same self capacitance.) When object 1 is large compared to object 2 and they are both far from ground, $C_{12} \gg C_{2G}$[43] and $c_{22} = C_{2G} + C_{12} \approx C_{12} = -c_{12}$. Referring to Figure 2.100, when object 2 is large compared to object 1 and they are both far from ground, $C_{12} \gg C_{1G}$ and $c_{11} = C_{1G} + C_{12} \approx C_{12} = -c_{12}$. Finally, a configuration of importance in electric shielding or screening is when a grounded conductor completely surrounds another conductor. In this case, the mutual coupling between the conductor inside the grounded shield and any conductors outside the grounded shield is zero. Referring to Figure 2.101, changing the charge or potential of object 1 does not affect the charge or potential on object 2 and vice

---

[42]If the closest distance to ground for either object is large compared to the largest dimension of either object and distance between the objects, then the two object's distance to ground is probably not important.

[43]In this case, object 1 might be viewed as a "local" reference for object 2.

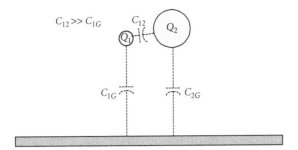

FIGURE 2.100   Distance between two objects, one much larger than the other, is small relative to their distance from the ground reference.

versa: the two objects are decoupled. The coefficients of capacitance for the situation depicted are

$$c_{11} = C_{1G} + C_{12} = C_{1G}, \ c_{22} = C_{2G} + C_{12} = C_{2G}, \ c_{12} = -C_{12} = 0 \tag{2.237}$$

Sometimes, a capacitance meter is used to measure these quantities instead of a voltage supply and an electrometer. For example, if the leads of the capacitance meter are connected to objects 1 and 2, shown in Figure 2.99, the capacitance measured is

$$C_{m12} = C_{12} + \frac{C_{1G} C_{2G}}{C_{1G} + C_{2G}} \tag{2.238}$$

Equation (2.238) is the capacitance between conductors 1 and 2 in the *presence of the ground plane*. As the distance between object 1 and the ground and object 2 and the ground increases, C_{m12} approaches C_{12} since C_{1G} and C_{2G} become small. If the leads of the capacitance meter are connected to object 1 and the ground, the capacitance measured is

$$C_{m1G} = C_{1G} + \frac{C_{12} C_{2G}}{C_{12} + C_{2G}} \tag{2.239}$$

Equation (2.239) is the capacitance between conductors 1 and ground in the *presence of conductor 2*. Finally, if the leads are connected to object 2 and the ground, the capacitance measured is

$$C_{m2G} = C_{2G} + \frac{C_{12} C_{1G}}{C_{12} + C_{1G}} \tag{2.240}$$

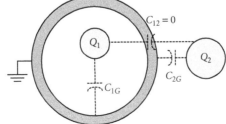

FIGURE 2.101   Object 1 is screened.

Equation (2.240) is the capacitance between conductors 2 and ground in *the presence of conductor 1*. From these three measurements, the three circuit-equivalent capacitances, C_{1G}, C_{2G}, and C_{12}, and the coefficients of self capacitance and induction, can be determined.

Before terminating this enlightening discussion, one final comment will be provided. Generally, unless the conducting objects are far from each other, the mutual *and* self coefficients of capacitance and the mutual *and* self circuit capacitances will be a function of the distance to and shape of their conducting neighbors. Of course, if the conductors are far from each other or isolated, then the mutual coupling terms are small. The mutual coupling terms can also be small when one or more conductors are shielded or screened.

2.34   Energy and Capacitance

Given the coefficients of capacitance for *n* conducting and charged objects, what is the total potential energy in the system? How can the capacitive energy be related to an "explosive-safe" voltage? [Greason, '02; Greason, '87; Greason, '95; Smythe; Jonassen]

The stored electrical energy in part determines the likelihood of an electrostatic discharge. Also, an estimate of the energy in a system before and after a discharge helps in the design of ESD-protection networks. From electromagnetic field theory, the electrical energy, *W*, contained in some volume, *V*, is given by

$$W = \frac{1}{2}\iiint_V \vec{D}\cdot\vec{E}\,dv \tag{2.241}$$

where $\vec{D}$ and $\vec{E}$ are the electric flux density and electric field, respectively. Although Ph.D.'s may appreciate the compact and elegant form of this energy expression, most engineers prefer to work with circuit concepts such as capacitance and voltage. A more amiable form for this expression begins by noting that for a given distribution of volume charge, ρ_V, and surface charge, ρ_S, (2.241) can be written in terms of the potential distribution, Φ:

$$W = \frac{1}{2}\iiint_V \rho_V\Phi\,dv + \frac{1}{2}\iint_S \rho_S\Phi\,ds \tag{2.242}$$

where *V* is the volume region containing the volume charge and *S* is the surface region containing the surface charge. Of course, terms can also be added for line and point charges. For electroquasistatic conditions, where charge exists only along the surface of conductors and the potential is constant along each of the conductors, the second integral can be written as a summation over the *n* conductors in the system:

$$W = \frac{1}{2}\iiint_V \rho_V\Phi\,dv + \frac{1}{2}\sum_{j=1}^{n} Q_j V_j \tag{2.243}$$

where Q_j is the total charge on and V_j is the voltage of each conductor *j*.[44] This expression allows for volume charge between the conductors. The volume integration, *V*, is between these conductors. Also, this expression assumes that surface charge can only exist along the *n* conducting surfaces. Therefore, surface charge along other surfaces such as dielectric interfaces are *not* included in this energy expression.

---

[44]If the reference conductor is included in this summation, it will have no effect since by definition its potential, *V*, is zero.

The summation can be written in terms of the coefficients of capacitance, which are related to the standard circuit capacitances. For n conducting objects, the charge Q_j is equal to

$$Q_j = c_{j1}V_1 + c_{j2}V_2 + \cdots + c_{jn}V_n = \sum_{k=1}^{n} c_{jk}V_k \qquad (2.244)$$

The variables c_{jk} are the coefficients of capacitance, and the voltages, V_j, are with respect to the reference conductor. Substituting this total charge expression for conductor j into the summation,

$$W = \frac{1}{2}\iiint_V \rho_V \Phi \, dv + \frac{1}{2}\sum_{j=1}^{n}\sum_{k=1}^{n} c_{jk}V_j V_k \qquad (2.245)$$

If no volume charge is present, the integral disappears.

As an introductory application of this energy expression, the total energy for two distant, charged, floating isolated conductors above a ground reference conductor will be determined. Then, the total energy for these two distant conductors will be determined after a thin wire conductively connects them together. The volume (line and surface) charge between the conductors is assumed zero. The initial charges on the conductors shown in Figure 2.102 are Q_1 and Q_2. In this figure, standard circuit capacitances, not coefficients of capacitance, are shown. Because the two conductors are far from each other, the capacitance between them, C_{12}, is negligible. The relationships between the circuit capacitances and coefficients of capacitance are (assuming C_{12} is negligible)

$$c_{11} = C_{1G}, \ c_{22} = C_{2G}, \ c_{12} = 0$$

Therefore, the total initial electrical energy is

$$W_i = \frac{1}{2}\sum_{j=1}^{2}\sum_{k=1}^{2} c_{jk}V_j V_k = \frac{1}{2}\sum_{j=1}^{2}(c_{j1}V_j V_1 + c_{j2}V_j V_2)$$

$$= \frac{1}{2}(c_{11}V_1 V_1 + c_{12}V_1 V_2 + c_{21}V_2 V_1 + c_{22}V_2 V_2) \qquad (2.246)$$

$$= \frac{C_{1G}V_1^2}{2} + \frac{C_{2G}V_2^2}{2} = \frac{Q_1^2}{2C_{1G}} + \frac{Q_2^2}{2C_{2G}} \quad (\text{since } Q_1 = C_{1G}V_1, Q_2 = C_{2G}V_2)$$

This is the classical result for the total energy of two isolated noninteracting capacitors. To determine the final energy of the system, after a wire connects the two conductors, two fundamental concepts must

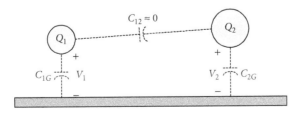

FIGURE 2.102 Two distant charged conductors.

be applied. First, the total charge must be conserved:

$$Q_1 + Q_2 = q_1 + q_2$$

where q_1 and q_2 are the respective charges on the two conductors after they are connected by the thin wire. Second, the final voltage of the two conductors (and conducting wire) must be the same, V_f. Using the basic charge-voltage relationship for the conductors

$$q_1 = C_{1G}V_f, \; q_2 = C_{2G}V_f$$

the final voltage is obtained:

$$Q_1 + Q_2 = C_{1G}V_f + C_{2G}V_f \;\; \Rightarrow V_f = \frac{Q_1 + Q_2}{C_{1G} + C_{2G}} \tag{2.247}$$

Because the wire is thin, it has little effect on the capacitances. The final energy is given by

$$W_f = \frac{C_{1G}V_f^2}{2} + \frac{C_{2G}V_f^2}{2} = \frac{C_{1G} + C_{2G}}{2}\left(\frac{Q_1 + Q_2}{C_{1G} + C_{2G}}\right)^2 = \frac{Q_1^2 + 2Q_1Q_2 + Q_2^2}{2(C_{1G} + C_{2G})} \tag{2.248}$$

What is most fascinating is that the initial energy is greater than or equal to the final energy:

$$W_i - W_f = \frac{Q_1^2}{2C_{1G}} + \frac{Q_2^2}{2C_{2G}} - \frac{Q_1^2 + 2Q_1Q_2 + Q_2^2}{2(C_{1G} + C_{2G})} = \frac{(C_{2G}Q_1 - C_{1G}Q_2)^2}{2C_{1G}C_{2G}(C_{1G} + C_{2G})} \geq 0 \tag{2.249}$$

Unless the initial voltage of the two objects is the same,

$$C_{2G}Q_1 - C_{1G}Q_2 = 0 \;\; \Rightarrow \frac{Q_2}{C_{2G}} = V_2 = \frac{Q_1}{C_{1G}} = V_1$$

the initial energy is greater than the final energy. If the two conductors and connecting wire are lossless, and the leakage to the surroundings is zero, this energy loss must be through some other mechanism such as radiation. Radiation is a source of electrical noise. Lastly, the final charge on each conductor, after the connection, is a function of the size of the object's capacitance:

$$q_1 = C_{1G}V_f = \frac{C_{1G}}{C_{1G} + C_{2G}}(Q_1 + Q_2), \; q_2 = C_{2G}V_f = \frac{C_{2G}}{C_{1G} + C_{2G}}(Q_1 + Q_2) \tag{2.250}$$

Since capacitance generally increases with surface area, when two objects touch, it is expected that the larger object will carry more of the total charge.

In the next situation, shown in Figure 2.103, the two conducting objects will not be assumed far away, which is a more realistic situation in ESD. Often, a charged object approaches another possibly charged

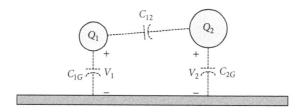

FIGURE 2.103   Two charged conductors.

object. Potentially, when the objects are close enough, an electrical discharge can occur between them. The initial energy in the system in terms of the coefficients of capacitance is

$$W_i = \frac{1}{2} \sum_{j=1}^{2} \sum_{k=1}^{2} c_{jk} V_j V_k = \frac{1}{2} \left(c_{11} V_1^2 + 2c_{12} V_1 V_2 + c_{22} V_2^2 \right)$$

$$= \frac{1}{2} \sum_{j=1}^{2} Q_j V_j = \frac{1}{2} (Q_1 V_1 + Q_2 V_2)$$

(2.251)

The relationships between the circuit and coefficient capacitances and between the charges and voltages are

$$c_{11} = C_{1G} + C_{12}, \ c_{22} = C_{2G} + C_{12}, \ c_{12} = -C_{12} = -C_{21}$$

$$Q_1 = c_{11} V_1 + c_{12} V_2, \ Q_2 = c_{21} V_1 + c_{22} V_2$$

The voltages can be written in terms of the charges:

$$V_1 = \frac{c_{22} Q_1 - c_{12} Q_2}{c_{11} c_{22} - c_{12}^2}, \ V_2 = \frac{c_{11} Q_2 - c_{12} Q_1}{c_{11} c_{22} - c_{12}^2}$$

(2.252)

Before the final energy is computed, note that the difference in potential is

$$V_1 - V_2 = \frac{c_{22} Q_1 - c_{12} Q_2}{c_{11} c_{22} - c_{12}^2} - \frac{c_{11} Q_2 - c_{12} Q_1}{c_{11} c_{22} - c_{12}^2} = \frac{Q_1 (c_{22} + c_{12}) - Q_2 (c_{11} + c_{12})}{c_{11} c_{22} - c_{12}^2}$$

(2.253)

or in terms of the circuit capacitances,

$$V_1 - V_2 = \frac{C_{2G} Q_1 - C_{1G} Q_2}{C_{1G} C_{2G} + C_{1G} C_{12} + C_{2G} C_{12}}$$

(2.254)

The electric field between two conductors is directly related to their difference in potential and "spacing." If the closest spacing, d, is small compared to the radii of curvature of the two conductors (near this closest spacing), then the electric field (near this closest spacing) is approximately

$$|\vec{E}| \approx \frac{\Delta \Phi}{\Delta x} \approx \frac{|V_1 - V_2|}{d} = \frac{1}{d} \frac{|C_{2G} Q_1 - C_{1G} Q_2|}{C_{1G} C_{2G} + C_{1G} C_{12} + C_{2G} C_{12}}$$

(2.255)

When this electric field exceeds the breakdown strength of the air, a spark can occur across the conductors. During this sparkover, charge is transferred between the conductors through one or more conducting channels. A single channel sparkover can be roughly modeled as a thin wire between the conductors. As before, even after the wire joins the two conductors, the charge must be conserved

$$Q_1 + Q_2 = q_1 + q_2$$

where q_1 and q_2 are the respective final charges on the two conductors.[45] Also, since the final voltage of the two conductors are the same

$$q_1 = c_{11} V_f + c_{12} V_f, \ q_2 = c_{21} V_f + c_{22} V_f$$

---

[45]This assumes that charge is not "lost" during the breakdown to the surroundings.

Solving for this final voltage,

$$V_f = \frac{Q_1 + Q_2}{c_{11} + 2c_{12} + c_{22}} \tag{2.256}$$

The final energy is then

$$W_f = \frac{1}{2}\sum_{j=1}^{2} Q_j V_j = \frac{1}{2}(Q_1 V_f + Q_2 V_f) = \frac{(Q_1 + Q_2)^2}{2(c_{11} + 2c_{12} + c_{22})} \tag{2.257}$$

or in terms of the circuit capacitances,

$$W_f = \frac{Q_1^2 + 2Q_1 Q_2 + Q_2^2}{2\,(C_{1G} + C_{2G})} \tag{2.258}$$

The final energy is not a function of C_{12} since it is eliminated by the connecting wire. The final charges on the two conductors are

$$q_1 = c_{11}V_f + c_{12}V_f = \frac{c_{11} + c_{12}}{c_{11} + 2c_{12} + c_{22}}(Q_1 + Q_2) = \frac{C_{1G}}{C_{1G} + C_{2G}}(Q_1 + Q_2)$$

$$q_2 = c_{21}V_f + c_{22}V_f = \frac{c_{22} + c_{12}}{c_{11} + 2c_{12} + c_{22}}(Q_1 + Q_2) = \frac{C_{2G}}{C_{1G} + C_{2G}}(Q_1 + Q_2) \tag{2.259}$$

In those situations where two conductors are involved (e.g., one conducting object and a ground reference), an "explosive-safe" voltage can be defined. The electrical energy contained in a capacitance, C, between a conducting object and ground reference conductor is

$$W = \frac{1}{2}CV^2 \tag{2.260}$$

where V is the voltage of the conductor relative to the ground reference. For many vapors and gases, the minimum ignition energy about 0.25 mJ. If a typical capacitance of 400 pF is selected, then the corresponding "explosion-safe" voltage is

$$V_{es} = \sqrt{\frac{2W}{C}} = \sqrt{\frac{2\,(0.25 \times 10^{-3})}{400 \times 10^{-12}}} \approx 1\,\text{kV} \tag{2.261}$$

Again, this "safe" voltage assumes a specific capacitance (or size and shape of the charged object) and a typical ignition energy for many vapors and gases. If other objects are in the vicinity of the charged conductor, then as was previously seen, the relationship is more complex. When protecting electrical devices from ESD, the maximum acceptable voltage level can be much lower than this "explosive-safe" voltage.

2.35   Capacitance Formula

List the expressions commonly seen for self and mutual capacitance. When can the capacitance be determined from the inductance or characteristic impedance? How can the "circuit capacitances" be determined, or at least, estimated if not specifically given in the table? [Paul, '92(b); Walker]

As throughout this book (unless noted otherwise), the SI system is used in Appendix B; that is, capacitances are in farads (F) and lengths and distances are in meters (m). Since capacitance is formally defined between two (or more) conductors (or objects), if the capacitance is listed for a single conductor, the second conductor is assumed at infinity (or at a great distance relative to the dimensions of the single conductor). In Appendix B, but not necessarily in other books, articles, or even elsewhere in this book, the following notation is used:

C_i capacitance of a single isolated conductor

C capacitance between two conductors with no other conductors nearby

C_{ab} "circuit capacitance" between conductors a and b with one or more other conductors nearby (also known as the equivalent circuit capacitance)

C_{mab} total capacitance between conductors a and b in the presence of one or more other conductors (i.e., the capacitance that would be measured between conductors a and b)

c_{ab} coefficient of self capacitance when $a = b$ and coefficient of induction (or coefficient of mutual capacitance) when $a \neq b$

A "single" conductor may actually consist of two or more conductors that are connected together with a wire or some other electrically conducting object.[46]

The capacitance expressions in Appendix B assume that the largest dimension, including the length, is electrically short (i.e., small compared to the wavelength of the signal in the medium surrounding the conductors). When an object is electrically small, its capacitance can be modeled as a lumped element. Unless otherwise denoted, the medium surrounding the conductor(s) is assumed free space ($\varepsilon_r = 1$). When dielectric material is present everywhere outside of a conductor (or at least where most of the electric fields and electric energy are located), the capacitance will increase proportionally to the relative permittivity (ε_r) of this dielectric material (assuming the dielectric material is linear and constant). The permeability or relative permeability is a magnetic field property that affects the inductance. It does not affect the capacitance, which is a function of the electric field. The conductivity of the surrounding medium does not affect the capacitance. It does, however, add resistance or loss to the system.

Unlike magnetic fields and inductance, the electric fields inside a conductor have virtually zero effect on the capacitance. Thus, most of the formula given are not a function of the thickness of the conductors; for example, the capacitance of a solid isolated sphere is the same as the capacitance of a thin-walled isolated sphere.[47] There is no difference between the low-frequency and high-frequency capacitance when free space surrounds the conductors. Of course, when real dielectric materials are used with frequency-dependent electrical properties, the capacitance can vary with frequency.

Many of the available simple analytical expressions for capacitance are provided in Appendix B. If the desired expression cannot be located in this specific table, the capacitance can possibly be determined from an inductance or a characteristic impedance expression. If the medium surrounding the conductors is homogeneous (i.e., the same everywhere), then the total capacitance can be determined from the expression for the characteristic impedance, Z_o. The capacitance of a long transmission line is related to the relative permittivity and permeability of the surrounding homogeneous medium (not of the conductors or objects) through the equation

$$C = \frac{l_{th}\sqrt{\mu_r \varepsilon_r}}{cZ_o} = \frac{l_{th}\sqrt{\mu_r \varepsilon_r}}{(3 \times 10^8)Z_o} \tag{2.262}$$

where l_{th} is the length of the transmission line, c is the speed of light, and C is the *total* capacitance. When using this expression, it is assumed that the length of the transmission line conductors is very large

---

[46]The potential difference between two conductors that are connected together via a perfect conductor is zero; therefore, the capacitance between these two connected conductors is also zero.

[47]There are situations where the thickness of one or both of the conductors does slightly affect the capacitance. For example, there is a small difference in the capacitance for a solid finite-length cylinder normal to a flat plane and for a thin finite-length cylinder normal to a flat plane.

compared to the other dimensions. The total capacitance between two transmission line-type conductors (that are long) can also be obtained from the *total* external or high-frequency inductance, L_e, if the surrounding medium is homogeneous:

$$C = \frac{\mu_r \varepsilon_r}{L_e c^2} l_{th}^2 = \frac{\mu_r \varepsilon_r l_{th}^2}{L_e (3 \times 10^8)^2} \tag{2.263}$$

Finally, these expressions assume that no other nearby conductors are present and the objects are good conductors. Also, if the surrounding dielectric medium is not homogeneous, then (2.262) and (2.263) should generally not be used. To obtain an estimate for the actual capacitance for nonhomogeneous medium, the free-space capacitance can be multiplied by an effective dielectric constant. This effective constant is based on a weighted average of the "total electric field" in each of the surrounding dielectrics.

For transmission-line structures involving more than two conductors, the circuit capacitances can be obtained from the mutual and self inductance terms if the medium surrounding the conductors is homogeneous or the same everywhere. In matrix form, the relationship between the inductance matrix for the line, **L**, and capacitance matrix for the line, **C**, is

$$\mathbf{LC} = \mu \varepsilon l_{th}^2 \mathbf{I}_n = \frac{\mu_r \varepsilon_r l_{th}^2}{c^2} \mathbf{I}_n = \frac{\mu_r \varepsilon_r l_{th}^2}{(3 \times 10^8)^2} \mathbf{I}_n \tag{2.264}$$

where l_{th} is the length of the conductors and $\mathbf{I}_n$ is the n by n identity matrix. The parameter matrix **L** contains the self and mutual inductances, and the parameter matrix **C** contains the various circuit capacitances. Notice that the capacitance terms can be obtained from the inductance terms (and vice versa):

$$\mathbf{C} = \frac{\mu_r \varepsilon_r l_{th}^2}{(3 \times 10^8)^2} \mathbf{L}^{-1} \mathbf{I}_n = \frac{\mu_r \varepsilon_r l_{th}^2}{(3 \times 10^8)^2} \mathbf{L}^{-1} \tag{2.265}$$

For a three-conductor line, a simple relationship exists between these inductances and capacitances using (2.264)

$$\begin{bmatrix} L_a & M \\ M & L_b \end{bmatrix} \begin{bmatrix} C_{ac} + C_{ab} & -C_{ab} \\ -C_{ab} & C_{bc} + C_{ab} \end{bmatrix} = \frac{\mu_r \varepsilon_r l_{th}^2}{(3 \times 10^8)^2} \begin{bmatrix} 1 & 0 \\ 0 & 1 \end{bmatrix} \tag{2.266}$$

or

$$\begin{bmatrix} C_{ac} + C_{ab} & -C_{ab} \\ -C_{ab} & C_{bc} + C_{ab} \end{bmatrix} = \frac{\mu_r \varepsilon_r l_{th}^2}{(3 \times 10^8)^2 (L_a L_b - M^2)} \begin{bmatrix} L_b & -M \\ -M & L_a \end{bmatrix} \tag{2.267}$$

where for conductors a and b, the reference conductor is c, as shown in Figure 2.104. (For this form of the expression, these variables correspond to the total inductances and total capacitances — not

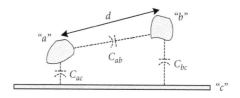

FIGURE 2.104 Circuit capacitances between all three conductors.

per-unit-length values.) Therefore, the circuit capacitances can be obtained from the self inductance of the loop generated by conductors a and c (L_a), from the self inductance of the loop generated by conductors b and c (L_b), and the mutual inductance between the loop generated by conductors a and c and the loop generated by conductors b and c (M):

$$C_{ab} = \frac{\mu_r \varepsilon_r l_{th}^2}{(3 \times 10^8)^2} \frac{M}{L_a L_b - M^2} \tag{2.268}$$

$$C_{ac} = \frac{\mu_r \varepsilon_r l_{th}^2}{(3 \times 10^8)^2} \frac{L_b}{L_a L_b - M^2} - C_{ab} \tag{2.269}$$

$$C_{bc} = \frac{\mu_r \varepsilon_r l_{th}^2}{(3 \times 10^8)^2} \frac{L_a}{L_a L_b - M^2} - C_{ab} \tag{2.270}$$

When the mutual inductance, M, is small compared to the product of the self inductances, L_a and L_b, these equations reduce to

$$C_{ab} \approx \frac{\mu_r \varepsilon_r l_{th}^2}{(3 \times 10^8)^2} \frac{M}{L_a L_b} \tag{2.271}$$

$$C_{ac} \approx \frac{\mu_r \varepsilon_r l_{th}^2}{(3 \times 10^8)^2} \frac{1}{L_a} - C_{ab} \tag{2.272}$$

$$C_{bc} \approx \frac{\mu_r \varepsilon_r l_{th}^2}{(3 \times 10^8)^2} \frac{1}{L_b} - C_{ab} \tag{2.273}$$

These expressions are sometimes referred to as the "weak" coupling version of the previous circuit capacitance equations. Furthermore, since for transmission-type conductors in homogeneous medium the external or high-frequency inductance and capacitance are related by Equation (2.263), these approximations can be written in terms of the isolated capacitances:

$$C_{ab} = \frac{M C_b}{L_a} = \frac{M C_a}{L_b} \tag{2.274}$$

$$C_{ac} = C_a - C_{ab} \tag{2.275}$$

$$C_{bc} = C_b - C_{ab} \tag{2.276}$$

where C_a is the capacitance between conductors a and c with conductor b not present and C_b is the capacitance between conductors b and c with conductor a not present.

If the surrounding dielectric medium is nonhomogeneous, then the *free-space* circuit capacitances can be obtained from these expressions since the inductances are not affected by dielectric materials that are nonmagnetic. The actual capacitances, corresponding to the nonhomogeneous medium, would have to be modified based on the effective dielectric constant(s).

As an example, the mutual circuit capacitance for the three-conductor line shown in Figure 2.105 will be determined. Since the free-space mutual and self inductances are obtained directly from the inductance

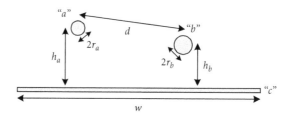

FIGURE 2.105   Three-conductor system involving two circular conductors above a ground plane.

table as

$$M = 10^{-7}l_{th}\ln\left(1+4\frac{h_a h_b}{d^2}\right),\ L_a = 2\times10^{-7}l_{th}\ln\left(\frac{2h_a}{r_a}\right),\ L_b = 2\times10^{-7}l_{th}\ln\left(\frac{2h_b}{r_b}\right)$$

the mutual capacitance is

$$C_{ab} = \frac{l_{th}^2}{(3\times10^8)^2}\frac{M}{L_a L_b - M^2}$$

$$= \frac{l_{th}^2}{(3\times10^8)^2}\frac{10^{-7}l_{th}\ln\left(1+4\dfrac{h_a h_b}{d^2}\right)}{\left[2\times10^{-7}l_{th}\ln\left(\dfrac{2h_a}{r_a}\right)\right]\left[2\times10^{-7}l_{th}\ln\left(\dfrac{2h_b}{r_b}\right)\right]-\left[10^{-7}l_{th}\ln\left(1+4\dfrac{h_a h_b}{d^2}\right)\right]^2}$$

$$= 4\pi\varepsilon_o l_{th}\frac{\ln\left(1+4\dfrac{h_a h_b}{d^2}\right)}{4\left[\ln\left(\dfrac{2h_a}{r_a}\right)\right]\left[\ln\left(\dfrac{2h_b}{r_b}\right)\right]-\left[\ln\left(1+4\dfrac{h_a h_b}{d^2}\right)\right]^2}$$

The "weak" coupled version of this equation is obtained by substituting the capacitance expression for a cylindrical wire above a large flat conductor,

$$C_b \approx \frac{2\pi\varepsilon_o l_{th}}{\ln\left(\dfrac{2h_b}{r_b}\right)}\quad\text{if }l_{th}\gg h_b,\, h_b\gg r_b$$

into Equation (2.274):

$$C_{ab} = \frac{10^{-7}l_{th}\ln\left(1+4\dfrac{h_a h_b}{d^2}\right)\left[\dfrac{2\pi\varepsilon_o l_{th}}{\ln\left(\dfrac{2h_b}{r_b}\right)}\right]}{2\times10^{-7}l_{th}\ln\left(\dfrac{2h_a}{r_a}\right)} = \frac{\pi\varepsilon_o l_{th}\ln\left(1+4\dfrac{h_a h_b}{d^2}\right)}{\ln\left(\dfrac{2h_a}{r_a}\right)\ln\left(\dfrac{2h_b}{r_b}\right)}$$

For transmission-line structures with more than three conductors, the relationships between the inductance and capacitance parameters are usually obtained numerically. For reference purposes, the matrix expression for a four-conductor system is

$$
\begin{bmatrix} L_a & M_{ab} & M_{ac} \\ M_{ab} & L_b & M_{bc} \\ M_{ac} & M_{bc} & L_c \end{bmatrix}\begin{bmatrix} C_{ad}+C_{ab}+C_{ac} & -C_{ab} & -C_{ac} \\ -C_{ab} & C_{bd}+C_{ab}+C_{bc} & -C_{bc} \\ -C_{ac} & -C_{bc} & C_{cd}+C_{ac}+C_{bc} \end{bmatrix}
$$

$$
= \frac{\mu_r \varepsilon_r l_{th}^2}{(3\times10^8)^2}\begin{bmatrix} 1 & 0 & 0 \\ 0 & 1 & 0 \\ 0 & 0 & 1 \end{bmatrix}
$$

(2.277)

where conductor d is the reference conductor. (Each total term of the capacitance matrix corresponds to a coefficient of capacitance, which can be negative.) The variable L_a is the self inductance of the loop generated by conductors a and d. The variable M_{ab} is the mutual inductance between the loop formed by conductors a and d and the loop formed by conductors b and d. Given the inductances, the capacitances can be determined by (i) multiplying both sides of Equation (2.277) by the inverse of the inductance matrix, (ii) equating the terms of the matrix elements on the two sides of the expression, and (iii) solving the set of equations for the capacitances.

There is another approach for estimating the circuit capacitances for a three-conductor system given the isolated capacitances. It is not necessary that the conductors be transmission-like with this approach. For two distant arbitrary shaped conductors, a and b, separated by a distance of d, an approximation for the circuit capacitances can be obtained from the isolated capacitances to ground:

$$
C_{ac} = \frac{4\pi\varepsilon_o dC_a(4\pi\varepsilon_o d - C_b)}{(4\pi\varepsilon_o d)^2 - C_a C_b} \approx C_a
$$

$$
C_{bc} = \frac{4\pi\varepsilon_o dC_b(4\pi\varepsilon_o d - C_a)}{(4\pi\varepsilon_o d)^2 - C_a C_b} \approx C_b
$$

$$
C_{ab} = \frac{4\pi\varepsilon_o dC_a C_b}{(4\pi\varepsilon_o d)^2 - C_a C_b} \approx \frac{C_a C_b}{4\pi\varepsilon_o d}
$$

where C_a is the capacitance between conductor a and the large reference conductor, c, when conductor b is not present, and C_b is the capacitance between conductor b and the large reference conductor, c, when conductor a is not present. The width and length of the reference conductor are large compared to the dimensions of a and b, their distance to c, and d. The distance d is large compared to the dimensions of conductors a and b.

The capacitance between two conductors generally increases with the surface area of the conductors and decreases with the distance between the conductors. Also, the capacitance between two objects increases as the relative permittivity or dielectric constant of the surrounding medium increases.

Sometimes the term common-mode capacitance is seen. For a three-conductor system, the capacitance between two of the conductors, when tied together, and the third conductor is a common-mode capacitance. For shielded twisted-pair cable, for example, the common-mode capacitance is the capacitance between the outer shield and the two twisted-pair conductors when tied together.

The capacitance can also be given for two objects that are not metallic such as between a human and tile floor. However, at the frequency, f, that the capacitance is to be used, it is more easily defined if both objects are good conductors:

$$
\frac{\sigma}{\omega\varepsilon} = \frac{\sigma}{2\pi f\varepsilon_r\varepsilon_o} = \frac{\sigma}{2\pi f\varepsilon_r(8.854\times10^{-12})} \gg 1
$$

where σ is the conductivity and ε_r is the relative permittivity of each object at the frequency f.

Appendix A   Charge Distributions and Their Electric Fields

Charge Distribution	Electric Field $(\vec{E})$
Point charge	$$\frac{Q}{4\pi\varepsilon_o r^2}\hat{a}_r$$ The field is radially outward from the point charge of total charge Q
Electric dipole [Cheng] 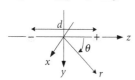	$$\frac{Qd\cos\theta}{2\pi\varepsilon_o r^3}\hat{a}_r + \frac{Qd\sin\theta}{4\pi\varepsilon_o r^3}\hat{a}_\theta \quad \text{if } r \gg d$$ $$= \frac{Qd}{2\pi\varepsilon_o r^3}\hat{a}_r \quad \text{if } r \gg d, \theta = 0° \text{ (along the } z \text{ axis)}$$ $$= \frac{Qd}{4\pi\varepsilon_o r^3}\hat{a}_\theta \quad \text{if } r \gg d, \theta = 90° \text{ (along the } z = 0 \text{ plane)}$$ An electric dipole consists of two point charges of equal magnitude (Q) but opposite sign separated by a distance of d, the charges are along the z axis and centered about the origin, the field decreases rapidly with r, the total charge is zero
Linear quadrupole [Jefimenko] 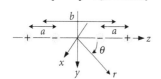	$$\frac{3Qab(3\cos^2\theta-1)}{4\pi\varepsilon_o r^4}\hat{a}_r + \frac{3Qab\cos\theta\sin\theta}{2\pi\varepsilon_o r^4}\hat{a}_\theta \quad \text{if } r \gg a, r \gg b$$ $$= \frac{3Qab}{2\pi\varepsilon_o r^4}\hat{a}_r \quad \text{if } r \gg a, r \gg b, \theta = 0° \text{ (along the } z \text{ axis)}$$ $$= -\frac{3Qab}{4\pi\varepsilon_o r^4}\hat{a}_r \quad \text{if } r \gg a, r \gg b, \theta = 90° \text{ (along the } z = 0 \text{ plane)}$$ A linear quadrupole consists of two pairs of point charges of equal magnitude (Q) but opposite sign, the charges are collinear and along the z axis, the distance between neighboring charges of opposite sign is a, the center-to-center distance between the two pairs is b, the field decreases very rapidly with r, the total charge is zero
Square quadrupole [Jefimenko] 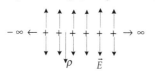	$$\frac{9Qa^2\cos\phi\sin\phi}{4\pi\varepsilon_o\rho^4}\hat{a}_\rho + \frac{3Qa^2(\sin^2\phi-\cos^2\phi)}{4\pi\varepsilon_o\rho^4}\hat{a}_\phi$$ in the $z = 0$ plane and if $\rho \gg a$ $$= -\frac{3Qa^2}{4\pi\varepsilon_o\rho^4}\hat{a}_\phi \quad \text{if } \rho \gg a, \phi = 0° \text{ (along the } x \text{ axis)}$$ $$= \frac{3Qa^2}{4\pi\varepsilon_o\rho^4}\hat{a}_\phi \quad \text{if } \rho \gg a, \phi = 90° \text{ (along the } y \text{ axis)}$$ Square quadrupole consists of two pairs of point charges of equal magnitude (Q) but opposite sign, the pair of charges located in the xy plane are centered about the origin in the corners of a square of side a, the distance between neighboring charges of opposite sign is a, the field decreases very rapidly with ρ, the total charge is zero
Uniformly charged, infinite-length, straight line charge [Cheng; Demarest]	$$\frac{\rho_L}{2\pi\varepsilon_o\rho}\hat{a}_\rho$$ The field outside the uniformly charged infinite line charge is radially outward

Charge Distribution	Electric Field $(\vec{E})$
Dipole infinite line charges [Zahn; Haus]	$\dfrac{\rho_L d\cos\phi}{2\pi\varepsilon_o\rho^2}\hat{a}_\rho + \dfrac{\rho_L d\sin\phi}{2\pi\varepsilon_o\rho^2}\hat{a}_\phi$ if $\rho \gg d$ $= \dfrac{\rho_L d}{2\pi\varepsilon_o\rho^2}\hat{a}_\rho$ if $\rho \gg d, \phi = 0°$ (along the $+x$ axis) $= \dfrac{\rho_L d}{2\pi\varepsilon_o\rho^2}\hat{a}_\phi$ if $\rho \gg d, \phi = 90°$ (along the $+y$ axis) A dipole line charge consists of two oppositely charged infinite line charges (in the z direction) each with a magnitude of ρ_L, they are centered about and parallel to the z axis, the distance from each line charge to the z axis is $d/2$, the total charge is zero
Quadrupole infinite line charges [Haus]	$-\dfrac{\rho_L d^2 \cos 2\phi}{2\pi\varepsilon_o\rho^3}\hat{a}_\rho - \dfrac{\rho_L d^2 \sin 2\phi}{2\pi\varepsilon_o\rho^3}\hat{a}_\phi$ $\rho \gg d$ Quadrupole line charges consists of two pairs of equally spaced positive and negative infinite line charges (in the z direction) each with a magnitude of ρ_L, they are centered about and parallel to the z axis, the distance from each line charge to the z axis is $d/2$, the total charge is zero
Uniformly charged, straight line charge [Marshall, '96; Zahn; Cheng] 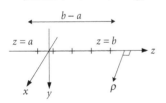	$\dfrac{\rho_L}{4\pi\varepsilon_o}\left[\dfrac{b-z}{\rho\sqrt{\rho^2+(b-z)^2}} - \dfrac{a-z}{\rho\sqrt{\rho^2+(a-z)^2}}\right]\hat{a}_\rho$ $+\dfrac{\rho_L}{4\pi\varepsilon_o}\left[\dfrac{1}{\sqrt{\rho^2+(b-z)^2}} - \dfrac{1}{\sqrt{\rho^2+(a-z)^2}}\right]\hat{a}_z$ $= \dfrac{\rho_L L}{4\pi\varepsilon_o\left(z^2 - \dfrac{L^2}{4}\right)}\hat{a}_z$ if $\rho = 0$, $a = -\dfrac{L}{2}, b = \dfrac{L}{2}, z > \dfrac{L}{2}$ (along the z axis) $= \dfrac{-\rho_L L}{4\pi\varepsilon_o\left(z^2 - \dfrac{L^2}{4}\right)}\hat{a}_z$ if $\rho = 0$, $a = -\dfrac{L}{2}, b = \dfrac{L}{2}, z < -\dfrac{L}{2}$ (along the z axis) $= \dfrac{\rho_L L}{4\pi\varepsilon_o\rho\sqrt{\rho^2 + \dfrac{L^2}{4}}}\hat{a}_\rho$ if $z = 0$, $a = -\dfrac{L}{2}, b = \dfrac{L}{2}$ (along the $z = 0$ plane)

The line charge exists along the z axis from $z = a$ to $z = b$, ρ is the distance from the z axis, the total charge on the line is $(b - a)\rho_L$

Charge Distribution	Electric Field $(\vec{E})$
Linearly charged straight line charge [Haus] 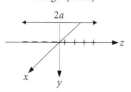	$\dfrac{A}{4\pi\varepsilon_o}\left[\ln\left(\dfrac{z-a}{z+a}\right)+\dfrac{2az}{z^2-a^2}\right]\hat{a}_z$ along the $+z$ axis $(0,0,z)$, $z>a$ The line charge varies according to the expression $\rho_L(z)=Az$, it exists along the z axis from $z=-a$ to $z=a$, the total charge on the line is zero
Uniformly charged circular ring [Guru; Zahn]	$\dfrac{\rho_L az}{2\varepsilon_o(a^2+z^2)^{\frac{3}{2}}}\hat{a}_z$ along the z axis $(0,0,z)$ The uniformly charged ring has a radius of a, it is located in the $z=0$ plane and centered about the z axis, the total charge on the ring is $2\pi a\rho_L$
Bipolar circular ring [Zahn]	$-\dfrac{\rho_L a^2}{\pi\varepsilon_o(a^2+z^2)^{\frac{3}{2}}}\hat{a}_y$ along the z axis $(0,0,z)$ The line charge for $y>0$ is $+\rho_L$ while the line charge for $y<0$ is $-\rho_L$, the ring has a radius of a, it is located in the $z=0$ plane and centered about the z axis, the total charge on the ring is zero
Sinusoidal-varying charged circular ring [Rao]	$-\dfrac{Aa^2}{4\varepsilon_o(a^2+z^2)^{\frac{3}{2}}}\hat{a}_y$ along the z axis $(0,0,z)$ The line charge on the round ring is varying according to the expression $\rho_L(\phi)=A\sin\phi$ where ϕ is the angle from the x axis (in the xy plane), the ring is in the $z=0$ plane and centered about the z axis, the radius of the ring is a, the total charge on the ring is zero
Sinusoidal-varying, charged, circular half ring [Lerner]	$-\dfrac{A}{8\varepsilon_o a}\hat{a}_y$ at the origin $(0,0,0)$ The line charge on the half ring is varying according to the expression $\rho_L(\phi)=A\sin\phi$ where ϕ is the angle from the x axis (in the xy plane), the half ring is in the $z=0$ plane and centered about the z axis, the radius of the ring is a, the total charge on the half ring is $2Aa$
Uniformly charged square loop [Zahn]	$\dfrac{\rho_L az}{\pi\varepsilon_o\left(\dfrac{a^2}{4}+z^2\right)\sqrt{\dfrac{a^2}{2}+z^2}}\hat{a}_z$ along the z axis $(0,0,z)$ The uniformly charged square loop has sides equal to a, the loop is located in the $z=0$ plane and centered about the z axis, the total charge on the loop is $4a\rho_L$
Uniformly charged, infinite flat sheet [Zahn; Cheng]	$\dfrac{\rho_S}{2\varepsilon_o}\hat{a}_z$ if $z>0$, $-\dfrac{\rho_S}{2\varepsilon_o}\hat{a}_z$ if $z<0$ The uniform field is normal to the sheet of charge, it is independent of the distance from the sheet, the sheet extends to infinity in both the x and y directions and is in the $z=0$ plane

Charge Distribution	Electric Field $(\vec{E})$
Uniformly charged circular disk [Demarest; Cheng]	$\dfrac{\rho_S}{2\varepsilon_o}\left(\pm1-\dfrac{z}{\sqrt{a^2+z^2}}\right)\hat{a}_z$ along the $\pm z$ axis $(0,0,z)$
	The uniformly charged disk has a radius of a, it is located in the $z=0$ plane and centered about the origin, the total charge on the ring is $\pi a^2 \rho_S$
Uniformly charged annular disk [Guru]	$\dfrac{\rho_S z}{2\varepsilon_o}\left(\dfrac{1}{\sqrt{b^2+z^2}}-\dfrac{1}{\sqrt{a^2+z^2}}\right)\hat{a}_z$ along the z axis $(0,0,z)$
	The annular cylindrical disk is uniformly charged, it is located in the $z=0$ plane and centered about the origin. the outer radius is a and inner radius is b, the total charge on the disk is $\pi(a^2-b^2)\rho_S$
Bipolar-charged circular disk [Zahn]	$-\dfrac{\rho_S}{\pi\varepsilon_o}\left[\ln\left(\dfrac{a+\sqrt{a^2+z^2}}{|z|}\right)-\dfrac{a}{\sqrt{a^2+z^2}}\right]\hat{a}_y$ along the z axis $(0,0,z)$
	The surface charge for $y>0$ is $+\rho_S$ while the surface charge for $y<0$ is $-\rho_S$, the disk has a radius of a, it is located in the $z=0$ plane and centered about the origin, the total charge on the disk is zero
Linearly varying, charged circular disk [Miner; Rao]	$\dfrac{Az}{2\varepsilon_o}\left[\ln\left(\dfrac{a+\sqrt{a^2+z^2}}{|z|}\right)-\dfrac{a}{\sqrt{a^2+z^2}}\right]\hat{a}_z$ along the z axis $(0,0,z)$
	The surface charge varies according to the expression $\rho_S(\rho)=A\rho$, the disk has a radius of a, it is located in the $z=0$ plane and centered about the origin, the total charge on the disk is $(2/3)A\pi a^3$
Uniformly charged square sheet [Zahn] 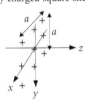	$\pm\dfrac{2\rho_S}{\pi\varepsilon_o}\left(\tan^{-1}\sqrt{\dfrac{\frac{a^2}{2}+z^2}{z^2}}-\dfrac{\pi}{4}\right)\hat{a}_z$ along the $\pm z$ axis $(0,0,z)$
	The uniformly charged square sheet has sides equal to a, it is located in the $z=0$ plane and centered about the origin, the total charge on the sheet is $a^2\rho_S$
Varying-charge square sheet [Haus]	$\dfrac{Az}{2\pi\varepsilon_o}\left(2\sqrt{a^2+4z^2}-\sqrt{2a^2+4z^2}-2z\right)\hat{a}_z$ along the $+z$ axis $(0,0,z)$
	The surface charge of the sheet varies according to the expression $\rho_S(x,y)=A|xy|$, it has sides equal to a, it is located in the $z=0$ plane and centered about the origin, the total charge on the sheet is $(A/16)a^4$

Charge Distribution	Electric Field $(\vec{E})$
Uniformly charged, infinite-length cylindrical surface [Demarest; Zahn] 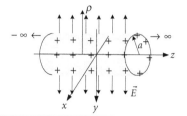	$\dfrac{\rho_S a}{\varepsilon_o \rho}\hat{a}_\rho$ if $\rho > a$ The field is radially outward from the cylinder's surface, the cylinder's axis is along the z axis and its radius is a, the charge is uniformly distributed at $\rho = a$ over its entire infinite length
Uniformly charged cylindrical surface [Zahn] 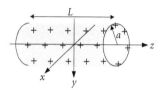	$\dfrac{\rho_S a}{2\varepsilon_o}\left[\dfrac{1}{\sqrt{a^2 + \left(z - \dfrac{L}{2}\right)^2}} - \dfrac{1}{\sqrt{a^2 + \left(z + \dfrac{L}{2}\right)^2}}\right]\hat{a}_z$ along the z axis $(0, 0, z)$ The cylinder of length, L, is charged along its outer curved surface at $\rho = a$, it is evenly spaced about the xy plane, the axis of the cylinder is along the z axis, the total charge on the cylinder is $2\pi a L \rho_S$
Uniformly charged, infinite-length semicylindrical surface [Zahn] 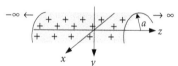	$\dfrac{\rho_S}{\pi\varepsilon_o}\hat{a}_y$ along the z axis $(0, 0, z)$ The semicylinder is infinitely long in the z direction and is uniformly charged along its outer surface at $\rho = a$, the axis of the cylinder is along the z axis
Parallel uniformly charged, infinite flat sheets of opposite sign [Zahn] 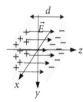	$\dfrac{\rho_S}{\varepsilon_o}\hat{a}_z$ if $|z| < \dfrac{d}{2}$, 0 if $|z| > \dfrac{d}{2}$ The field between the sheets is uniform, the sheets are separated by a distance of d and are centered about the $z = 0$ plane, the sheets are located at $z = \pm d/2$, the sheet extends to infinity in both the x and y directions
Uniformly charged infinite-length strip surface [Zahn] 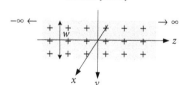	$\dfrac{\rho_S}{2\pi\varepsilon_o}\left[\dfrac{\pi}{2} + \sin^{-1}\left(\dfrac{\dfrac{w^2}{4} - x^2}{\dfrac{w^2}{4} + x^2}\right)\right]\hat{a}_x$ if $x > 0$ and in the xz plane $(x, 0, z)$ The field in the xz plane $(y = 0)$ is normal to the strip surface, the strip of width w is centered about $y = 0$ and infinite in the z direction, it has zero depth in the x direction
Uniformly charged spherical surface [Zahn]	$\dfrac{\rho_S a^2}{\varepsilon_o r^2}\hat{a}_r$ if $r > a$ The field outside the spherical surface charge distribution of radius a is identical to a point charge equal to $Q = 4\pi a^2 \rho_S$

Charge Distribution	Electric Field $(\vec{E})$
Uniformly charged hemispherical surface [Zahn; Jefimenko]	$\dfrac{\rho_S}{4\varepsilon_o}\hat{a}_z$ at the origin $(0,0,0)$ The total charge on the surface of the hemisphere of radius a is $2\pi a^2 \rho_S$
Sinusoidal-varying charged spherical surface [Haus] 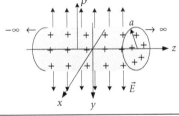	$\dfrac{2Aa^3}{3\varepsilon_o z^3}\hat{a}_z$ along the $+z$ axis $(0,0,z)$ The charge on the surface of the sphere of radius a is varying according to the expression $\rho_S(\theta)=A\cos\theta$ where θ is the angle from the z axis, the total charge on the surface of the sphere is zero
Uniformly charged, infinite-length cylindrical volume [Demarest] 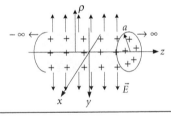	$\dfrac{\rho_V a^2}{2\varepsilon_o \rho}\hat{a}_\rho$ if $\rho \geq a$ The field is radially outward from the cylinder's surface of radius a, cylinder's axis is along the z axis, the charge is uniformly distributed throughout its infinite length, the field is continuous at $\rho=a$ since no surface charge is present
Cylindrically symmetric, infinite-length volume charge 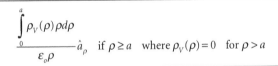	$\dfrac{\displaystyle\int_0^a \rho_V(\rho)\rho\,d\rho}{\varepsilon_o \rho}\hat{a}_\rho$ if $\rho \geq a$ where $\rho_V(\rho)=0$ for $\rho > a$ The field outside this symmetrical cylindrical charge distribution of radius a is only a function of the total charge per unit length, the volume charge is only a function of the radial (ρ) distance, the field is continuous at $\rho=a$ since no surface charge is present
Uniformly charged cylindrical volume [Zahn; Jefimenko] 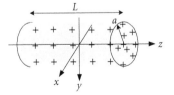	$\dfrac{\rho_V}{2\varepsilon_o}\left[\sqrt{a^2+\left(z-\dfrac{L}{2}\right)^2}-\left|z-\dfrac{L}{2}\right| \right.$ $\left. -\sqrt{a^2+\left(z+\dfrac{L}{2}\right)^2}+\left|z+\dfrac{L}{2}\right|\right]\hat{a}_z$ along the z axis $(0,0,z)$ $\approx \dfrac{\rho_V}{2\varepsilon_o}\left[\sqrt{a^2+(\Delta z)^2}-|\Delta z|\right]\hat{a}_z$ if $L \gg \Delta z, L \gg a$ along the z axis near the $+$ face $\left(0,0,\Delta z+\dfrac{L}{2}\right)$ The cylinder of length, L, and radius a is uniformly charged throughout its volume, it is evenly spaced about the xy plane, the axis of the cylinder is along the z axis, the total charge in the cylinder is $\pi a^2 L\rho_V$

Charge Distribution	Electric Field ($\vec{E}$)
Uniformly charged, infinite-length semicylindrical volume [Zahn] 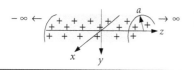	$\dfrac{\rho_V a}{\pi \varepsilon_o}\hat{a}_y$ along the z axis $(0,0,z)$ The semicylinder of radius a is infinitely long in the z direction and is uniformly charged throughout its volume, the axis of the cylinder is along the z axis
Uniformly charged, infinite-length thin-walled cylindrical shell [Jefimenko] 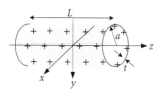	$\dfrac{\rho_V a t}{2\varepsilon_o \sqrt{a^2+(\Delta z)^2}}\hat{a}_z$ if $L \gg \Delta z, L \gg a, a \gg t$ along the z axis near $+$ face $\left(0,0,\Delta z + \dfrac{L}{2}\right)$ The thin-walled cylindrical shell of length, L, is uniformly charged, it is evenly spaced about the xy plane, the axis of the cylinder is along the z axis, the radius of the cylinder is a and the shell's thickness is t, the total charge in the cylinder is $[\pi a^2 - \pi(a-t)^2]L\rho_V$
Uniformly charged spherical volume [Demarest]	$\dfrac{\rho_V a^3}{3\varepsilon_o r^2}\hat{a}_r$ if $r \geq a$ The field outside the spherical charge distribution of radius a is identical to a point charge with a charge equal to $(4/3)\pi a^3 \rho_V$, the field is continuous at $r = a$ since no surface charge is present
Spherically symmetrical volume charge	$\dfrac{Q}{4\pi\varepsilon_o r^2}\hat{a}_r$ if $r \geq a$ where $Q = \iiint\limits_V \rho_V(r)\,dv, \rho_V(r)=0$ for $r > a$ The field outside the spherical charge distribution of radius a is identical to a point charge equal to Q, the volume charge is only a function of the radial (r) distance, Q is the total charge contained within the volume, the field is continuous at $r = a$ since no surface charge is present
Uniformly charged hemispherical volume [Zahn]	$\dfrac{\rho_V a}{4\varepsilon_o}\hat{a}_z$ at the origin $(0,0,0)$ The total charge in the hemisphere of radius a is $(2/3)\pi a^3 \rho_V$
Uniformly charged infinite slab [Zahn] 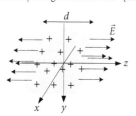	$\dfrac{\rho_V d}{2\varepsilon_o}\hat{a}_z$ if $z \geq \dfrac{d}{2}$, $-\dfrac{\rho_V d}{2\varepsilon_o}\hat{a}_z$ if $z \leq -\dfrac{d}{2}$ The uniform field is normal to the two sides of the slab, the slab of charge extends from $z = -d/2$ to $z = d/2$, it extends to infinity in both the x and y directions, the field is continuous at $z = \pm d/2$ since no surface charge is present

Charge Distribution	Electric Field ($\vec{E}$)

Nonuniformly charged infinite slab

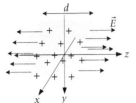

$$\pm \frac{\displaystyle\int_{-\frac{d}{2}}^{\frac{d}{2}} \rho_V(z)\,dz}{2\varepsilon_o}\,\hat{a}_z \quad \text{if } |z| \ge \pm\frac{d}{2} \quad \text{where } \rho_V(z) = 0 \quad \text{for } |z| > \frac{d}{2}$$

The uniform field is normal to the two sides of the slab, the slab of charge extends from $z = -d/2$ to $z = d/2$, it extends to infinity in both the x and y directions, the field outside this nonuniform charge distribution is only a function of the total charge per unit area, the volume charge is only a function of z, the field is continuous at $z = \pm d/2$ since no surface charge is present

Appendix B   Capacitance Formula

Spherical Conductor   [Zahn]

$$C_i = 4\pi\varepsilon_o R$$

Coated Spherical Conductor   [Jefimenko]

$$C_i = 4\pi\varepsilon_o R \left[1 + \frac{t(\varepsilon_{r1} - 1)}{t + \varepsilon_{r1} R} \right]$$

$$C_i \approx 4\pi\varepsilon_{r1}\varepsilon_o R \quad \text{if } t \gg R$$

where t is the thickness of the dielectric coating with a relative permittivity of ε_r.

Prolate Spheroid Conductor   [Weber; Gray]

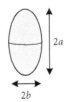

$$C_i = 4\pi\varepsilon_o \frac{\sqrt{a^2 - b^2}}{\tanh^{-1}\sqrt{1 - \frac{b^2}{a^2}}} = 4\pi\varepsilon_o \frac{\sqrt{a^2 - b^2}}{\frac{1}{2}\ln\left(\frac{1 + \sqrt{1 - \frac{b^2}{a^2}}}{1 - \sqrt{1 - \frac{b^2}{a^2}}}\right)} \quad \text{if } a > b$$

$$C_i \approx 4\pi\varepsilon_o a \quad \text{if } a \approx b$$

$$C_i \approx 4\pi\varepsilon_o \frac{a}{\ln\left(\frac{2a}{b}\right)} \quad \text{if } a \gg b$$

Oblate Spheroid Conductor [Weber; Moon, '61]

$$C_i = 4\pi\varepsilon_o \frac{\sqrt{a^2 - c^2}}{\tan^{-1}\sqrt{\frac{a^2}{c^2} - 1}} \quad \text{if } a > c$$

$$C_i \approx 4\pi\varepsilon_o c \quad \text{if } a \approx c$$

$$C_i \approx 4\pi\varepsilon_o \frac{a}{\tan^{-1}\left(\frac{a}{c}\right)} \approx 8\varepsilon_o a \quad \text{if } a \gg c$$

Ellipsoid Conductor [Smythe; Morse]

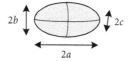

$$C_i = \frac{4\pi\varepsilon_o \sqrt{a^2 - c^2}}{\displaystyle\int_0^{\sqrt{1-\left(\frac{c}{a}\right)^2}} \frac{dx}{\sqrt{(1-x^2)\left[1 - \left(\frac{a^2 - b^2}{a^2 - c^2}\right)x^2\right]}}} \quad \text{if } a > b > c$$

$$C_i = 4\pi\varepsilon_o \frac{\sqrt{a^2 - b^2}}{\tanh^{-1}\sqrt{1 - \frac{b^2}{a^2}}} \quad \text{if } a > b, \ b = c$$

$$C_i = 4\pi\varepsilon_o \frac{\sqrt{a^2 - c^2}}{\tan^{-1}\sqrt{\frac{a^2}{c^2} - 1}} \quad \text{if } a = b, \ a > c$$

$$C_i \approx 8\varepsilon_o a \quad \text{if } a = b, \ a \gg c$$

Circular Disk Conductor [Weber; Morse; Dwight, '36]

$$C_i = 8\varepsilon_o R$$

The disk is thin compared to its radius, R.

Elliptical Disk Conductor [Gray; Morse]

$2b$ $2a$

$$C_i = \cfrac{4\pi\varepsilon_o a}{\displaystyle\int_0^1 \cfrac{dx}{\sqrt{(1-x^2)\left[1-\left(1-\dfrac{b^2}{a^2}\right)x^2\right]}}} \quad \text{if } a > b$$

$$= \cfrac{4\pi\varepsilon_o a}{\dfrac{\pi}{2}\left[1+\left(\dfrac{1}{2}\right)^2\left(1-\dfrac{b^2}{a^2}\right)+\left(\dfrac{1\times 3}{2\times 4}\right)^2\left(1-\dfrac{b^2}{a^2}\right)^2+\left(\dfrac{1\times 3\times 5}{2\times 4\times 6}\right)^2\left(1-\dfrac{b^2}{a^2}\right)^4+\cdots\right]}$$

$$C_i \approx \cfrac{8\varepsilon_o a}{\left[1+\left(\dfrac{1}{2}\right)^2\left(1-\dfrac{b^2}{a^2}\right)\right]} \quad \text{if } a \approx b$$

The disk is thin compared to its semiaxes a and b.

Circular Flat Ring Conductor [Gray]

a b

$$C_i = \frac{16\varepsilon_o}{\pi}a\left[\cos^{-1}\left(\frac{b}{a}\right)+\sqrt{1-\frac{b^2}{a^2}}\tanh^{-1}\left(\frac{b}{a}\right)\right]\left[1+\frac{0.0143a}{b}\tan^3\left(\frac{1.28b}{a}\right)\right] \quad \text{if } a \geq 1.1b$$

$$C_i \approx \frac{16\varepsilon_o}{\pi}a\left(\frac{\pi}{2}+\frac{b}{a}\right) \approx 8\varepsilon_o a \quad \text{if } a \gg b$$

$$C_i = \frac{5\pi\varepsilon_o}{8}\frac{a+b}{\ln\left(16\dfrac{a+b}{a-b}\right)} \quad \text{if } b \leq a < 1.1b$$

This ring, flat circular annulus, or "washer," is thin compared to a and b.

Rectangular Plate Conductor [Dwight, '36]

b

a

$$C_i = 2\pi\varepsilon_o \cfrac{1}{\dfrac{1}{a}\ln\left(\dfrac{a+\sqrt{a^2+b^2}}{b}\right)+\dfrac{1}{b}\ln\left(\dfrac{b+\sqrt{a^2+b^2}}{a}\right)+\dfrac{a}{3b^2}+\dfrac{b}{3a^2}-\dfrac{(a^2+b^2)\sqrt{a^2+b^2}}{3a^2b^2}}$$

$$C_i = \pi\varepsilon_o \frac{a}{\ln\left(1+\sqrt{2}\right)+\dfrac{1}{3}-\dfrac{\sqrt{2}}{3}} \approx 4.2\varepsilon_o a \quad \text{if } b=a$$

$$C_i \approx 2\pi\varepsilon_o \frac{b}{\ln\left(\dfrac{2b}{a}\right)} \quad \text{if } b \gg a$$

The rectangular plate is thin compared to a and b.

Circular Ring Conductor   [Weber; Gray; Dwight, '36]

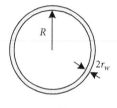

$$C_i = 4\pi\varepsilon_o \frac{\pi R}{\ln\left(\dfrac{8R}{r_w}\right)} \quad \text{if } R \gg r_w$$

Spherical Bowl Conductor   [Weber; Gray]

$$C_i = 4\varepsilon_o R\left[\sin(\alpha)+\alpha\right]$$

$$C_i \approx 8\varepsilon_o R\alpha \quad \text{if } \alpha \ll 1$$

$$C_i \approx 4\pi\varepsilon_o R \quad \text{if } 2\alpha = 2\pi \,(= 360°)$$

where a is in radians and the thickness of the bowl is thin compared to R. The figure shown is a cross-sectional view.

Solid Cylindrical Conductor   [Butler, '80; Weber; Gray; Dwight, '36]

$$C_i = 8\varepsilon_o R + 6.95\varepsilon_o R\left(\frac{l_{th}}{2R}\right)^{0.76} \quad \text{if } 4 \le \frac{l_{th}}{R} < 16$$

Using a long, prolate spheroid approximation (inscribed inside or outside the cylindrical rod)

$$C_i \approx 2\pi\varepsilon_o \frac{l_{th}}{\ln\left(\dfrac{l_{th}}{R}\right)} \quad \text{or} \quad C_i \approx 2\sqrt{2}\pi\varepsilon_o \frac{l_{th}}{\ln\left(\dfrac{l_{th}}{R}\right)} \quad \text{if } l_{th} \gg R$$

where l_{th} is the length of the cylinder.

Cylindrical Tube Conductor [Butler, '80; Weber]

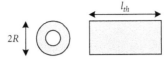

For thick tubes,

$$C_i = 4\pi\varepsilon_o \frac{\pi R}{\ln\left(\frac{32R}{l_{th}}\right)} \quad \text{if } \frac{l_{th}}{R} \le 8$$

For thin tubes,

$$C_i = 2\pi\varepsilon_o \frac{l_{th}}{\ln\left(\frac{2l_{th}}{R}\right) - 1} \approx 2\pi\varepsilon_o \frac{l_{th}}{\ln\left(\frac{2l_{th}}{R}\right)} \quad \text{if } \frac{l_{th}}{R} \ge 8$$

where l_{th} is the length of the cylindrical tube.

Strip Conductor [Dwight, '36]

$$C_i = 2\pi\varepsilon_o \frac{l_{th}}{\ln\left(\frac{2l_{th}}{a}\right) + \frac{a^2 - \pi ab}{2(a+b)^2}} \quad \text{if } l_{th} \gg \max(a,b), \ a > 8b$$

$$C_i \approx 2\pi\varepsilon_o \frac{l_{th}}{\ln\left(\frac{2l_{th}}{a}\right)} \quad \text{if } l_{th} \ggg \max(a,b), \ a > 8b$$

where a and b are the width and thickness of the strip of length l_{th}.

Figures of Rotation Conductor [Gray]

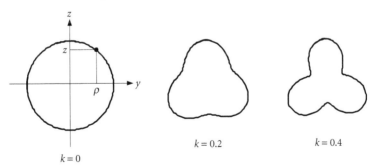

$k = 0.2$ $k = 0.4$

$k = 0$

$$C_i = 4\pi\varepsilon_o a \left(1 - 0.06857k^2 - 0.00559k^4\right) \quad \text{if } 0 \le k \le \frac{1}{2}$$

where the figure of rotation is described by

$$z = a\left[\cos(u) + k\cos(2u)\right], \quad \rho = a\left[\sin(u) - k\sin(2u)\right]$$

and u varies from 0 to 2π. The closed path described by this set of equations (and shown in the given figures) is rotated 360° about the z axis to generate the solid body. For a sphere,

$$z = a\cos(u), \quad \rho = a\sin(u) \quad \text{if } k = 0$$

$$C = 4\pi\varepsilon_o a \quad \text{if } k = 0$$

Two Spherical Conductors in Contact [Smythe; Zahn]

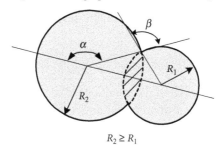

$$C_i = -4\pi\varepsilon_o \frac{ab}{a+b}\left[1.154 + \psi\left(\frac{a}{a+b}\right) + \psi\left(\frac{b}{a+b}\right)\right]$$

$$C_i = -4\pi\varepsilon_o \frac{a}{2}\left[1.154 + \psi\left(\frac{1}{2}\right) + \psi\left(\frac{1}{2}\right)\right] = 8\pi\varepsilon_o a\ln(2) \quad \text{if } a = b$$

$$C_i \approx -4\pi\varepsilon_o a\left[1.154 + \psi\left(\frac{a}{b}\right) + \psi(1)\right] \approx 4\pi\varepsilon_o b \quad \text{if } b \gg a$$

where

$$\psi(x) = \frac{\Gamma'(x)}{\Gamma(x)} = \frac{\displaystyle\int_0^\infty e^{-\lambda}\ln(\lambda)\,\lambda^{x-1}d\lambda}{\displaystyle\int_0^\infty e^{-\lambda}\lambda^{x-1}d\lambda} = -0.5772 - \frac{1}{x} + \sum_{n=1}^\infty\left(\frac{1}{n} - \frac{1}{n+x}\right)$$

Two Orthogonally Intersecting Spherical Conductors [Weber; Smythe]

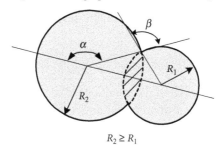

$$R_2 \geq R_1$$

$$C_i = 4\pi\varepsilon_o R_2\left[\frac{1 + \dfrac{R_2}{R_1}}{\dfrac{R_2}{R_1}} - \frac{1}{\sqrt{1 + \left(\dfrac{R_2}{R_1}\right)^2}}\right] \quad \text{if } \beta = \frac{\pi}{2}(=90°)$$

$$C_i \approx 4\pi\varepsilon_o R_2 \quad \text{if } \beta = \frac{\pi}{2}, R_2 \gg R_1$$

$$C_i \approx 4\pi\varepsilon_o R_1 \quad \text{if } \beta = \frac{\pi}{2}, R_1 \gg R_2$$

Two Spherical Conductors Connected By Thin Wire [Gray; Smythe]

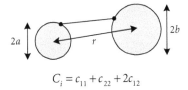

$$C_i = c_{11} + c_{22} + 2c_{12}$$

where

$$c_{11} = 4\pi\varepsilon_o ab \sinh(\alpha) \sum_{m=1}^{\infty} \frac{1}{b\sinh(m\alpha) + a\sinh[(m-1)\alpha]} \quad \text{if } r > a+b$$

$$c_{22} = 4\pi\varepsilon_o ab \sinh(\alpha) \sum_{m=1}^{\infty} \frac{1}{a\sinh(m\alpha) + b\sinh[(m-1)\alpha]} \quad \text{if } r > a+b$$

$$c_{12} = -4\pi\varepsilon_o ab \sinh(\alpha) \sum_{m=1}^{\infty} \frac{1}{r\sinh(m\alpha)} \quad \text{if } r > a+b$$

$$\alpha = \cosh^{-1}\left(\frac{r^2 - a^2 - b^2}{2ab}\right)$$

$$C_i \approx 4\pi\varepsilon_o ab \sum_{m=1}^{\infty} \left[\frac{1}{bm + a(m-1)} + \frac{1}{am + b(m-1)} - 2\frac{1}{(a+b)m} \right] \quad \text{if } r \approx a+b$$

$$C_i \approx 4\pi\varepsilon_o ab \left\{ \frac{1}{b} + \frac{1}{a} - \frac{2}{a+b} + \sum_{m=2}^{\infty} \left[\frac{1}{bm + a(m-1)} + \frac{1}{am + b(m-1)} - 2\frac{1}{(a+b)m} \right] \right\}$$

$$\approx 4\pi\varepsilon_o ab\left(\frac{1}{a} + \frac{1}{b}\right) \approx 4\pi\varepsilon_o b \quad \text{if } r \approx a+b, b \gg a$$

$$C_i \approx 4\pi\varepsilon_o a \sum_{m=1}^{\infty} \left(\frac{2}{2m-1} - \frac{1}{m}\right) \approx 8\pi\varepsilon_o a \ln(2) \quad \text{if } r \approx a+b, a=b$$

$$C_i = 8\pi\varepsilon_o a \sinh\left[\cosh^{-1}\left(\frac{r}{2a}\right)\right] \sum_{m=1}^{\infty} (-1)^{m+1} \frac{1}{\sinh\left[m\cosh^{-1}\left(\frac{r}{2a}\right)\right]} \quad \text{if } a=b$$

where

$$\alpha = \cosh^{-1}\left(\frac{r^2 - a^2 - b^2}{2ab}\right)$$

$$\alpha \approx \cosh^{-1}\left(\frac{r^2}{2ab}\right) \approx \ln\left(\frac{r^2}{ab}\right) \quad \text{if } r \gg \max(a,b)$$

$$\alpha \approx 0 \quad \text{if } r \approx a+b$$

The wire is thin relative to the radii of the spheres.

Arbitrary Shaped Isolated Conductor [Chow; Gray]

"1" "2"

If the capacitance and surface area of object "1" are C_{i1} and S_1, respectively, then the isolated capacitance of a somewhat similar shaped object "2" of surface area S_2 is approximately

$$C_{i2} \approx C_{i1}\sqrt{\frac{S_2}{S_1}}$$

The relationship is exact if the shapes of objects "1" and "2" are identical. The capacitance of a nearly spherical object of surface area S_2 is

$$C_{i2} \approx 4\pi\varepsilon_o R\sqrt{\frac{S_2}{4\pi R^2}} = 2\sqrt{\pi}\varepsilon_o\sqrt{S_2} \approx 31\times 10^{-12}\sqrt{S_2}$$

If the known capacitance of the shape most similar to a cube is a sphere, then an approximation for the isolated capacitance of a cube of side a is

$$C_{i2} \approx 2\sqrt{\pi}\varepsilon_o\sqrt{6a^2} = 2\sqrt{6\pi}\varepsilon_o a \approx 77\times 10^{-12}a$$

A lower limit given for a cube of side a is

$$C_{i2} \approx 73\times 10^{-12}a$$

Two Spherical Conductors [Smythe; Weber; Gray; Greason, '87]

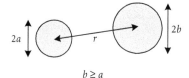

$2a$ r $2b$

$$b \geq a$$

$$C = \frac{c_{11}c_{22}-c_{12}^2}{c_{11}+c_{22}+2c_{12}} \quad {}^{48}$$

where the *coefficients* of capacitance are equal to

$$c_{11} = 4\pi\varepsilon_o ab\sinh(\alpha)\sum_{m=1}^{\infty}\frac{1}{b\sinh(m\alpha)+a\sinh[(m-1)\alpha]} \quad \text{if } r>a+b$$

$$c_{22} = 4\pi\varepsilon_o ab\sinh(\alpha)\sum_{m=1}^{\infty}\frac{1}{a\sinh(m\alpha)+b\sinh[(m-1)\alpha]} \quad \text{if } r>a+b$$

---

[48]In terms of the typical circuit capacitances, and assuming that the voltage reference is far away,

$$C = \frac{C_{1G}C_{2G}}{C_{1G}+C_{2G}}+C_{12}$$

That is, the total capacitance between the two spheres is the capacitance C_{12} in parallel with the series combination of the capacitance between one sphere and the voltage reference (G) and the capacitances between the other sphere and the voltage reference (G). See the coefficient of capacitances discussion for more details.

$$c_{12} = -4\pi\varepsilon_o ab\sinh(\alpha)\sum_{m=1}^{\infty}\frac{1}{r\sinh(m\alpha)} \quad \text{if } r > a+b$$

$$\alpha = \cosh^{-1}\left(\frac{r^2 - a^2 - b^2}{2ab}\right)$$

$$C \approx \frac{4\pi\varepsilon_o}{\frac{a+b}{ab} - \frac{1}{r}} \approx 4\pi\varepsilon_o\frac{ab}{a+b} \quad \text{if } r \gg a+b \;\; [49]$$

$$C \approx \frac{4\pi\varepsilon_o}{\frac{2}{a} - \frac{1}{r}} \approx 2\pi\varepsilon_o a \quad \text{if } r \gg a+b, a = b$$

$$C = 2\pi\varepsilon_o a\sinh(\beta)\sum_{m=1}^{\infty}\left\{\frac{1}{\sinh[(2m-1)\beta]} + \frac{1}{\sinh(2m\beta)}\right\} \quad \text{if } a = b$$

$$\beta = \cosh^{-1}\left(\frac{r}{2a}\right)$$

These results assume that the charge on each spherical conductor is equal but opposite. When a thin wire connects two spheres, so that their potential is the same, the total charge available would be distributed based on the relative size of the spheres. In this case, the set of expressions for two spheres connected by a thin wire provided in this table should be used.

Spherical Conductor and Large Flat Conductor [Smythe; Deno, '75]

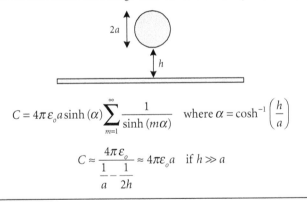

$$C = 4\pi\varepsilon_o a\sinh(\alpha)\sum_{m=1}^{\infty}\frac{1}{\sinh(m\alpha)} \quad \text{where } \alpha = \cosh^{-1}\left(\frac{h}{a}\right)$$

$$C \approx \frac{4\pi\varepsilon_o}{\frac{1}{a} - \frac{1}{2h}} \approx 4\pi\varepsilon_o a \quad \text{if } h \gg a$$

Cylindrical Conductor Perpendicular to Large Flat Conductor [Weber]

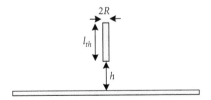

---

[49]This interesting approximation that is independent of the distance between the two spheres can be interpreted as the *series* combination of the capacitance of one sphere to a ground plane at infinity and the capacitance from this ground plane at infinity to the other sphere.

$$C = \frac{2\pi\varepsilon_o l_{th}}{\ln\left(\frac{l_{th}}{R}\sqrt{\frac{2h+\frac{l_{th}}{2}}{2h+\frac{3l_{th}}{2}}}\right)} \quad \text{if } l_{th} \gg R$$

$$C \approx \frac{2\pi\varepsilon_o l_{th}}{\ln\left(\frac{l_{th}}{R}\right)} \quad \text{if } l_{th} \gg R, h \gg l_{th}$$

$$C \approx \frac{2\pi\varepsilon_o l_{th}}{\ln\left(\frac{l_{th}}{R\sqrt{3}}\right)} \quad \text{if } l_{th} \gg h$$

where R is the radius of the solid cylinder. The width of the flat plane is large compared to h and l_{th}.

Cylindrical Conductor Parallel to Large Flat Conductor [Weber; Smythe]

$$C = \frac{2\pi\varepsilon_o l_{th}}{\ln\left\{\frac{l_{th}^2}{4hR}\left[\sqrt{1+\left(\frac{4h}{l_{th}}\right)^2}-1\right]\right\}} \quad \text{if } l_{th} \gg R$$

$$C \approx \frac{2\pi\varepsilon_o l_{th}}{\ln\left(\frac{2h}{R}\right)} \quad \text{if } l_{th} \gg R, l_{th} \gg h$$

where R is the radius of the solid cylinder. The width of the flat plane is large compared to h and l_{th}.

Circular Ring Conductor Parallel to Large Flat Conductor [Weber]

$$C = 4\pi\varepsilon_o \frac{\pi}{\frac{1}{R}\ln\left(\frac{8R}{r_w}\right) - \frac{2K(k)}{\sqrt{(2R)^2+(2h)^2}}} \quad \text{if } R \gg r_w \quad \text{where } k = \frac{1}{1+\left(\frac{h}{R}\right)^2}$$

$$C \approx 4\pi\varepsilon_o \frac{\pi}{\frac{1}{R}\ln\left(\frac{8R}{r_w}\right)-\frac{\pi}{2h}} \approx 4\pi\varepsilon_o \frac{R\pi}{\ln\left(\frac{8R}{r_w}\right)} \quad \text{if } R \gg r_w,\ h \gg R$$

$$C \approx 4\pi\varepsilon_o \frac{R\pi}{\ln\left(\frac{2\sqrt{2}h}{r_w}\right)} \quad \text{if } R \gg \max(r_w, h)$$

where $K(k)$ is the complete elliptical integral of the first kind

$$K(k) = \int_0^{\frac{\pi}{2}} \frac{d\theta}{\sqrt{1 - k^2 \sin^2 \theta}}$$

$$K(k) \approx \frac{\pi}{2} \quad \text{if } k \ll 1, \quad K(k) \approx \ln\left(\frac{4}{\sqrt{1-k^2}}\right) \quad \text{if } k \approx 1$$

R is the radius of the ring and r_w is the radius of the solid circular conductor of the ring. The width of the flat plane is large compared to h and R.

Two Identical, Parallel, Circular Ring Conductors [Weber]

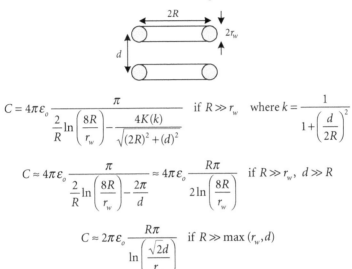

$$C = 4\pi\varepsilon_o \frac{\pi}{\frac{2}{R}\ln\left(\frac{8R}{r_w}\right) - \frac{4K(k)}{\sqrt{(2R)^2 + (d)^2}}} \quad \text{if } R \gg r_w \quad \text{where } k = \frac{1}{1 + \left(\frac{d}{2R}\right)^2}$$

$$C \approx 4\pi\varepsilon_o \frac{\pi}{\frac{2}{R}\ln\left(\frac{8R}{r_w}\right) - \frac{2\pi}{d}} \approx 4\pi\varepsilon_o \frac{R\pi}{2\ln\left(\frac{8R}{r_w}\right)} \quad \text{if } R \gg r_w, \ d \gg R$$

$$C \approx 2\pi\varepsilon_o \frac{R\pi}{\ln\left(\frac{\sqrt{2d}}{r_w}\right)} \quad \text{if } R \gg \max(r_w, d)$$

where $K(k)$ is the complete elliptical integral of the first kind

$$K(k) = \int_0^{\frac{\pi}{2}} \frac{d\theta}{\sqrt{1 - k^2 \sin^2 \theta}}$$

$$K(k) \approx \frac{\pi}{2} \quad \text{if } k \ll 1, \quad K(k) \approx \ln\left(\frac{4}{\sqrt{1-k^2}}\right) \quad \text{if } k \approx 1$$

R is the radius of the identical rings and r_w is the radii of the solid circular conductors.

Two Concentric Cylindrical Conductors [Zahn]

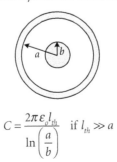

$$C = \frac{2\pi\varepsilon_o l_{th}}{\ln\left(\frac{a}{b}\right)} \quad \text{if } l_{th} \gg a$$

$$C \approx \varepsilon_o \frac{S}{d} \quad \text{if } l_{th} \gg a, \ a \approx b, \ S = 2\pi a l_{th}, \ d = a - b$$

where l_{th} is the length of the coaxial cable.

Two Concentric Cylindrical Conductors with Varying Dielectric [Haus]

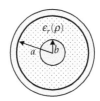

$$C = \frac{2\pi \varepsilon_o \beta l_{th}}{\ln\left(\dfrac{\alpha + \beta a}{\alpha + \beta b}\right)} \quad \text{if } l_{th} \gg a$$

$$C = \frac{2\pi \varepsilon_r \varepsilon_o l_{th}}{\ln\left(\dfrac{a}{b}\right)} \quad \text{if } l_{th} \gg a, \ \alpha = 0, \ \beta = \varepsilon_r$$

where the relative permittivity varies uniformly from one conductor to the other conductor only in the ρ direction as

$$\varepsilon_r(\rho) = \frac{\alpha}{\rho} + \beta$$

and l_{th} is the length of the coaxial cable. When the relative permittivity varies as

$$\varepsilon_r(r) = \frac{\chi}{\rho^2} + \xi$$

then the capacitance is

$$C = \frac{4\pi \varepsilon_o \xi l_{th}}{\ln\left(\dfrac{\chi + \xi a^2}{\chi + \xi b^2}\right)} \quad \text{if } l_{th} \gg a$$

When the relative permittivity varies only in the ϕ direction (around the center conductor) as

$$\varepsilon_r(\phi) = \varepsilon_{ra} + \varepsilon_{rb} \cos^2 \phi$$

then

$$C = \frac{(2\varepsilon_{ra} + \varepsilon_{rb}) \pi l_{th}}{\ln\left(\dfrac{a}{b}\right)} \quad \text{if } l_{th} \gg a$$

Two Concentric Cylindrical Conductors with Layered Dielectrics [Zahn; Jefimenko]

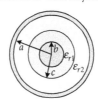

$$C = \frac{2\pi \varepsilon_{r1}\varepsilon_{r2}\varepsilon_o l_{th}}{\varepsilon_{r2}\ln\left(\frac{c}{b}\right) - \varepsilon_{r1}\ln\left(\frac{c}{a}\right)} \quad \text{if } l_{th} \gg a$$

$$C \approx \frac{\varepsilon_{r1}\varepsilon_{r2}\varepsilon_o S}{\varepsilon_{r2}s + \varepsilon_{r1}d} \quad \text{if } l_{th} \gg a,\ a \approx b \approx c,\ S = 2\pi a l_{th},\ d = a - c,\ s = c - b$$

$$\varepsilon_r = \begin{cases} \varepsilon_{r1} & \text{if } b < r < c \\ \varepsilon_{r2} & \text{if } c < r < a \end{cases}$$

where l_{th} is the length of the coaxial cable.

Two Concentric Cylindrical Conductors with Two Parallel Dielectrics [Zahn; Haus]

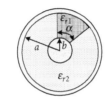

$$C = \frac{[\alpha\varepsilon_{r1} + (2\pi - \alpha)\varepsilon_{r2}]\varepsilon_o l_{th}}{\ln\left(\frac{a}{b}\right)} \quad \text{if } l_{th} \gg a$$

$$C = \frac{\pi(\varepsilon_{r1} + 1)\varepsilon_o l_{th}}{\ln\left(\frac{a}{b}\right)} \quad \text{if } l_{th} \gg a,\ \alpha = \pi\,(=180°),\ \varepsilon_{r2} = 1$$

where α is in radians and l_{th} is the length of the coaxial cable.

Two Partial Concentric Cylindrical Conductors

$$C = \frac{\varepsilon_o \alpha l_{th}}{\ln\left(\frac{a}{b}\right)} \quad \text{if } l_{th} \gg a,\ \alpha a \gg (a - b)$$

$$C \approx \frac{\varepsilon_o \alpha l_{th} a}{a - b} \quad \text{if } l_{th} \gg a,\ \alpha a \gg (a - b),\ a \approx b$$

where α is in radians and l_{th} is the length of the conductors located at $r = b$ and $r = a$.

Two Slanted Conductors [Jefimenko]

$$C = \frac{\varepsilon_o l_{th}}{2\sin^{-1}\left(\frac{d - s}{2w}\right)}\ln\left(\frac{d}{s}\right) \quad \text{if } l_{th} \gg w,\ w \gg d$$

$$C \approx \frac{\varepsilon_o l_{th} w}{s} \quad \text{if } l_{th} \gg w, \ w \gg d, \ d \approx s$$

where s and d are the respective distances between the two ends of the electrodes (not the arc lengths) and l_{th} is the length of the two conductors.

Two Slanted Conductors with Layered Dielectrics [Zahn]

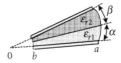

$$C = \frac{\varepsilon_{r1}\varepsilon_{r2}\varepsilon_o l_{th}}{\alpha \varepsilon_{r2} + \beta \varepsilon_{r1}} \ln\left(\frac{a}{b}\right) \quad \text{if } l_{th} \gg a, \frac{\pi}{2} \gg (\alpha + \beta)$$

$$C = \frac{\varepsilon_{r1}\varepsilon_o l_{th}}{\alpha + \beta} \ln\left(\frac{a}{b}\right) \quad \text{if } l_{th} \gg a, \ \frac{\pi}{2} \gg (\alpha + \beta), \ \varepsilon_{r1} = \varepsilon_{r2}$$

where α and β are in radians and l_{th} is the length. The two electrodes are in the radial direction and extend from $r = b$ to a (and $b \neq 0$).

Two Cylindrical Conductors with Displaced Inner Conductor [Zahn]

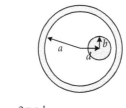

$$C = \frac{2\pi \varepsilon_o l_{th}}{\cosh^{-1}\left(\frac{a^2 + b^2 - d^2}{2ab}\right)} \quad \text{if } l_{th} \gg a, \ (a - b) > d$$

$$C \approx \frac{2\pi \varepsilon_o l_{th}}{\ln\left(\frac{a^2 + b^2 - d^2}{ab}\right)} \quad \text{if } l_{th} \gg a, \ (a - b) \gg d$$

$$C = \frac{2\pi \varepsilon_o l_{th}}{\ln\left(\frac{a}{b}\right)} \quad \text{if } l_{th} \gg a, \ d = 0$$

where l_{th} is the length of the cable and d is the distance between the axis of the outer conductor of radius a and axis of the inner conductor of radius b.

Three Concentric Cylindrical Conductors

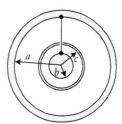

A fine insulated wire passes through a hole in the intermediate cylindrical conductor, connecting the inner and outer cylindrical conductors. The capacitance between the intermediate cylinder and the outer and inner cylinder is

$$C = 2\pi\varepsilon_o l_{th} \left[\frac{1}{\ln\left(\dfrac{a}{c}\right)} + \frac{1}{\ln\left(\dfrac{c}{b}\right)} \right] \quad \text{if } l_{th} \gg a$$

$$C \approx \varepsilon_o S \left(\frac{1}{d_1} + \frac{1}{d_2} \right) \quad \text{if } l_{th} \gg a, \; a \approx b \approx c, \; S = 2\pi a l_{th}, \; d_1 = a - c, \; d_2 = c - b$$

where l_{th} is the length of the three conductors.

Two Concentric Conductors with Slotted Outer Conductor [Howard W. Sams]

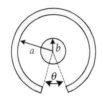

$$C \geq \frac{2\pi\varepsilon_o l_{th}}{\ln\left(\dfrac{a}{b}\right) + 5\times 10^{-4}\theta^2} \quad \text{if } l_{th} \gg a$$

where θ, in radians, corresponds to the opening of the slot and l_{th} is the length of the slotted coax.

Confocal Elliptical Conductors [Hilberg; Weber]

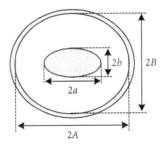

If the focal lengths of the inner and outer elliptical-shaped conductors are the same

$$\sqrt{A^2 - B^2} = \sqrt{a^2 - b^2}$$

and $A > B$ and $a > b$, then the capacitance is

$$C = \frac{2\pi\varepsilon_o l_{th}}{\cosh^{-1}\left(\dfrac{A}{\sqrt{A^2-B^2}}\right) - \cosh^{-1}\left(\dfrac{a}{\sqrt{a^2-b^2}}\right)} = \frac{2\pi\varepsilon_o l_{th}}{\ln\left(\dfrac{A+B}{a+b}\right)} \quad \text{if } l_{th} \gg A$$

$$C \approx \frac{2\pi\varepsilon_o l_{th}}{\ln\left(\dfrac{A}{a}\right)} \quad \text{if } l_{th} \gg A, \; A \approx B, \; a \approx b$$

where l_{th} is the length of this confocal elliptical coax.

Cylindrical Conductor in Square Enclosure [Lo; Howard W. Sams]

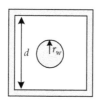

$$C \approx \frac{2\pi\varepsilon_o l_{th}}{\left\{ \ln\left(\dfrac{d}{2r_w}\right) + 0.108 - 0.039\left[\dfrac{1+0.405\left(\dfrac{d}{2r_w}\right)^{-4}}{1-0.405\left(\dfrac{d}{2r_w}\right)^{-4}}\right] - 0.008\left[\dfrac{1+0.163\left(\dfrac{d}{2r_w}\right)^{-8}}{1-0.163\left(\dfrac{d}{2r_w}\right)^{-8}}\right] \atop \qquad - 0.002\left[\dfrac{1+0.067\left(\dfrac{d}{2r_w}\right)^{-12}}{1-0.067\left(\dfrac{d}{2r_w}\right)^{-12}}\right] \right\}} \qquad \text{if } l_{th} \gg d$$

$$C \approx \frac{2\pi\varepsilon_o l_{th}}{\ln\left(0.54\dfrac{d}{r_w}\right)} \quad \text{if } l_{th} \gg d, d \gg r_w$$

where l_{th} is the length of these conductors.

Square Conductor in Square Enclosure [Lo]

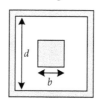

$$C = \begin{cases} \dfrac{8\varepsilon_o l_{th}\left(0.279 + 0.721\dfrac{b}{d}\right)}{1 - \dfrac{b}{d}} & \text{if } l_{th} \gg d, \dfrac{d}{4} \le b \le \dfrac{d}{2} \\[4ex] \dfrac{6.33 l_{th}\varepsilon_o}{\ln\left(0.956\dfrac{d}{b}\right)} & \text{if } l_{th} \gg d, b \le \dfrac{d}{2} \end{cases}$$

$$C \approx \frac{8\varepsilon_o d l_{th}}{d-b} = 4\frac{\varepsilon_o d l_{th}}{\left(\dfrac{d-b}{2}\right)} \quad \text{if } l_{th} \gg d, \ b \approx d$$

where l_{th} is the length of this square coaxial line.

Cylindrical Conductor in Trough [Lo; Howard W. Sams]

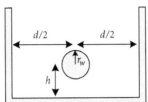

$$C \approx \frac{2\pi \varepsilon_o l_{th}}{\ln\left[\tanh\left(\dfrac{\pi h}{d}\right)\coth\left(\dfrac{\pi r_w}{2d}\right)\right]} \quad \text{if } l_{th} \gg \max(d,h),\, d > 8r_w,\, h > 3r_w$$

$$C \approx \frac{2\pi \varepsilon_o l_{th}}{\ln\left[\dfrac{2d}{\pi r_w}\tanh\left(\dfrac{\pi h}{d}\right)\right]} \quad \text{if } l_{th} \gg \max(d,h),\, d \gg r_w,\, h \gg r_w$$

$$C \approx \frac{2\pi \varepsilon_o l_{th}}{\ln\left(\dfrac{2h}{r_w}\right)} \quad \text{if } l_{th} \gg \max(d,h),\, d \gg r_w,\, h \gg r_w,\, d \gg h$$

where l_{th} is the length of these conductors.

Rectangular Conductor in Rectangular Enclosure [Lo; Tippet]

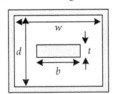

$$C = \frac{2\pi \varepsilon_o l_{th}}{\ln\left(\dfrac{1+\dfrac{w}{d}}{\dfrac{b}{d}+\dfrac{t}{d}}\right)} \quad \text{if } l_{th} \gg \max(w,d),\, d > 3t,\, w > 1.25b$$

$$C = \frac{2\pi \varepsilon_o l_{th}}{\ln\left(\dfrac{d}{b}\right)} \quad \text{if } l_{th} \gg \max(w,d),\, d > 3t,\, w > 1.25b,\, w = d,\, b = t$$

$$C = \frac{2\pi \varepsilon_o l_{th}}{\ln\left(\dfrac{d+w}{b}\right)} \quad \text{if } l_{th} \gg \max(w,d),\, d \gg t,\, w > 1.25b$$

$$C = 4\varepsilon_o l_{th}\left\{\frac{b}{d}+\frac{2}{\pi}\ln\left[1+\coth\left(\frac{\pi w}{2d}\right)\right]\right\} \quad \text{if } l_{th} \gg \max(w,d),\, d \gg t,\, w > 2.86d \quad [50]$$

$$C \approx 2\frac{2\varepsilon_o l_{th} b}{d} \quad \text{if } l_{th} \gg \max(w,d),\, d \gg t,\, w \gg 2.86d,\, b \gg d$$

where l_{th} is the length of this rectangular coaxial line.

---

[50]A more accurate expression, used for TEM cells, is available when the distance $w-b$ is small.

Cylindrical Conductor in Corner Trough [Hilberg; Johnson, '61]

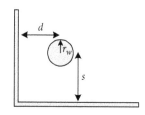

$$C = \frac{2\pi \varepsilon_o l_{th}}{\cosh^{-1}\left(\dfrac{ds}{r_w \sqrt{s^2+d^2}}\right)} \quad \text{if } l_{th} \gg \max(d,s),\ \min(d,s) \gg r_w$$

$$= \frac{2\pi \varepsilon_o l_{th}}{\ln\left[\dfrac{ds}{r_w \sqrt{s^2+d^2}} + \sqrt{\left(\dfrac{ds}{r_w \sqrt{s^2+d^2}}\right)^2 - 1}\ \right]} \quad \text{if } l_{th} \gg \max(d,s),\ \min(d,s) \gg r_w$$

$$C = \frac{2\pi \varepsilon_o l_{th}}{\cosh^{-1}\left(\dfrac{d}{r_w \sqrt{2}}\right)} \approx \frac{2\pi \varepsilon_o l_{th}}{\ln\left(\dfrac{\sqrt{2}d}{r_w}\right)} \quad \text{if } l_{th} \gg \max(d,s),\ \min(d,s) \gg r_w,\ s = d$$

$$C \approx \frac{2\pi \varepsilon_o l_{th}}{\ln\left[\dfrac{d}{r_w} + \sqrt{\left(\dfrac{d}{r_w}\right)^2 - 1}\ \right]} \approx \frac{2\pi \varepsilon_o l_{th}}{\ln\left(\dfrac{2d}{r_w}\right)} \quad \text{if } l_{th} \gg \max(d,s),\ \min(d,s) \gg r_w,\ s \gg d$$

The width and height of the flat walls are large compared to d and s. The length of the conductors is l_{th}.

Hyperbolic Conductor in Corner Trough [Zahn]

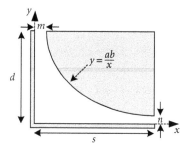

$$C \approx \frac{\varepsilon_o l_{th}}{2ab}(s^2+d^2) \quad \text{if } l_{th} \gg \max(d,s)$$

So that the fringing fields are negligible, $d \gg m$ and $s \gg n$. The length of the conductors is l_{th}.

Cylindrical Conductor near Concave Cylindrical Conductor [Hilberg]

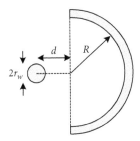

For free space between the two conductors,

$$C = \cfrac{2\pi\varepsilon_o l_{th}}{\cosh^{-1}\left[\cfrac{\dfrac{\pi}{4}+\dfrac{1}{2}\tan^{-1}\left(\dfrac{2Rd}{R^2-d^2+r_w^2}\right)}{\dfrac{1}{2}\tan^{-1}\left(\dfrac{2Rr_w}{R^2+d^2-r_w^2}\right)}\right]} \qquad \text{if } l_{th} \gg (d+R),\ R \gg r_w$$

$$C \approx \cfrac{2\pi\varepsilon_o l_{th}}{\cosh^{-1}\left(\dfrac{\pi}{4}\dfrac{R}{r_w}+\dfrac{d}{r_w}\right)} \approx \cfrac{2\pi\varepsilon_o l_{th}}{\ln\left(\dfrac{\pi}{2}\dfrac{R}{r_w}+\dfrac{2d}{r_w}\right)}$$

$$\approx \cfrac{2\pi\varepsilon_o l_{th}}{\ln\left(\dfrac{R}{r_w}\right)+0.45+1.27\dfrac{d}{R}} \qquad \text{if } l_{th} \gg R,\ R \gg \max{(r_w,d)}$$

where l_{th} is the length of the conductors.

Two Concentric Spherical Conductors [Zahn]

$$C = \frac{4\pi\varepsilon_o}{\dfrac{1}{b}-\dfrac{1}{a}} = \frac{4\pi\varepsilon_o ab}{a-b}$$

$$C \approx \varepsilon_o \frac{S}{d} \quad \text{if } a \approx b,\ S = 4\pi a^2,\ d = a-b$$

$$C \approx 4\pi\varepsilon_o b \quad \text{if } a \gg b$$

Two Concentric Spherical Conductors with Varying Dielectric

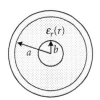

$$C = \frac{4\pi \varepsilon_o \alpha}{\ln \left[\dfrac{a\,(\alpha + \beta b)}{b\,(\alpha + \beta a)} \right]}$$

$$C = \frac{4\pi \varepsilon_o \varepsilon_r}{\dfrac{1}{b} - \dfrac{1}{a}} \quad \text{if } \alpha = 0,\ \beta = \varepsilon_r$$

where the relative permittivity varies uniformly from one conductor to the other conductor only in the r direction as

$$\varepsilon_r(r) = \frac{\alpha}{r} + \beta$$

When the relative permittivity varies as

$$\varepsilon_r(r) = \frac{\chi}{r^2} + \xi$$

then the capacitance is

$$C = \frac{4\pi \varepsilon_o \sqrt{\chi \xi}}{\tan^{-1} \left(\dfrac{a}{\sqrt{\dfrac{\chi}{\xi}}} \right) - \tan^{-1} \left(\dfrac{b}{\sqrt{\dfrac{\chi}{\xi}}} \right)}$$

$$C = \frac{4\pi \varepsilon_o \varepsilon_r}{\dfrac{1}{b} - \dfrac{1}{a}} \quad \text{if } \chi = 0,\ \xi = \varepsilon_r$$

Two Concentric Spherical Conductors with Layered Dielectrics [Zahn; Jefimenko]

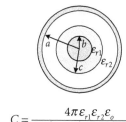

$$C = \frac{4\pi \varepsilon_{r1} \varepsilon_{r2} \varepsilon_o}{\varepsilon_{r2} \left(\dfrac{1}{b} - \dfrac{1}{c} \right) - \varepsilon_{r1} \left(\dfrac{1}{a} - \dfrac{1}{c} \right)}$$

$$C \approx \frac{4\pi \varepsilon_o ab}{a - b} \left[1 + \frac{ta\,(\varepsilon_{r1} - 1)}{\varepsilon_{r1} b\,(a - b)} \right] \quad \text{if } \varepsilon_{r2} = 1,\ b \gg (c - b) = t$$

$$C \approx \frac{\varepsilon_{r1}\varepsilon_{r2}\varepsilon_o S}{\varepsilon_{r2}s + \varepsilon_{r1}d} \quad \text{if } a \approx b \approx c, S = 4\pi a^2, d = a - c, s = c - b$$

$$\varepsilon_r = \begin{cases} \varepsilon_{r1} & \text{if } b < r < c \\ \varepsilon_{r2} & \text{if } c < r < a \end{cases}$$

Two Concentric Spherical Conductors with Two Parallel Dielectrics [Zahn]

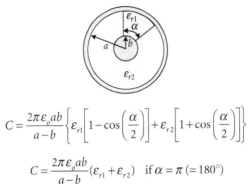

$$C = \frac{2\pi\varepsilon_o ab}{a - b} \left\{ \varepsilon_{r1} \left[1 - \cos\left(\frac{\alpha}{2}\right) \right] + \varepsilon_{r2} \left[1 + \cos\left(\frac{\alpha}{2}\right) \right] \right\}$$

$$C = \frac{2\pi\varepsilon_o ab}{a - b} (\varepsilon_{r1} + \varepsilon_{r2}) \quad \text{if } \alpha = \pi\, (= 180°)$$

where α, in radians, is the angle of the wedge dielectric (in the θ direction).

Two Concentric Spherical Cap Conductors

$$C = \frac{2\pi\varepsilon_o ab}{a - b} \left[1 - \cos\left(\frac{\alpha}{2}\right) \right] \quad \text{if } \alpha a \gg (a - b)$$

$$C = \frac{4\pi\varepsilon_o ab}{a - b} \quad \text{if } \alpha a \gg (a - b), \alpha = 2\pi\, (= 360°)$$

where α is in radians, and the spherical caps are located at $r = b$ and $r = a$.

Two Conducting Hemispheres [Moon, '61]

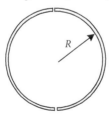

$$C = 2.72 R\varepsilon_o$$

where the two hemispheres are separated by a narrow gap.

Two Spherical Conductors with Displaced Inner Conductor [Smythe; Weber]

$$C = 4\pi\varepsilon_o ab\sinh(\alpha)\sum_{m=1}^{\infty}\frac{1}{a\sinh(m\alpha)-b\sinh[(m-1)\alpha]} \quad \text{if } d < a-b$$

$$C \approx 4\pi\varepsilon_o b \quad \text{if } d = 0, a \gg b$$

where

$$\alpha = \cosh^{-1}\left(\frac{a^2+b^2-d^2}{2ab}\right)$$

Three Concentric Spherical Conductors [Jefimenko]

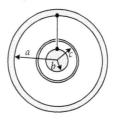

A fine insulated wire passes through a hole in the intermediate spherical conductor, connecting the inner and outer spherical conductors. The capacitance between the intermediate sphere and the outer and inner sphere is

$$C = 4\pi\varepsilon_o\left(\frac{bc}{c-b}+\frac{ac}{a-c}\right)$$

$$C \approx \varepsilon_o S\left(\frac{1}{d_1}+\frac{1}{d_2}\right) \quad \text{if } a \approx b \approx c, S = 4\pi a^2, d_1 = a-c, d_2 = c-b$$

Spherical Conductor between Two Large Flat Conductors [Weber; Gray]

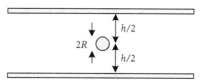

The capacitance between the two plane conductors, which are connected together, and spherical conductor of radius R is

$$C = \frac{4\pi\varepsilon_o}{\dfrac{1}{R}-\dfrac{2}{h}\ln(2)} \approx 4\pi\varepsilon_o R \quad \text{if } h \gg R$$

The width of the plane conductors is large compared to h.

Spherical Conductor inside Cylindrical Conductor [Gray; Smythe]

$$C = R \left[\begin{array}{l} 1.11285 - 0.9277 \left(\dfrac{R}{a}\right) - 0.114 \left(\dfrac{R}{a}\right)^2 \\ -0.1955 \left(\dfrac{R}{a}\right)^3 + 1.8858 \left(\dfrac{R}{a}\right)\left(1 - \dfrac{R}{a}\right)^{-0.5463} \end{array} \right] \times 10^{-10} \quad \text{if } l_{th} \gg a, 0 \le R \le 0.95a$$

$$C \approx 1.11285 \times 10^{-10} R = 4\pi\varepsilon_o R \quad \text{if } l_{th} \gg a, a \gg R$$

The length of the cylinder conductor is l_{th}, and the center of the sphere is on the axis of the cylinder.

Two Conical Conductors [Weeks, '68; Hilberg]

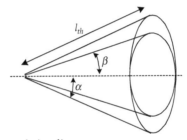

For this coaxial, circular cone transmission line,

$$C = \frac{2\pi\varepsilon_o l_{th}}{\ln\left[\dfrac{\tan\left(\dfrac{\alpha}{2}\right)}{\tan\left(\dfrac{\beta}{2}\right)}\right]} \quad \text{if } 1 \gg \alpha - \beta$$

$$C = \frac{2\pi\varepsilon_o l_{th}}{\ln\left[\cot\left(\dfrac{\beta}{2}\right)\right]} \quad \text{if } 1 \gg \alpha - \beta, \alpha = \frac{\pi}{2} (= 90°)$$

$$C = \frac{2\pi\varepsilon_o l_{th}}{\ln\left[\cot^2\left(\dfrac{\beta}{2}\right)\right]} \quad \text{if } 1 \gg \alpha - \beta, \alpha = \pi - \beta$$

where l_{th} is the length of the cone-shaped conductors.

Two Parallel Conductors [Zahn; Weber]

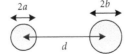

$$C = \frac{2\pi\varepsilon_o l_{th}}{\cosh^{-1}\left(\dfrac{d^2 - a^2 - b^2}{2ab}\right)} \quad \text{if } l_{th} \gg d$$

$$C \approx \frac{2\pi\varepsilon_o l_{th}}{\cosh^{-1}\left(\dfrac{d^2}{2ab}\right)} \approx \frac{2\pi\varepsilon_o l_{th}}{\ln\left(\dfrac{d^2}{ab}\right)} = \frac{\pi\varepsilon_o l_{th}}{\ln\left(\dfrac{d}{\sqrt{ab}}\right)} \quad \text{if } l_{th} \gg d,\, d \gg \max(a,b)$$

$$C = \frac{\pi\varepsilon_o l_{th}}{\cosh^{-1}\left(\dfrac{d}{2a}\right)} = \frac{\pi\varepsilon_o l_{th}}{\ln\left[\dfrac{d}{2a} + \sqrt{\left(\dfrac{d}{2a}\right)^2 - 1}\right]} \quad \text{if } l_{th} \gg d,\, a = b$$

$$C \approx \frac{\pi\varepsilon_o l_{th}}{\ln\left(\dfrac{d}{a}\right)} \quad \text{if } l_{th} \gg d,\, a = b,\, d \gg a$$

where d is the distance between the centers of the two conductors and l_{th} is the length of the conductors.

Two Thinly Insulated Parallel Conductors

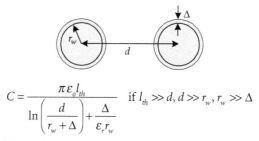

$$C = \frac{\pi\varepsilon_o l_{th}}{\ln\left(\dfrac{d}{r_w + \Delta}\right) + \dfrac{\Delta}{\varepsilon_r r_w}} \quad \text{if } l_{th} \gg d,\, d \gg r_w,\, r_w \gg \Delta$$

where Δ is the thickness of the dielectric coating with a relative permittivity of ε_r, and l_{th} is the length of the conductors.

Two Rectangular Conductors [Paul, '92(b); Lo]

$$C = \begin{cases} \dfrac{\pi\varepsilon_o l_{th}}{\ln\left[2\dfrac{(1+\sqrt{k})}{(1-\sqrt{k})}\right]} & \text{if } l_{th} \gg (d+2w),\, \dfrac{1}{\sqrt{2}} \le k \le 1,\, w \gg t \\[2em] \dfrac{\varepsilon_o l_{th}}{\pi}\ln\left\{2\dfrac{\left[1+(1-k^2)^{\frac{1}{4}}\right]}{\left[1-(1-k^2)^{\frac{1}{4}}\right]}\right\} & \text{if } l_{th} \gg (d+2w),\, 0 \le k \le \dfrac{1}{\sqrt{2}},\, w \gg t \end{cases}$$

$$C \approx \frac{\pi\varepsilon_o l_{th}}{\ln\left(\dfrac{4d}{w} - 2\right)} \approx \frac{\pi\varepsilon_o l_{th}}{\ln\left(\dfrac{4d}{w}\right)} \quad \text{if } l_{th} \gg d,\, d \gg w$$

$$C \approx 1.27 \varepsilon_o l_{th} \ln\left(2\sqrt{\frac{2w+d}{d}}\right) \approx 1.27 \varepsilon_o l_{th} \ln\left(2\sqrt{\frac{2w}{d}}\right) \quad \text{if } l_{th} \gg d, w \gg \max(d,t)$$

where $k = d/(d+2w)$ and l_{th} is the length of the conductors.

Two Equal Rectangular Conductors [Walker]

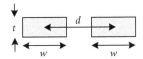

$$C = \frac{\pi \varepsilon_o l_{th}}{\ln\left[\frac{\pi(d-w)}{w+t}+1\right]} \approx \frac{\pi \varepsilon_o l_{th}}{\ln\left(\frac{\pi d}{w+t}+1\right)} \quad \text{if } l_{th} \gg d, d \gg \max(w,t)$$

where l_{th} is the length of the conductors.

Two Unequal Rectangular Conductors [Walker]

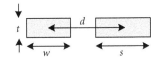

$$C \approx \frac{\pi \varepsilon_o l_{th}}{\ln\left[\frac{\pi d}{\sqrt{(w+t)(s+t)}}\right]} \quad \text{if } l_{th} \gg d, d \gg \max(w,t)$$

where l_{th} is the length of the conductors.

Two Parallel Conductors [Lo; Montrose, '99]

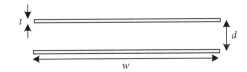

$$C = \frac{\pi \varepsilon_o l_{th}}{\ln\left[\frac{6}{u}+\frac{2\pi-6}{u}e^{-\left(\frac{30.7}{u}\right)^{0.7528}}+\sqrt{1+\left(\frac{2}{u}\right)^2}\right]} \quad \text{if } l_{th} \gg w, w > 3d$$

$$C \approx \varepsilon_o \frac{w l_{th}}{d} \quad \text{if } l_{th} \gg w, w \gg d, d > 3t$$

where

$$u = \frac{2w}{d}+\frac{2t}{\pi d}\ln\left(1+\frac{d}{2t}\frac{4e}{\coth^2\sqrt{13\frac{w}{d}}}\right)$$

where l_{th} is the length of the conductors.

Two Thin, Parallel Plane Conductors [Chipman]

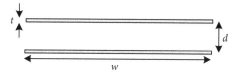

$$C = \frac{\pi \varepsilon_o l_{th}}{2} \frac{w}{d \tan^{-1}\left(\frac{w}{2d}\right) + \frac{w}{4}\ln(4d^2 + w^2) - \frac{w}{2}\ln(w)} \quad \text{if } l_{th} \gg d, d > 3t$$

$$C \approx \frac{\pi \varepsilon_o l_{th}}{2} \frac{w}{d \tan^{-1}\left(\frac{w}{2d}\right)} \approx \varepsilon_o l_{th}\frac{w}{d} \quad \text{if } l_{th} \gg d, d > 3t, w \gg d$$

where t is small compared to all other dimensions and l_{th} is the length of the conductors.

Two Close, Parallel Plane Conductors with Varying Dielectric [Haus; Jefimenko; Boast]

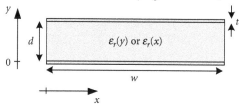

$$C = \frac{\varepsilon_o l_{th} w}{d} \frac{\varepsilon_{r2} - \varepsilon_{r1}}{\ln\left(\frac{\varepsilon_{r2}}{\varepsilon_{r1}}\right)} \quad \text{if } l_{th} \gg d, d > 3t, w \gg d$$

where the relative permittivity varies uniformly from one conductor to the other conductor (in only the y direction) as

$$\varepsilon_r(y) = \left(\frac{\varepsilon_{r2} - \varepsilon_{r1}}{d}\right)y + \varepsilon_{r1}$$

and where l_{th} is the length of the conductors. When the relative permittivity varies exponentially in the y direction as

$$\varepsilon_r(y) = \alpha e^{-\frac{y}{\beta}}$$

the capacitance is

$$C = \frac{\varepsilon_o l_{th} w \alpha}{\beta\left(e^{\frac{d}{\beta}} - 1\right)} \quad \text{if } l_{th} \gg d, d > 3t, w \gg d$$

If the relative permittivity varies only in the x direction as

$$\varepsilon_r(x) = \left(\frac{\varepsilon_{r4} - \varepsilon_{r3}}{w}\right)x + \varepsilon_{r3}$$

then

$$C = \frac{(\varepsilon_{r4} + \varepsilon_{r3})\varepsilon_o l_{th} w}{2d} \quad \text{if } l_{th} \gg d, d > 3t, w \gg d$$

If the relative permittivity varies only in the x direction as (and $0 < A < 1$)

$$\varepsilon_r(x) = \varepsilon_{r5}\left[1 + A\cos(Bx)\right]$$

then

$$C = \frac{\varepsilon_{r5}\varepsilon_o l_{th}}{d}\left[w + \frac{A}{B}\sin(Bw)\right] \quad \text{if } l_{th} \gg d, d > 3t, w \gg d$$

$$C = \frac{\varepsilon_{r5}\varepsilon_o l_{th}w}{d} \quad \text{if } l_{th} \gg d, d > 3t, w \gg d, Bw = \pi$$

Two Close, Parallel Plane Conductors with Layered Dielectrics [Jefimenko; Zahn]

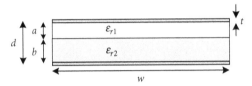

$$C = \frac{1}{\dfrac{1}{\dfrac{\varepsilon_{r1}\varepsilon_o l_{th}w}{a}} + \dfrac{1}{\dfrac{\varepsilon_{r2}\varepsilon_o l_{th}w}{b}}} = \frac{\varepsilon_{r1}\varepsilon_{r2}\varepsilon_o l_{th}w}{\varepsilon_{r1}b + \varepsilon_{r2}a} \quad \text{if } l_{th} \gg d, d = a+b > 3t, w \gg d$$

where l_{th} is the length of the conductors.

Two Close, Parallel Plane Conductors with Parallel Dielectrics [Jefimenko; Zahn]

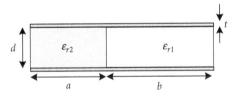

$$C = \frac{\varepsilon_{r2}\varepsilon_o l_{th}a}{d} + \frac{\varepsilon_{r1}\varepsilon_o l_{th}b}{d} = \frac{\varepsilon_o l_{th}(\varepsilon_{r2}a + \varepsilon_{r1}b)}{d} \quad \text{if } l_{th} \gg d, d > 3t, a+b \gg d$$

where l_{th} is the length of the conductors.

Two Parallel Disk Conductors [Nishiyama; Gray]

$$C = \frac{\varepsilon_o \pi R^2}{d} + \varepsilon_o R\left\{-1 + \ln\left[\frac{16\pi R}{d}\left(1 + \frac{t}{d}\right)\right] + \frac{4\pi t}{d}\ln\left(1 + \frac{d}{t}\right)\right\} \quad \text{if } R \gg \max(d,t)$$

$$C \approx \frac{\varepsilon_o \pi R^2}{d}\left[1 + \frac{d}{\pi R}\ln\left(\frac{16\pi R}{d}\right) - \frac{d}{\pi R}\right] \approx \frac{\varepsilon_o \pi R^2}{d} \quad \text{if } R \gg d, d \gg t$$

$$C = \frac{\varepsilon_o \pi R^2}{d}\left[1 + 1.298\left(\frac{d}{R}\right)^{0.876}\right] \quad \text{if } \frac{R}{100} \le d \le R, d \gg t$$

$$C = \frac{\varepsilon_o \pi R^2}{d}\left[1+1.298\left(\frac{d}{R}\right)^{0.982}\right] \quad \text{if } R \le d \le 10R, d \gg t$$

where R is the radius of each circular conductor.

Disk Conductor above Large Flat Conductor [Shen]

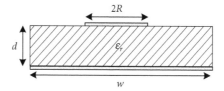

$$C \approx \frac{\varepsilon_r \varepsilon_o \pi R^2}{d}\left\{1+\frac{2d}{\varepsilon_r \pi R}\left[\ln\left(\frac{\pi R}{2d}\right)+1.77\right]\right\} \quad \text{if } w \gg 2R, R \gg d$$

where R is the radius of the circular disk conductor.

Cylindrical Conductor above Large Flat Conductor [Zahn; Weber]

$$C = \frac{2\pi \varepsilon_o l_{th}}{\cosh^{-1}\left(\frac{h+r_w}{r_w}\right)} = \frac{2\pi \varepsilon_o l_{th}}{\ln\left[\frac{h+r_w}{r_w}+\sqrt{\left(\frac{h+r_w}{r_w}\right)^2-1}\right]} \quad \text{if } l_{th} \gg h$$

$$C \approx \frac{2\pi \varepsilon_o l_{th}}{\ln\left(\frac{2h}{r_w}\right)} \quad \text{if } l_{th} \gg h, h \gg r_w$$

where l_{th} is the length of the conductors. The width of the flat conductor is large compared to h and r_w.

Flat Conductor above Large Flat Conductor [Pucel; Barke; Walker]

$$C = \frac{4\pi \varepsilon_o l_{th}}{\ln\left\{1+\frac{32h^2}{w^2}\left[1+\sqrt{1+\left(\frac{\pi w^2}{8h^2}\right)^2}\right]\right\}} \quad \text{if } l_{th} \gg \max(h,w), 2 \ge \frac{w}{h}$$

If the thickness is significant, then

$$C \approx \frac{2\pi \varepsilon_o l_{th}}{\ln\left(\frac{8h}{w'}\right)+\frac{1}{32}\left(\frac{w'}{h}\right)^2} \quad \text{if } l_{th} \gg \max(h,w), 2 \ge \frac{w}{h}$$

$$\approx 2\varepsilon_o l_{th}\left\{\frac{w'}{2h}+\frac{1}{\pi}\ln\left[2\pi e\left(\frac{w'}{2h}+0.94\right)\right]\right\} \quad \text{if } l_{th} \gg \max(h,w), \frac{w}{h} \ge 2$$

where

$$w' = w + \frac{t}{\pi} \ln\left(\frac{4\pi w}{t} + 1\right) \quad \text{if } \frac{1}{2\pi} \ge \frac{w}{h}$$

$$= w + \frac{t}{\pi} \ln\left(\frac{2h}{t} + 1\right) \quad \text{if } \frac{w}{h} \ge \frac{1}{2\pi}$$

and l_{th} is the length of the conductors. A simpler numerical-based expression for significant thicknesses is

$$C = \varepsilon_o l_{th} \left[\frac{w}{h} + 0.77 + 1.06 \left(\frac{w}{h}\right)^{0.25} + 1.06 \left(\frac{t}{h}\right)^{0.5}\right] \quad \text{if } l_{th} \gg \max(h, w), \frac{w}{h} \ge 0.3, \frac{h}{t} \ge 0.1$$

The width of the larger "ground" plane is large compared to h, w, and t.

Dielectric Sandwiched between Strip and Large Flat Conductor   [Janssen; Schneider; Cahill; White, '82]

ground plane

$$C = \begin{cases} \dfrac{2\pi \varepsilon_{reff} \varepsilon_o l_{th}}{\ln\left(\dfrac{8h}{w} + 0.25\dfrac{w}{h}\right)} & \text{if } l_{th} \gg h, \dfrac{w}{h} \le 1 \\[4mm] \varepsilon_{reff} \varepsilon_o l_{th} \left[\dfrac{w}{h} + 1.393 + 0.667\ln\left(\dfrac{w}{h} + 1.444\right)\right] & \text{if } l_{th} \gg h, \dfrac{w}{h} \ge 1 \end{cases}$$

$$C \approx \begin{cases} \dfrac{2\pi \varepsilon_{reff} \varepsilon_o l_{th}}{\ln\left(\dfrac{8h}{w}\right) + 0.0313\left(\dfrac{w}{h}\right)^2} \approx \dfrac{2\pi \varepsilon_{reff} \varepsilon_o l_{th}}{\ln\left(\dfrac{8h}{w}\right)} & \text{if } l_{th} \gg h, h \gg w \\[4mm] \varepsilon_{reff} \varepsilon_o l_{th}\left[\dfrac{w}{h} + 2.42 - 0.44\dfrac{h}{w} + \left(1 - \dfrac{h}{w}\right)^6\right] \approx \varepsilon_{reff} \varepsilon_o l_{th}\left(\dfrac{w}{h} + 1.393\right) & \text{if } l_{th} \gg h, w \gg h \end{cases}$$

where l_{th} is the length of the conductors and the width of the larger "ground" plane is large compared to h, w, and t. Also,

$$\varepsilon_{reff} = \begin{cases} \dfrac{\varepsilon_r + 1}{2} + \dfrac{\varepsilon_r - 1}{2}\left[\dfrac{1}{\sqrt{1 + 12\dfrac{h}{w}}} + 0.04\left(1 - \dfrac{w}{h}\right)^2\right] & \text{if } \dfrac{w}{h} \le 1 \\[6mm] \dfrac{\varepsilon_r + 1}{2} + \dfrac{\varepsilon_r - 1}{2\sqrt{1 + 12\dfrac{h}{w}}} & \text{if } \dfrac{w}{h} > 1 \end{cases}$$

$$\varepsilon_{reff} = 1 \quad \text{if } \varepsilon_r = 1$$

$$\varepsilon_{reff} \approx \frac{1}{2}(\varepsilon_r + 1) \quad \text{if } h \gg w$$

The thickness of the strip is considered small compared to its width, w. If the thickness is not negligible, it can be incorporated into these expressions by replacing the real width w with an effective width, w_{eff}:

$$w_{eff} = \begin{cases} w + \dfrac{t}{\pi}\left[1+\ln\left(\dfrac{4\pi w}{t}\right)\right] & \text{if } \dfrac{1}{2\pi} > \dfrac{w}{h} > 2\dfrac{t}{h} \\[3ex] w + \dfrac{t}{\pi}\left[1+\ln\left(\dfrac{2h}{t}\right)\right] & \text{if } \dfrac{w}{h} > \dfrac{1}{2\pi} > 2\dfrac{t}{h} \end{cases}$$

Actually, these correction terms for finite thickness straps assume that $\varepsilon_r = 1$. It has been recommended that one-half of the *correction* term added to w be used when $\varepsilon_r = 10$ and with a proportional adjustment for $1 < \varepsilon_r < 10$.

Two Coplanar Flat Conductors on Dielectric [Paul, '92(b); Gupta, '79; Walker]

$$C = \begin{cases} \dfrac{\pi \varepsilon_{reff}\varepsilon_o l_{th}}{\ln\left[2\dfrac{1+\sqrt{k}}{1-\sqrt{k}}\right]} & \text{if } l_{th} \gg h, \dfrac{1}{\sqrt{2}} \le k \le 1 \\[5ex] \dfrac{\varepsilon_{reff}\varepsilon_o l_{th}}{\pi}\ln\left[2\dfrac{1+(1-k^2)^{\frac{1}{4}}}{1-(1-k^2)^{\frac{1}{4}}}\right] & \text{if } l_{th} \gg h, 0 \le k \le \dfrac{1}{\sqrt{2}} \end{cases}$$

where l_{th} is the length of the conductors and $w \gg t$ (thin conductors)

$$k = \frac{d}{d+2w}$$

$$\varepsilon_{reff} = \frac{\varepsilon_r + 1}{2}\left\{\tanh\left[0.775\ln\left(\frac{h}{w}\right)+1.75\right]+\frac{kw}{h}[0.04-0.7k+0.01(1-0.1\varepsilon_r)(0.25+k)]\right\}$$

If the thickness, t, is not small, then the k should be replaced with k_e only in the C expression where

$$k_e \approx k - (1-k^2)\left\{\frac{1.25t}{2\pi w}\left[1+\ln\left(\frac{4\pi w}{t}\right)\right]\right\}$$

Furthermore, the effective relative permittivity ε_{reff} should be replaced with ε_{ereff}:

$$\varepsilon_{ereff} = \begin{cases} \varepsilon_{reff} - \dfrac{1.4(\varepsilon_{reff}-1)\dfrac{t}{d}}{\dfrac{\pi}{\ln\left[2\dfrac{1+\sqrt{k}}{1-\sqrt{k}}\right]}+\dfrac{1.4t}{d}} & \text{if } \dfrac{1}{\sqrt{2}} \le k \le 1 \\[8ex] \varepsilon_{reff} - \dfrac{1.4(\varepsilon_{reff}-1)\dfrac{t}{d}}{\dfrac{\ln\left[2\dfrac{1+(1-k^2)^{\frac{1}{4}}}{1-(1-k^2)^{\frac{1}{4}}}\right]}{\pi}+\dfrac{1.4t}{d}} & \text{if } 0 \le k \le \dfrac{1}{\sqrt{2}} \end{cases}$$

A much simpler expression for the capacitance that yields very good results for reasonable sized microstrips is

$$C \approx \frac{\pi \varepsilon_{ereff} \varepsilon_o l_{th}}{\ln \left[\frac{\pi (d-w)}{w+t} + 1 \right]} \approx \frac{\pi \varepsilon_{ereff} \varepsilon_o l_{th}}{\ln \left(\frac{\pi d}{w+t} \right)} \quad \text{if } l_{th} \gg w+d, w \geq t, d \gg w$$

Cylindrical Conductor above Slotted Large Flat Conductor [Hilberg]

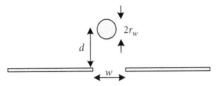

The capacitance between the circular conductor and large planes is

$$C = \frac{2\pi \varepsilon_o l_{th}}{\cosh^{-1} \left\{ \frac{\sin \left[\frac{\pi}{2} - \frac{1}{2} \tan^{-1} \left(\frac{4wd}{w^2 - 4d^2 + 4r_w^2} \right) \right]}{\sin \left[\frac{1}{2} \tan^{-1} \left(\frac{4wr_w}{w^2 + 4d^2 - 4r_w^2} \right) \right]} \right\}} \quad \text{if } l_{th} \gg \max (d,w), d \gg r_w$$

$$C \approx \frac{2\pi \varepsilon_o l_{th}}{\cosh^{-1} \left(\frac{w}{2r_w} \right)} \approx \frac{2\pi \varepsilon_o l_{th}}{\ln \left(\frac{w}{r_w} \right)} \quad \text{if } l_{th} \gg w, d \gg r_w, w \gg d$$

$$C \approx \frac{2\pi \varepsilon_o l_{th}}{\cosh^{-1} \left(\frac{d}{r_w} \right)} \approx \frac{2\pi \varepsilon_o l_{th}}{\ln \left(\frac{2d}{r_w} \right)} \quad \text{if } l_{th} \gg d, d \gg r_w, d \gg w$$

The circular conductor is centered over the slot in a very wide flat plane conductor (compared to d and w). The length of the conductors is l_{th}.

Embedded Strip above Large Flat Conductor [Montrose, '99]

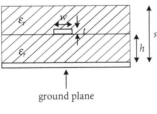

ground plane

$$C \approx \frac{7.17 \varepsilon_{reff} \varepsilon_o l_{th}}{\ln \left(\frac{5.98h}{0.8w + t} \right)} \quad \text{if } l_{th} \gg s$$

where l_{th} is the length of the conductors and

$$\varepsilon_{reff} = \varepsilon_r \left[1 - e^{-\left(1.55 \frac{s}{h} \right)} \right]$$

The thickness of the upper coating, $s - (t + h)$, should be comparable or large compared to h. When the upper coating is thin ($s < 1.2h$), the uncoated microstrip expressions can be used. The width of the larger "ground" plane is large compared to h, w, and t.

Cylindrical Conductor between Two Large Flat Conductors [Howard W. Sams; Frankel; Johnson, '61]

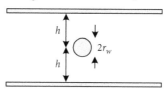

The capacitance between the circular conductor and two large planar conductors is

$$C = \frac{25\varepsilon_o l_{th}}{\ln\left[1+1.314u+\sqrt{(1.314u)^2+2u}\right]} \approx \frac{2\pi\varepsilon_o l_{th}}{\ln\left(\dfrac{4h}{\pi r_w}\right) - \dfrac{0.0843\left(\dfrac{r_w}{h}\right)^4}{1-0.355\left(\dfrac{r_w}{h}\right)^4}} \quad \text{if } l_{th} \gg h$$

$$C \approx \frac{2\pi\varepsilon_o l_{th}}{\ln\left(\dfrac{4h}{\pi r_w}\right)} \quad \text{if } l_{th} \gg h, h \gg r_w$$

where l_{th} is the length of the conductors and

$$u = \left(\frac{h}{r_w}\right)^4 - 1$$

The width of the flat conductors is large compared to h.

Flat Conductor between Two Large Flat Conductors [Lo; Schneider; Hilberg; White, '82; Lee, '98; Montrose, '00]

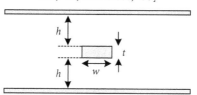

The capacitance between the narrow center conductor and two large flat conductors is

$$C = \frac{4\pi\varepsilon_o l_{th}}{\ln\left\{1+\dfrac{8h}{\pi u}\left[\dfrac{16h}{\pi u}+\sqrt{\left(\dfrac{16h}{\pi u}\right)^2+6.27}\right]\right\}} \quad \text{if } l_{th} \gg h$$

$$C \approx \frac{4\pi\varepsilon_o l_{th}}{\ln\left\{1+\dfrac{8h}{\pi w}\left[\dfrac{16h}{\pi w}+\sqrt{\left(\dfrac{16h}{\pi w}\right)^2+6.27}\right]\right\}} \quad \text{if } l_{th} \gg h, w \gg t$$

$$C \approx \frac{2\pi\varepsilon_o l_{th}}{\ln\left(\dfrac{3.8h}{0.8w+t}\right)} \approx \frac{2\pi\varepsilon_o l_{th}}{\ln\left(\dfrac{16h}{\pi w}\right)} \quad \text{if } l \gg h, h \gg w \gg t$$

$$C \approx \frac{4.93\varepsilon_o w l_{th}}{h\left(2.5 + \dfrac{16h}{\pi w}\right)} \approx 1.97\varepsilon_o l_{th}\frac{w}{h} \quad \text{if } l_{th} \gg w, w \gg \max(h,t)$$

where l_{th} is the length of the conductors and

$$u = w + \frac{t}{\pi} - \frac{t}{2\pi}\ln\left[\left(\frac{1}{\dfrac{4h}{t}+1}\right)^2 + \left(\frac{\dfrac{\pi}{4}}{\dfrac{w}{t}+1.1}\right)^{\frac{2}{1+\frac{t}{3h}}}\right]$$

The width of the larger "ground" planes is large compared to h, w, and t.

When the center flat conductor of width w is not centered between the two large flat conductors, an order-of-magnitude approximation for the off-center capacitance is

$$C = \varepsilon_o l_{th}\left\{w\left(\frac{1}{h_1}+\frac{1}{h_2}\right) + 0.77 + 0.891\left[w\left(\frac{1}{h_1}+\frac{1}{h_2}\right)\right]^{0.25} + 0.891\sqrt{t}\left(\frac{1}{h_1^2}+\frac{1}{h_2^2}\right)^{0.25}\right\}$$

$$\text{if } l_{th} \gg \max(h,w), \frac{w}{h} \ge 0.3, \frac{h}{t} \ge 0.1$$

where h_1 and h_2 are the distances between the center flat conductor and large conducting planes (in the given figure $h = h_1 = h_2$).

Three Coplanar Flat Conductors on a Substrate [Lo; Gupta, '81]

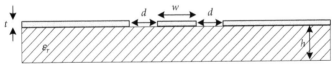

The capacitance between the two wide conductors, which are connected together, and center conductor of width w is

$$C = \begin{cases} 1.27\varepsilon_{reff}\varepsilon_o l_{th}\ln\left(2\dfrac{1+\sqrt{k}}{1-\sqrt{k}}\right) & \text{if } l_{th} \gg \max(w+d,h), \dfrac{1}{\sqrt{2}} \le k \le 1 \\[4mm] \dfrac{4\pi\varepsilon_{reff}\varepsilon_o l_{th}}{\ln\left[2\dfrac{1+(1-k^2)^{\frac{1}{4}}}{1-(1-k^2)^{\frac{1}{4}}}\right]} & \text{if } l_{th} \gg \max(w+d,h), 0 \le k \le \dfrac{1}{\sqrt{2}} \end{cases}$$

where l_{th} is the length of the conductors and

$$k = \frac{w + \dfrac{1.25t}{\pi}\left[1 + \ln\left(\dfrac{4\pi w}{t}\right)\right]}{w + 2d - \dfrac{1.25t}{\pi}\left[1 + \ln\left(\dfrac{4\pi w}{t}\right)\right]}$$

$$\varepsilon_{reff} = \varepsilon_m - \frac{0.7(\varepsilon_m - 1)\dfrac{t}{d}}{\varepsilon_n + 0.7\dfrac{t}{d}}$$

$$\varepsilon_m = \frac{\varepsilon_r + 1}{2} \left\{ \begin{array}{l} \tanh\left[0.775\ln\left(\frac{h}{d}\right)+1.75\right]+ \\ \\ \frac{d}{h}\left(\frac{w}{w+2d}\right)\left[0.04-0.7\frac{w}{w+2d}+0.01\left(1-0.1\varepsilon_r\right)\left(0.25+\frac{w}{w+2d}\right)\right] \end{array} \right\}$$

$$\varepsilon_n = \left\{ \begin{array}{ll} 0.318\ln\left(2\dfrac{1+\sqrt{\dfrac{w}{w+2d}}}{1-\sqrt{\dfrac{w}{w+2d}}}\right) & \text{if } \dfrac{1}{\sqrt{2}} \le \dfrac{w}{w+2d} \le 1 \\ \\ \dfrac{3.14}{\ln\left\{2\dfrac{1+\left[1-\left(\dfrac{w}{w+2d}\right)^2\right]^{\frac{1}{4}}}{1-\left[1-\left(\dfrac{w}{w+2d}\right)^2\right]^{\frac{1}{4}}}\right\}} & \text{if } 0 \le \dfrac{w}{w+2d} \le \dfrac{1}{\sqrt{2}} \end{array} \right.$$

The two outer conductors are wide in comparsion to w and d.

Three Equal Cylindrical Conductors [Johnson, '61; Hilberg]

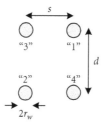

The capacitance between the two outer conductors, "b" and "c," which are connected together, and center conductor "a" is

$$C = \frac{4.18\varepsilon_o l_{th}}{\ln\left(0.398\dfrac{d}{r_w}\right)} \quad \text{if } l_{th} \gg d, d \gg r_w$$

where l_{th} is the length of the conductors.

Four Cylindrical Conductors [Johnson, '61; Lo; Hilberg]

The capacitance between conductors "1" and "2," which are connected together, and conductors "3" and "4," which are connected together, is

$$C = \frac{2\pi\varepsilon_o l_{th}}{\ln\left[\dfrac{d}{r_w\sqrt{1+\left(\dfrac{d}{s}\right)^2}}\right]} \quad \text{if } l_{th} \gg \max(d,s),\ \min(d,s) \gg r_w$$

$$C \approx \frac{2\pi\varepsilon_o l_{th}}{\ln\left(\dfrac{d}{r_w}\right)} \quad \text{if } l_{th} \gg \max(d,s),\ \min(d,s) \gg r_w,\ s \gg d$$

The capacitance between conductors "1" and "3" and conductors "2" and "4" is

$$C = \frac{2\pi\varepsilon_o l_{th}}{\ln\left[\dfrac{d}{r_w}\sqrt{1+\left(\dfrac{d}{s}\right)^2}\right]} \quad \text{if } l_{th} \gg \max(d,s),\ \min(d,s) \gg r_w$$

$$C \approx \frac{2\pi\varepsilon_o l_{th}}{\ln\left(\dfrac{d}{r_w}\right)} \quad \text{if } l_{th} \gg \max(d,s),\ \min(d,s) \gg r_w,\ s \gg d$$

where l_{th} is the length of the conductors.

Five Cylindrical Conductors   [Johnson, '61; Howard W. Sams; Lo]

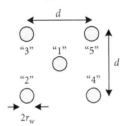

The capacitance between conductors "2" through "5," which are connected together, and conductor "1" is

$$C = \frac{5\varepsilon_o l_{th}}{\ln\left(\dfrac{d}{1.9r_w}\right)} \quad \text{if } l_{th} \gg d,\ d \gg r_w$$

where l_{th} is the length of the conductors.

2N Cylindrical Conductors   [Hilberg]

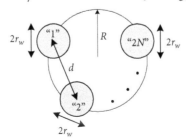

For 2N parallel, identical round conductors uniformly spaced on the perimeter of a circle of radius R, the capacitance between conductors "1," "3,"..."$2N − 1$," which are connected together, and conductors "2," "4,"..."$2N$," which are connected together, is

$$C = \cfrac{\pi \varepsilon_o N l_{th}}{\cosh^{-1}\left\{\cfrac{1}{\sin\left[N \sin^{-1}\left(\cfrac{r_w}{R}\right)\right]}\right\}} \quad \text{if } l_{th} \gg R, R \gg r_w$$

$$C \approx \cfrac{\pi \varepsilon_o N l_{th}}{\cosh^{-1}\left(\cfrac{R}{N r_w}\right)} \approx \cfrac{\pi \varepsilon_o N l_{th}}{\ln\left(\cfrac{2R}{N r_w}\right)} \quad \text{if } l_{th} \gg R, R \gg N r_w$$

$$C \approx \cfrac{\pi \varepsilon_o l_{th}}{\ln\left(\cfrac{2R}{r_w}\right)} \quad \text{if } l_{th} \gg R, R \gg r_w, N = 1$$

$$C \approx \cfrac{2\pi \varepsilon_o l_{th}}{\ln\left(\cfrac{R}{r_w}\right)} \quad \text{if } l_{th} \gg R, R \gg 2r_w, N = 2$$

where l_{th} is the length of the conductors.

Three Cylindrical Conductors   [Hilberg]

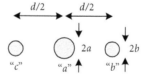

The capacitance between conductors "b" and "c" in the presence of conductor "a" is

$$C_{mbc} = \cfrac{\pi \varepsilon_o l_{th}}{\cosh^{-1}\left\{\cfrac{\sin\left[2\tan^{-1}\left(\cfrac{ad}{a^2 - \cfrac{d^2}{4} + b^2}\right)\right]}{\sin\left[2\tan^{-1}\left(\cfrac{-2ab}{a^2 + \cfrac{d^2}{4} - b^2}\right)\right]}\right\}} \quad \text{if } l_{th} \gg d, d \gg \max(a,b)$$

$$C_{mbc} \approx \cfrac{\pi \varepsilon_o l_{th}}{\ln\left(\cfrac{d}{b}\right)} \quad \text{if } l_{th} \gg d, d \gg \max(a,b)$$

where l_{th} is the length of the conductors. The radii of conductors "b" and "c" are the same.

Two Cylindrical Conductors in Conducting Trough [Lo; Howard W. Sams]

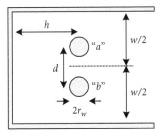

The capacitance between conductors "a" and "b" in the presence of the trough is

$$C_{mab} = \frac{\pi \varepsilon_o l_{th}}{\ln\left[\dfrac{w}{\pi r_w \sqrt{\csc^2\left(\dfrac{\pi d}{w}\right) + \operatorname{csch}^2\left(\dfrac{2\pi h}{w}\right)}}\right]} \qquad \text{if } l_{th} \gg \max(w,h),\ \min(d,h,w) \gg r_w$$

$$C_{mab} \approx \frac{\pi \varepsilon_o l_{th}}{\ln\left[\dfrac{d}{r_w\sqrt{1+\left(\dfrac{d}{2h}\right)^2}}\right]} \qquad \text{if } l_{th} \gg \max(h,w),\ \min(d,h,w) \gg r_w,\ w \gg \max(h,d)$$

$$C_{mab} \approx \frac{\pi \varepsilon_o l_{th}}{\ln\left[\dfrac{w}{\pi r_w \csc\left(\dfrac{\pi d}{w}\right)}\right]} \qquad \text{if } l_{th} \gg \max(h,w),\ \min(d,h,w) \gg r_w,\ h \gg w$$

where l_{th} is the length of the conductors.

Two Cylindrical Conductors between Two Large Flat Conductors [Frankel]

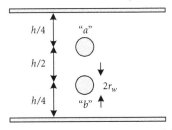

The capacitance between conductors "a" and "b" in the presence of the two large planar conductors is

$$C_{mab} = \frac{\pi \varepsilon_o l_{th}}{\ln\left(\dfrac{h}{\pi r_w}\right)} \qquad \text{if } l_{th} \gg h,\ h \gg r_w$$

where l_{th} is the length of the conductors. The width of the flat conductors is large compared to h.

Two Coplanar Cylindrical Conductors between Two Large Plane Conductors [Lo; Frankel]

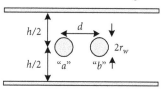

The capacitance between conductors "a" and "b" in the presence of the flat conducting planes is

$$C_{mab} = \frac{\pi \varepsilon_o l_{th}}{\ln \left[\tanh\left(\frac{\pi d}{2h}\right) \coth\left(\frac{\pi r_w}{2h}\right) \right]} \quad \text{if } l_{th} \gg h, \min(d,h) \gg r_w$$

$$C_{mab} \approx \frac{\pi \varepsilon_o l_{th}}{\ln \left[\frac{2h \tanh\left(\frac{\pi d}{2h}\right)}{\pi r_w} \right]} \approx \frac{\pi \varepsilon_o l_{th}}{\ln\left(\frac{d}{r_w}\right)} \quad \text{if } l_{th} \gg h, \min(d,h) \gg r_w, h \gg d$$

The capacitance between conductors "a" and "b," which are connected together, and the two large flat conducting planes is

$$C = \frac{4\pi \varepsilon_o l_{th}}{\ln \left[\coth\left(\frac{\pi d}{2h}\right) \coth\left(\frac{\pi r_w}{2h}\right) \right]} \quad \text{if } l_{th} \gg h, \min(d,h) \gg r_w \ ^{51}$$

$$C \approx \frac{4\pi \varepsilon_o l_{th}}{\ln \left[\frac{2h \coth\left(\frac{\pi d}{2h}\right)}{\pi r_w} \right]} \approx \frac{2\pi \varepsilon_o l_{th}}{\ln\left(\frac{2h}{\pi\sqrt{r_w d}}\right)} \quad \text{if } l_{th} \gg h, \min(d,h) \gg r_w, h \gg d$$

where l_{th} is the length of the conductors. The width of the flat conductors is large compared to h.

Three Cylindrical Conductors between Two Large Plane Conductors [Frankel]

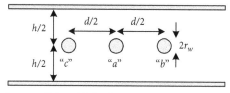

The capacitance between conductors "b" and "c," which are connected together, and conductor "a" in the presence of the two large planar conductors is

---

[51]The expression from the reference was modified. Now, when the two conductors are far apart, the total capacitance is larger than the capacitance of a single conductor between two planes.

$$C_{mab-c} = \frac{4.18\varepsilon_o l_{th}}{\ln\left\{\dfrac{2h\tanh\left(\dfrac{\pi d}{2h}\right)}{\pi r_w\left[1+\mathrm{sech}\left(\dfrac{\pi d}{2h}\right)\right]^{\frac{4}{3}}}\right\}} \qquad \text{if } l_{th} \gg h,\ \min(d,h) \gg r_w$$

$$C \approx \frac{4.18\varepsilon_o l_{th}}{\ln\left(0.397\dfrac{d}{r_w}\right)} \qquad \text{if } l_{th} \gg h,\ h \gg d,\ d \gg r_w$$

where l_{th} is the length of the conductors. The width of the flat conductors is large compared to h.

Three Cylindrical Conductors in Rectangular Conducting Enclosure [Frankel]

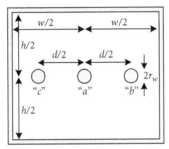

The capacitance between conductors "b" and "c," which are connected together, and conductor "a" in the presence of the enclosure is

$$C_{mab-c} = \frac{4.18\varepsilon_o l_{th}}{\left[\ln\left\{\dfrac{2h\tanh\left(\dfrac{\pi d}{2h}\right)}{\pi r_w\left[1+\mathrm{sech}\left(\dfrac{\pi d}{2h}\right)\right]^{\frac{4}{3}}}\right\} -\dfrac{2}{3}\displaystyle\sum_{m=1}^{\infty}(-1)^m \ln\left[\dfrac{1-\dfrac{\sinh^2\left(\dfrac{\pi d}{2h}\right)}{\sinh^2\left(\dfrac{m\pi w}{2h}\right)}}{1+\dfrac{\sinh^2\left(\dfrac{\pi d}{2h}\right)}{\cosh^2\left(\dfrac{m\pi w}{2h}\right)}}\right]\left[\dfrac{1+\dfrac{\sinh^2\left(\dfrac{\pi d}{4h}\right)}{\cosh^2\left(\dfrac{m\pi w}{2h}\right)}}{1-\dfrac{\sinh^2\left(\dfrac{\pi d}{4h}\right)}{\sinh^2\left(\dfrac{m\pi w}{2h}\right)}}\right]\right]}$$

$$\text{if } l_{th} \gg \max(w,h),\ \min(d,h,w) \gg r_w$$

$$C_{mab-c} \approx \frac{4.18\varepsilon_o l_{th}}{\ln\left\{\dfrac{2h\tanh\left(\dfrac{\pi d}{2h}\right)}{\pi r_w\left[1+\operatorname{sech}\left(\dfrac{\pi d}{2h}\right)\right]^{\frac{4}{3}}}\right\}} \quad \text{if } l_{th} \gg \max(w,h),\ \min(d,h,w) \gg r_w,\ w \gg d$$

where l_{th} is the length of the conductors.

Two Cylindrical Conductors Surrounded by Four Cylindrical Conductors [Howard W. Sams]

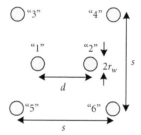

The capacitance between conductors "1" and "2" in the presence of conductors "3" through "6," which are connected together, is

$$C_{m12} = \frac{\pi \varepsilon_o l_{th}}{\ln\left(\dfrac{d}{r_w}\right) - \dfrac{\ln\dfrac{\left[1+\left(1+\dfrac{d}{s}\right)^2\right]^2}{1+\left(1-\dfrac{d}{s}\right)^2}}{\ln\left(\dfrac{s\sqrt{2}}{r_w}\right)}} \quad \text{if } l_{th} \gg s$$

$$C_{m12} \approx \frac{\pi \varepsilon_o l_{th}}{\ln\left(\dfrac{d}{r_w}\right)} \quad \text{if } l_{th} \gg s,\ s \gg d$$

$$C_{m12} \approx \frac{\pi \varepsilon_o l_{th}}{\ln\left(\dfrac{d}{r_w}\right)} \quad \text{if } l_{th} \gg s,\ d \gg s$$

where l_{th} is the length of the conductors.

Two Cylindrical Conductors in Rectangular Enclosure [Frankel; King]

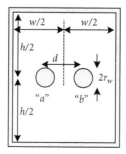

The capacitance between conductors "*a*" and "*b*" in the presence of the enclosure is

$$C_{mab} = \frac{\pi \varepsilon_o l_{th}}{\ln\left[\frac{2h\tanh\left(\frac{\pi d}{2h}\right)}{\pi r_w}\right] - \sum_{m=1}^{\infty} \ln\left[\frac{1 + \frac{\sinh^2\left(\frac{\pi d}{2h}\right)}{\cosh^2\left(\frac{m\pi w}{2h}\right)}}{1 - \frac{\sinh^2\left(\frac{\pi d}{2h}\right)}{\sinh^2\left(\frac{m\pi w}{2h}\right)}}\right]} \qquad \text{if } l_{th} \gg \max(h,w), \min(d,h,w) \gg r_w$$

$$C_{mab} \approx \frac{\pi \varepsilon_o l_{th}}{\ln\left[\frac{2h\tanh\left(\frac{\pi d}{2h}\right)}{\pi r_w}\right] - \ln\left[\frac{1 + \frac{\sinh^2\left(\frac{\pi d}{2h}\right)}{\cosh^2\left(\frac{\pi w}{2h}\right)}}{1 - \frac{\sinh^2\left(\frac{\pi d}{2h}\right)}{\sinh^2\left(\frac{\pi w}{2h}\right)}}\right]} \qquad \text{if } l_{th} \gg \max(h,w), \min(d,h,w) \gg r_w$$

$$C_{mab} \approx \frac{\pi \varepsilon_o l_{th}}{\ln\left[\frac{2h\tanh\left(\frac{\pi d}{2h}\right)}{\pi r_w}\right]} \qquad \text{if } l_{th} \gg \max(h,w), \min(d,h,w) \gg r_w, w \gg \max(h,d)$$

The capacitance between conductors "*a*" and "*b*," which are connected together, and the rectangular enclosure is

$$C = \frac{4\pi \varepsilon_o l_{th}}{\ln\left[\frac{2h\coth\left(\frac{\pi d}{2h}\right)}{\pi r_w}\right] + \sum_{m=1}^{\infty}(-1)^m \ln\left[\frac{1 + \frac{\cosh^2\left(\frac{\pi d}{2h}\right)}{\sinh^2\left(\frac{m\pi w}{2h}\right)}}{1 - \frac{\cosh^2\left(\frac{\pi d}{2h}\right)}{\cosh^2\left(\frac{m\pi w}{2h}\right)}}\right]} \qquad \text{if } l_{th} \gg \max(h,w), \min(d,h,w) \gg r_w$$

$$C \approx \frac{4\pi\varepsilon_o l_{th}}{\ln\left[\frac{2h\coth\left(\frac{\pi d}{2h}\right)}{\pi r_w}\right] - \ln\left[\frac{1 + \frac{\cosh^2\left(\frac{\pi d}{2h}\right)}{\sinh^2\left(\frac{\pi w}{2h}\right)}}{1 - \frac{\cosh^2\left(\frac{\pi d}{2h}\right)}{\cosh^2\left(\frac{\pi w}{2h}\right)}}\right]} \quad \text{if } l_{th} \gg \max(h,w),\ \min(d,h,w) \gg r_w$$

where l_{th} is the length of the conductors.

Two Broadside Flat Conductors between Two Large Flat Conductors [Cohn, '60; Lo]

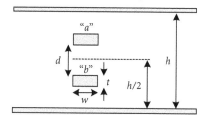

The capacitance between conductors "*a*" and "*b*" in the presence of the two large flat conductors, assuming that the dual strips are thin (e.g., $h \gg t$), is

$$C_{mab} = 2\varepsilon_o l_{th}\left(\frac{w}{h-d} + \frac{w}{d} + k\right) \quad \text{if } l_{th} \gg h,\ w > 0.35d,\ w > 0.35(h-d)$$

where

$$k = \frac{h}{\pi d}\left[\ln\left(\frac{h}{h-d}\right) + \frac{d}{h-d}\ln\left(\frac{h}{d}\right)\right]$$

The capacitance between conductors "*a*" and "*b*," which are connected together, and the large flat conductors is

$$C = 2\varepsilon_o l_{th}\left(\frac{w}{h-d} + 0.4413 + k\frac{d}{h}\right) \quad \text{if } l_{th} \gg h,\ w > 0.35d,\ w > 0.35(h-d)$$

where k is as previously defined and l_{th} is the length of the conductors. The width of the outer flat conductors is large compared to w and h.

Two Cylindrical Conductors above Large Flat Conductor [Paul, '92(b); Weber; Walker]

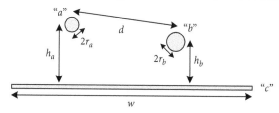

The "circuit capacitances" are

$$C_{ab} = 4\pi \varepsilon_o l_{th} \frac{\ln\left(1 + 4\frac{h_a h_b}{d^2}\right)}{4\ln\left(\frac{2h_a}{r_a}\right)\ln\left(\frac{2h_b}{r_b}\right) - \left[\ln\left(1 + 4\frac{h_a h_b}{d^2}\right)\right]^2} \quad \text{if } l_{th} \gg \max(d, h_a, h_b), \min(d, h_a, h_b) \gg \max(r_a, r_b)$$

$$C_{ab} \approx \frac{\pi \varepsilon_o l_{th} \ln\left(1 + \frac{4h_a h_b}{d^2}\right)}{\ln\left(\frac{2h_a}{r_a}\right)\ln\left(\frac{2h_b}{r_b}\right)} \quad \text{if } l_{th} \gg \max(d, h_a, h_b), \min(d, h_a, h_b) \gg \max(r_a, r_b), d \gg \max(h_a, h_b)$$

$$C_{ac} = 4\pi\varepsilon_o l_{th} \frac{2\ln\left(\frac{2h_b}{r_b}\right)}{4\ln\left(\frac{2h_a}{r_a}\right)\ln\left(\frac{2h_b}{r_b}\right) - \left[\ln\left(1 + 4\frac{h_a h_b}{d^2}\right)\right]^2} - C_{ab}$$

$$C_{ac} \approx \frac{2\pi\varepsilon_o l_{th}}{\ln\left(\frac{2h_a}{r_a}\right)} - C_{ab} \quad \text{if } d \gg \max(h_a, h_b)$$

$$C_{bc} = 4\pi\varepsilon_o l_{th} \frac{2\ln\left(\frac{2h_a}{r_a}\right)}{4\ln\left(\frac{2h_a}{r_a}\right)\ln\left(\frac{2h_b}{r_b}\right) - \left[\ln\left(1 + 4\frac{h_a h_b}{d^2}\right)\right]^2} - C_{ab}$$

$$C_{bc} \approx \frac{2\pi\varepsilon_o l_{th}}{\ln\left(\frac{2h_b}{r_b}\right)} - C_{ab} \quad \text{if } d \gg \max(h_a, h_b)$$

The capacitance between conductors "a" and "b" in the presence of the large flat conducting plane is

$$C_{mab} = \frac{2\pi\varepsilon_o l_{th}}{\ln\left(\frac{d^2}{r_a r_b}\frac{4h_a h_b}{4h_a h_b + d^2}\right)} = C_{ab} + \frac{C_{ac} C_{bc}}{C_{ac} + C_{bc}} \quad \text{if } l_{th} \gg \max(d, h_a, h_b), \min(d, h_a, h_b) \gg \max(r_a, r_b)$$

$$C_{mab} = \frac{2\pi\varepsilon_o l_{th}}{\ln\left(\frac{d^2}{r_w^2}\frac{4h_a h_b}{4h_a h_b + d^2}\right)} = \frac{\pi\varepsilon_o l_{th}}{\ln\left(\frac{d}{r_w\sqrt{1 + \frac{d^2}{4h_a h_b}}}\right)} \quad \text{if } l_{th} \gg \max(d, h_a, h_b), \min(d, h_a, h_b) \gg r_w = r_a = r_b$$

$$C_{mab} = \frac{\pi\varepsilon_o l_{th}}{\ln\left(\frac{d}{r_w\sqrt{1 + \left(\frac{d}{2h}\right)^2}}\right)} = \frac{\pi\varepsilon_o l_{th}}{\ln\left(\frac{2h}{r_w\sqrt{1 + \left(\frac{2h}{d}\right)^2}}\right)}$$

$$\text{if } l_{th} \gg \max(d, h_a, h_b), \min(d, h_a, h_b) \gg r_w = r_a = r_b, h = h_a = h_b$$

$$C_{mab} \approx \frac{\pi \varepsilon_o l_{th}}{\ln\left(\frac{2h}{r_w}\right)} \quad \text{if } l_{th} \gg \max(d, h_a, h_b), \min(d, h_a, h_b) \gg r_w = r_a = r_b, h = h_a = h_b, d \gg h$$

$$C_{mab} \approx \frac{\pi \varepsilon_o l_{th}}{\ln\left(\frac{d}{r_w}\right)} \quad \text{if } l_{th} \gg (d, h_a, h_b), \min(d, h_a, h_b) \gg r_w = r_a = r_b, h = h_a = h_b, h \gg d$$

The capacitance between conductors "a" and "b," which are connected together, and the large flat conducting plane is

$$C = \frac{4\pi \varepsilon_o l_{th}}{\ln\left[\frac{2h}{r_w}\sqrt{1+\left(\frac{2h}{d}\right)^2}\right]} \quad \text{if } l_{th} \gg \max(d, h_a, h_b), \min(d, h_a, h_b) \gg r_w = r_a = r_b, h = h_a = h_b$$

$$C \approx \frac{4\pi \varepsilon_o l_{th}}{\ln\left(\frac{2h}{r_w}\right)} \quad \text{if } l_{th} \gg \max(d, h_a, h_b), \min(d, h_a, h_b) \gg r_w = r_a = r_b, h = h_a = h_b, d \gg h$$

$$C \approx \frac{2\pi \varepsilon_o l_{th}}{\ln\left(\frac{2h}{\sqrt{r_w d}}\right)} \quad \text{if } l_{th} \gg \max(d, h_a, h_b), \min(d, h_a, h_b) \gg r_w = r_a = r_b, h = h_a = h_b, h \gg d$$

where l_{th} is the length of the conductors. The width, w, of the large flat conductors is large compared to the h_a, h_b, and d.

Two Thin Rectangular Conductors above Large Flat Conductor [Holloway, '98; Pucel; Paul, '92(b)]

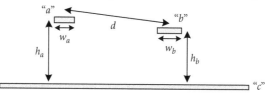

The "circuit capacitances," assuming that "a" and "b" are not too close to "c" or to each other (≈ 3 to 5 cross-sectional dimensions away), are

$$C_{ab} = 4\pi \varepsilon_o l_{th} \frac{\ln\left(1 + 4\frac{h_a h_b}{d^2}\right)}{\left(\left[\ln\left\{1 + \frac{32 h_a^2}{w_a^2}\left[1 + \sqrt{1 + \left(\frac{\pi w_a^2}{8 h_a^2}\right)^2}\right]\right\}\ln\left\{1 + \frac{32 h_b^2}{w_b^2}\left[1 + \sqrt{1 + \left(\frac{\pi w_b^2}{8 h_b^2}\right)^2}\right]\right\}\right] - \left[\ln\left(1 + 4\frac{h_a h_b}{d^2}\right)\right]^2\right)} \quad \text{if } l_{th} \gg \max(h_a, h_b, d)$$

$$C_{ab} \approx 4\pi \varepsilon_o l_{th} \frac{\frac{h^2}{d^2}}{\left[\ln\left(\frac{8h}{w}\right)\right]^2} \quad \text{if } l_{th} \gg \max(h_a, h_b, d), d \gg \sqrt{h_a h_b}, h_a = h_b = h, h \gg w_a = w_b = w$$

$$C_{ac} = 4\pi\varepsilon_o l_{th} \frac{\ln\left\{1+\frac{32h_b^2}{w_b^2}\left[1+\sqrt{1+\left(\frac{\pi w_b^2}{8h_b^2}\right)^2}\right]\right\}}{\left(\left[\ln\left\{1+\frac{32h_a^2}{w_a^2}\left[1+\sqrt{1+\left(\frac{\pi w_a^2}{8h_a^2}\right)^2}\right]\right\}\ln\left\{1+\frac{32h_b^2}{w_b^2}\left[1+\sqrt{1+\left(\frac{\pi w_b^2}{8h_b^2}\right)^2}\right]\right\}\right]-\left[\ln\left(1+4\frac{h_a h_b}{d^2}\right)\right]^2\right)} - C_{ab}$$

$$C_{ac} \approx 2\pi\varepsilon_o l_{th} \frac{1}{\ln\left(\frac{8h}{w}\right)} \quad \text{if } l_{th} \gg \max(h_a, h_b, d), d \gg \sqrt{h_a h_b}, h_a = h_b = h, h \gg w_a = w_b = w$$

$$C_{bc} = 4\pi\varepsilon_o l_{th} \frac{\ln\left\{1+\frac{32h_a^2}{w_a^2}\left[1+\sqrt{1+\left(\frac{\pi w_a^2}{8h_a^2}\right)^2}\right]\right\}}{\left(\left[\ln\left\{1+\frac{32h_a^2}{w_a^2}\left[1+\sqrt{1+\left(\frac{\pi w_a^2}{8h_a^2}\right)^2}\right]\right\}\ln\left\{1+\frac{32h_b^2}{w_b^2}\left[1+\sqrt{1+\left(\frac{\pi w_b^2}{8h_b^2}\right)^2}\right]\right\}\right]-\left[\ln\left(1+4\frac{h_a h_b}{d^2}\right)\right]^2\right)} - C_{ab}$$

$$C_{bc} \approx 2\pi\varepsilon_o l_{th} \frac{1}{\ln\left(\frac{8h}{w}\right)} \quad \text{if } l_{th} \gg \max(h_a, h_b, d), d \gg \sqrt{h_a h_b}, h_a = h_b = h, h \gg w_a = w_b = w$$

where the thicknesses of "*a*" and "*b*" are small compared to h_a and h_b, and l_{th} is the length of the conductors. The width of the "ground" plane is large compared to the h's, w's, and d.

Two Distant Arbitrary Shaped Conductors above Large Flat Conductor   [Smythe; Jefimenko]

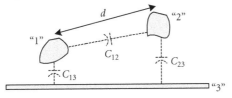

The "circuit capacitances" are

$$C_{13} = c_{11} + c_{12} = \frac{4\pi\varepsilon_o d C_1 (4\pi\varepsilon_o d - C_2)}{(4\pi\varepsilon_o d)^2 - C_1 C_2} \approx C_1$$

$$C_{23} = c_{22} + c_{12} = \frac{4\pi\varepsilon_o d C_2 (4\pi\varepsilon_o d - C_1)}{(4\pi\varepsilon_o d)^2 - C_1 C_2} \approx C_2$$

$$C_{12} = -c_{12} = \frac{4\pi\varepsilon_o d C_1 C_2}{(4\pi\varepsilon_o d)^2 - C_1 C_2} \approx \frac{C_1 C_2}{4\pi\varepsilon_o d}$$

where

$$c_{11} = \frac{(4\pi\varepsilon_o d)^2 C_1}{(4\pi\varepsilon_o d)^2 - C_1 C_2}, \quad c_{22} = \frac{(4\pi\varepsilon_o d)^2 C_2}{(4\pi\varepsilon_o d)^2 - C_1 C_2}, \quad c_{12} = -\frac{4\pi\varepsilon_o d C_1 C_2}{(4\pi\varepsilon_o d)^2 - C_1 C_2} \approx -\frac{C_1 C_2}{4\pi\varepsilon_o d}$$

where C_1 is the capacitance between the large flat conductor ("3") and object "1" when object "2" is not present and C_2 is the capacitance between the large flat conductor ("3") and object "2" when object "1" is not present. The width and length of the flat reference conductor are large compared to the dimensions of "1" and "2," their distance to "3," and d. The distance d is large compared to *all* of the dimensions of objects "1" and "2."

Two Cylindrical Conductors and Cylindrical Return   [Paul, '92(b)]

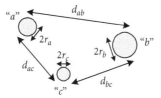

The "circuit capacitances," assuming the conductors are not too close to each other ($\approx$ 3 to 5 radii away), are

$$C_{ab} = 2\pi\varepsilon_o l_{th} \frac{\ln\left(\dfrac{d_{ac}d_{bc}}{r_c d_{ab}}\right)}{\ln\left(\dfrac{d_{ac}^2}{r_a r_c}\right)\ln\left(\dfrac{d_{bc}^2}{r_b r_c}\right) - \left[\ln\left(\dfrac{d_{ac}d_{bc}}{r_c d_{ab}}\right)\right]^2} \quad \text{if } l_{th} \gg \max\left(d_{ab}, d_{ac}, d_{bc}\right)$$

$$C_{ac} = 2\pi\varepsilon_o l_{th} \frac{\ln\left(\dfrac{d_{bc}^2}{r_b r_c}\right)}{\ln\left(\dfrac{d_{ac}^2}{r_a r_c}\right)\ln\left(\dfrac{d_{bc}^2}{r_b r_c}\right) - \left[\ln\left(\dfrac{d_{ac}d_{bc}}{r_c d_{ab}}\right)\right]^2} - C_{ab}$$

$$C_{bc} = 2\pi\varepsilon_o l_{th} \frac{\ln\left(\dfrac{d_{ac}^2}{r_a r_c}\right)}{\ln\left(\dfrac{d_{ac}^2}{r_a r_c}\right)\ln\left(\dfrac{d_{bc}^2}{r_b r_c}\right) - \left[\ln\left(\dfrac{d_{ac}d_{bc}}{r_c d_{ab}}\right)\right]^2} - C_{ab}$$

The capacitance between conductors "a" and "b" in the presence of conductor "c" is

$$C_{mab} = C_{ab} + \frac{C_{ac}C_{bc}}{C_{ac} + C_{bc}}$$

where the d's are the distances between the center of the conductors and l_{th} is the length of the conductors.

Two Cylindrical Conductors inside Cylindrical Shield   [Paul, '92(b); Lo; Attwood; Howard W. Sams]

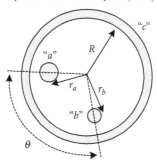

The "circuit capacitances," assuming the conductors are not too close to each other, are

$$C_{ab} = 2\pi\varepsilon_o l_{th} \frac{\ln\left[\frac{r_b}{R}\sqrt{\frac{(r_a r_b)^2 + R^4 - 2r_a r_b R^2 \cos\theta}{(r_a r_b)^2 + r_b^4 - 2r_a r_b^3 \cos\theta}}\right]}{\ln\left(\frac{R^2 - r_a^2}{aR}\right)\ln\left(\frac{R^2 - r_b^2}{bR}\right) - \left\{\ln\left[\frac{r_b}{R}\sqrt{\frac{(r_a r_b)^2 + R^4 - 2r_a r_b R^2 \cos\theta}{(r_a r_b)^2 + r_b^4 - 2r_a r_b^3 \cos\theta}}\right]\right\}^2} \quad \text{if } l_{th} \gg \max(r_a, r_b, R)$$

$$C_{ac} = 2\pi\varepsilon_o l_{th} \frac{\ln\left(\frac{R^2 - r_b^2}{bR}\right)}{\ln\left(\frac{R^2 - r_a^2}{aR}\right)\ln\left(\frac{R^2 - r_b^2}{bR}\right) - \left\{\ln\left[\frac{r_b}{R}\sqrt{\frac{(r_a r_b)^2 + R^4 - 2r_a r_b R^2 \cos\theta}{(r_a r_b)^2 + r_b^4 - 2r_a r_b^3 \cos\theta}}\right]\right\}^2} - C_{ab}$$

$$C_{bc} = 2\pi\varepsilon_o l_{th} \frac{\ln\left(\frac{R^2 - r_a^2}{aR}\right)}{\ln\left(\frac{R^2 - r_a^2}{aR}\right)\ln\left(\frac{R^2 - r_b^2}{bR}\right) - \left\{\ln\left[\frac{r_b}{R}\sqrt{\frac{(r_a r_b)^2 + R^4 - 2r_a r_b R^2 \cos\theta}{(r_a r_b)^2 + r_b^4 - 2r_a r_b^3 \cos\theta}}\right]\right\}^2} - C_{ab}$$

where the r's are the distances between the center of the conductors and the center of the cylindrical shield, and a and b are the radii of conductors "a" and "b," respectively. If the radii of both conductors are equal to a, the capacitance between conductors "a" and "b" in the presence of the cylindrical enclosure is

$$C_{mab} \approx \frac{\pi\varepsilon_o l_{th}}{\ln\left[\frac{d}{a}\frac{1-\left(\frac{d}{2R}\right)^2}{1+\left(\frac{d}{2R}\right)^2}\right]} = C_{ab} + \frac{C_{ac} C_{bc}}{C_{ac} + C_{bc}} \quad \text{if } l_{th} \gg R, \min\left(d, R-\frac{d}{2}\right) \gg a, a = b, r_a = r_b = \frac{d}{2}, \theta = 180°$$

$$C_{mab} \approx \frac{\pi\varepsilon_o l_{th}}{\ln\left(\frac{d}{a}\right)} \quad \text{if } l_{th} \gg R, \min\left(d, R-\frac{d}{2}\right) \gg a, R \gg d, a = b, r_a = r_b = \frac{d}{2}, \theta = 180°$$

If insulating beads are used at frequent intervals (i.e., at electrically-short intervals) between the inner conductors and outer conductor, then the capacitance will increase by

$$\varepsilon_{reff} = 1 + (\varepsilon_r - 1)\frac{w}{s}$$

where ε_r is the relative permittivity of the beads, w is the width of the beads, and $s - w$ is the distance between the beads. The capacitance between conductors "a" and "b," which are connected together, and the cylindrical enclosure "c" is

$$C = \frac{4\pi\varepsilon_o l_{th}}{\ln\left\{\frac{R^2}{da}\left[1-\left(\frac{d}{2R}\right)^4\right]\right\}} = C_{ac} + C_{bc} \quad \text{if } l_{th} \gg R, \min\left(d, R-\frac{d}{2}\right) \gg a, a = b, r_a = r_b = \frac{d}{2}, \theta = 180°$$

where l_{th} is the length of the conductors.

Two Flat Conductors between Two Large Flat Conductors [Walker; Lee, '98]

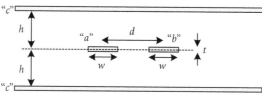

The "circuit capacitances" are

$$C_{ab} = 4\pi\varepsilon_o l_{th} \frac{\left(\dfrac{h}{d}\right)^2}{\left[2\ln\left(\dfrac{16h}{\pi w}\right)\right]^2 - \left(\dfrac{h}{d}\right)^4} \quad \text{if } l_{th} \gg \max(h,d,w),\, h \gg \max(w,t)$$

$$C_{ac} = 4\pi\varepsilon_o l_{th} \frac{2\ln\left(\dfrac{16h}{\pi w}\right)}{\left[2\ln\left(\dfrac{16h}{\pi w}\right)\right]^2 - \left[\ln\left(\dfrac{h}{d}\right)^2\right]^2} - C_{ab} \quad \text{if } l_{th} \gg \max(h,d,w),\, h \gg \max(w,t)$$

$$C_{bc} = C_{ac} \quad \text{if } l_{th} \gg \max(h,d,w),\, h \gg \max(w,t)$$

This assumes that the width of the flat planes, "c," is much greater than h, d, or w, and the length of all of the conductors, l_{th}, is large compared to these dimensions.

Multiple Cylindrical Conductors between Two Large Flat Conductors [Ramo; Weber]

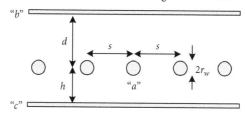

The "circuit capacitances" are

$$C_{bc} = -\frac{l_{th}\ln\left[2\sin\left(\dfrac{\pi r_w}{s}\right)\right]}{2\pi\varepsilon_o\Delta} \quad \text{if } l_{th} \gg h,\, \min(d,s,h) \gg r_w,\, 2\pi\min(h,d) \gg s$$

$$C_{bc} \approx \frac{l_{th}\varepsilon_o s \ln\left(\dfrac{s}{2\pi r_w}\right)}{2h\left[\dfrac{h\pi}{s} + \ln\left(\dfrac{s}{2\pi r_w}\right)\right]} \quad \text{if } l_{th} \gg h,\, \min(d,s,h) \gg r_w,\, 2\pi\min(h,d) \gg s,\, h = d$$

$$C_{bc} \approx \frac{l_{th}\varepsilon_o s}{2h} \quad \text{if } l_{th} \gg h,\, \min(d,s,h) \gg r_w,\, 2\pi\min(h,d) \gg s,\, h = d,\, s \gg h$$

$$C_{ac} = \frac{l_{th}d}{\varepsilon_o s\Delta} \quad \text{if } l_{th} \gg h,\, \min(d,s,h) \gg r_w,\, 2\pi\min(h,d) \gg s$$

$$C_{ab} = \frac{l_{th}h}{\varepsilon_o s\Delta} \quad \text{if } l_{th} \gg h,\, \min(d,s,h) \gg r_w,\, 2\pi\min(h,d) \gg s$$

$$C_{ac} = C_{ab} \approx \frac{l_{th}\varepsilon_o\pi}{\dfrac{h\pi}{s} + \ln\left(\dfrac{s}{2\pi r_w}\right)} \quad \text{if } l_{th} \gg h, \min(d, s, h) \gg r_w, 2\pi\min(h, d) \gg s, h = d$$

where

$$\Delta = \frac{h+d}{\varepsilon_o s}\left\{\frac{h}{\varepsilon_o s} - \frac{1}{2\pi\varepsilon_o}\ln\left[2\sin\left(\frac{\pi r_w}{s}\right)\right]\right\} - \left(\frac{h}{\varepsilon_o s}\right)^2$$

$$\Delta \approx \frac{h}{\varepsilon_o^2\pi s}\left[\frac{h\pi}{s} + \ln\left(\frac{s}{2\pi r_w}\right)\right] \quad \text{if } h = d$$

and l_{th} is the length of the conductors. The cylindrical conductors "*a*" of radius r_w are connected together (i.e., they are at the same potential).

Index

References

Adamczewski, Ignacy, *Ionization, Conductivity, and Breakdown in Dielectric Liquids*, Barnes & Noble, 1969.

Anderson, Dan, *Electrostatic Discharge*, Richmond Technology, Videocassette, 1994.

ANSI/IEEE Grounding of Industrial and Commercial Power Systems, Standard 142, *Green Book*, 1982.

Asano, Kazutoshi, *On the Electrostatic Problem Consisting of a Charged Nonconducting Disk and a Metal Sphere*, Journal of Applied Physics, Vol. 48, No. 3, March 1977.

ASM International, *Electronic Materials Handbook, Volume 1: Packaging*, 1989.

Attwood, Stephen S., *Electric and Magnetic Fields*, Dover, 1967.

Auger, Raymond W., *The Relay Guide*, Reinhold, 1960.

Barke, Erich, *Line-to-Ground Capacitance Calculation for VLSI: A Comparison*, IEEE Transactions on Computer-Aided Design, Vol. 7, No. 2, February 1988.

Bazelyan, E.M. and Yu P. Raizer, *Spark Discharge*, CRC Press, 1998.

Bell Telephone Laboratories, *Integrated Device and Connection Technology*, Vol. 3, Prentice-Hall, 1971.

Bendjamin, J., R. Thottappillil, and V. Scuka, *Time Varying Magnetic Fields Generated by Human Metal (ESD) Electrostatic Discharges*, Journal of Electrostatics, Vol. 46, 1999.

Benner, Linda S., T. Suzuki, K. Meguro, and S. Tanaka, ed., *Precious Metals: Science and Technology*, International Precious Metals Institute, 1991.

Boast, Warren B., *Vector Fields*, Harper & Row, 1964.

Bosich, Joseph F., *Corrosion Prevention for Practicing Engineers*, Barnes & Noble, 1970.

Bouwers, A. and P.G. Cath, *The Maximum Electric Field Strength For Several Simple Electrode Configurations*, Philips Technical Review, Vol. 6, No. 9, September 1941.

Boxleitner, Warren, *Electrostatic discharge and electronic equipment: a practical guide for designing to prevent ESD problems*, IEEE Press, 1989.

Browne, Thomas E., Jr., ed., *Circuit Interruption: Theory and Techniques*, Marcel Dekker, 1984.

Butler, Chalmers M., *Capacitance of a Finite-Length Conducting Cylindrical Tube*, Journal of Applied Physics, Vol. 51, No. 11, November 1980.

Cahill, L.W., *Approximate Formulae for Microstrip Transmission Lines*, Proceedings of the I.R.E., Australia 35, No. 10, October 1974.

Cheng, David K., *Field and Wave Electromagnetics*, Addison-Wesley, 1989.

Chipman, Robert A., *Theory and Problems of Transmission Lines*, Schaum's Outline Series, McGraw-Hill, 1968.

Chow, Y.L. and M.M. Yovanovich, *The Shape Factor of the Capacitance of a Conductor*, Journal of Applied Physics, Vol. 53, No. 12, December 1982.

Chubb, J.N. and G.J. Butterworth, *Charge Transfer and Current Flow Measurements in Electrostatic Discharges*, Journal of Electrostatics, Vol. 13, 1982.

Cohn, Seymour B., *Characteristic Impedances of Broadside-Coupled Strip Transmission Lines*, IRE Transactions on Microwave Theory and Techniques, Vol. MTT-8, No. 6, November 1960.

Cooperman, Gene, *A New Current-Voltage Relation for Duct Precipitators Valid for Low and High Current Densities*, IEEE Transactions on Industry Applications, Vol. 1A-17, No. 2, March/April 1981.

Cooperman, P., *A Theory for Space-Charge-Limited Currents with Application to Electrical Precipitation*, Transactions of the AIEE, Vol. 79, Part 1, 1960.

Copson, David A., *Microwave Heating*, AVI, 1975.

Cross, Jean, *Electrostatics: Principles, Problems and Applications*, Adam Hilger, 1987.

Crowley, Joseph M., *Fundamentals of Applied Electrostatics*, John Wiley & Sons, 1986.

Crowley, Joseph M., *The Electrostatics of Static-Dissipative Worksurfaces*, Journal of Electrostatics, Vol. 24, 1990.

Cruft Laboratory Electronics Training Staff, *Electronic Circuits and Tubes*, McGraw-Hill, 1947.

Dascalescu, Lucian, Patrick Ribardière, Claude Duvanaud, and Jean-Marie Paillot, *Electrostatic Discharges from Charged Spheres Approaching a Grounded Surface*, Journal of Electrostatics, Vol. 47, 1999.

Davidson, J.L., T.J. Williams, A.G. Bailey, and G.L. Hearn, *Characterisation of Electrostatic Discharges from insulating surfaces*, Journal of Electrostatics, Vol. 51–52, 2001.

Demarest, Kenneth R., *Engineering Electromagnetics*, Prentice-Hall, 1998.

Deno, Don W., *Electrostatic Effect Induction Formulae*, IEEE Transactions on Power Apparatus and Systems, Vol. PAS-94, No. 5, September/October 1975.

Donovan, John E., *Triboelectric Noise Generation in Some Cables Commonly Used with Underwater Electroacoustic Transducers*, The Journal of the Acoustical Society of America, Vol. 48, No. 3 (part 2), 1970.

Dorf, Richard D., ed., *The Electrical Engineering Handbook*, CRC Press, 1993.

Dorsett, Henry G., Jr. and John Nagy, *Dust Explosibility of Chemicals, Drugs, Dyes, and Pesticides*, Bureau of Mines, Report of Investigations 7132, 1968.

Dummer, G.W.A., *Materials for Conductive and Resistive Functions*, Hayden, 1970.

Dwight, H.B., *Calculation of Resistances to Ground*, AIEE Transactions, Vol. 55, December 1936.

Eaton, Robert J., *Electric Power Transmission Systems*, Prentice-Hall, 1972.

Felici, N.J., *Electrostatics and Electrostatic Engineering*, Proceedings of the Static Electrification Conference, Institute of Physics and the Physical Society, 1967.

Feng, James Q., *An analysis of corona currents between two concentric cylindrical electrodes*, Journal of Electrostatics, Vol. 46, 1999.

Fenn, John B., *Lean Flammability Limit and Minimum Spark Ignition Energy*, Industrial and Engineering Chemistry, Vol. 43, No. 12, December 1951.

Fowler, E.P., *Microphony of Coaxial Cables*, Proceedings of the IEE, Vol. 123, No. 10, October 1976.

Frankel, Sidney, *Characteristic Impedance of Parallel Wires in Rectangular Troughs*, Proceedings of the I.R.E., Vol. 30, No. 4, April 1942.

Gaertner, Reinhold, Karl-Heinz Helling, Gerhard Biermann, Erich Brazda, Roland Haberhauer, Wilfried Koehl, Richard Mueller, Werner Niggemeier, and Bernhard Soder, *Grounding Personnel via the Floor/Footwear System*, Electrostatic Overstress/Electrostatic Discharge Symposium Proceedings, 1997.

Gandhi, O.P. and I. Chatterjee, *Radio-Frequency Hazards in the VLF and MF Band*, Proceedings of the IEEE, Vol. 70, No. 12, December 1982.

Gibson, Norbert, *Static Electricity — An Industrial Hazard Under Control?*, Journal of Electrostatics, Vol. 40–41, 1997.

Glor, Martin, *Electrostatic Hazards in Powder Handling*, Research Studies Press, 1988.

Gray, Dwight E., ed., *American Institute of Physics Handbook*, McGraw-Hill, 1972.

Greason, William D., *Electrostatic Damage in Electronics: Devices and Systems*, Research Studies Press, 1987.

Greason, William D., *Generalized Model of Electrostatic Discharge (ESD) for Bodies in Approach: Analyses of Multiple Discharges and Speed of Approach*, Journal of Electrostatics, Vol. 54, 2002.

Greason, William D., *Quasi-static Analysis of Electrostatic Discharge (ESD) and the Human Body Using a Capacitance Model*, Journal of Electrostatics, Vol. 35, 1995.

Greenwald, E.K., ed., *Electrical Hazards and Accidents: Their Cause and Prevention*, Van Nostrand Reinhold, 1991.

Grigsby, L.L., ed., *The Electric Power Engineering Handbook*, CRC Press LLC, 2001.

Gunn, Ross, *The Electrification of Precipitation and Thunderstorms*, Proceedings of the IRE, Vol. 45, No. 10, 1957.

Gupta, K.C., Ramesh Garg, and I.J. Bahl, *Microstrip Lines and Slotlines*, Artech House, 1979.

Gupta, K.C., Ramesh Garg, and Rakesh Chadha, *Computer-Aided Design of Microwave Circuits*, Artech House, 1981.

Guru, Bhag Singh and Hüseyin R. Hiziroğlu, *Electromagnetic Field Theory Fundamentals*, PWS, 1998.

Haase, Heinz, *Electrostatic Hazards: Their Evaluation and Control*, Veriag Chemie-Weinheim, 1977.

Hara, Masanori and Masanori Akazaki, *A Method for Prediction of Gaseous Discharge Threshold Voltage in the Presence of a Conducting Particle*, Journal of Electrostatics, Vol. 2, 1976/1977.

Harper, Charles A., ed., *Handbook of Wiring, Cabling, and Interconnecting for Electronics*, McGraw-Hill, 1972.

Harper, W.R., *Contact and Frictional Electrification*, Oxford University Press, 1967.

Hartmann, Irving, John Nagy, and Murray Jacobson, *Explosive Characteristics of Titanium, Zirconium, Thorium, Uranium and Their Hyrides*, Bureau of Mines, Report of Investigations 4835, 1951.

Hartmann, Irving, *Recent Research on the Explosibility of Dust Dispersion*, Industrial and Engineering Chemistry, Vol. 40, 1948.

Haus, Herman A. and James R. Melcher, *Electromagnetic Fields and Energy*, Prentice-Hall, 1989.

Hayt, William H., Jr. and John A. Buck, *Engineering Electromagnetics*, McGraw-Hill, 2001.

Heidelberg, E., *Generation of Igniting Brush Discharges by Charged Layers on Earthed Conductors*, Proceedings of the Static Electrification Conference, Institute of Physics and the Physical Society, 1967.

Hilberg, Wolfgang, *Electrical Characteristics of Transmission Lines*, Artech House, 1979.

Holloway, Christopher L. and Edward F. Kuester, *Net and Partial Inductance of a Microstrip Ground Plane*, IEEE Transactions on Electromagnetic Compatibility, Vol. 40, No. 1, February 1998.

Holm, Ragnar, *Electric Contacts: Theory and Application*, Springer-Verlag, 1967.

Horowitz, Paul and Winfield Hill, *The Art of Electronics*, Cambridge University Press, 1989.

Howard W. Sams, *Reference Data for Radio Engineers*, 1975.

Hughes, J.F., *Electrostatic Powder Coating*, Research Studies Press, 1984.

Jacobson, Murray, John Nagy, and Austin R. Cooper, *Explosibililty of Dusts Used in the Plastics Industry*, Bureau of Mines, Report of Investigations 5971, 1962.

Jacobson, Murray, John Nagy, Austin R. Cooper, and Frank J. Ball, *Explosibililty of Agricultural Dusts*, Bureau of Mines, Report of Investigations 5753, 1961.

Janssen W. and F. Nilber, *High-Frequency Circuit Engineering*, IEE, 1996.

Jefimenko, Oleg D., *Electricity and Magnetism*, Electret Scientific, 1966.

Johnson, Richard C. and Henry Jasik, ed., *Antenna Engineering Handbook*, McGraw-Hill, 1961.

Jonassen, Niels, *Electrostatics*, Chapman & Hall, 1998.

Jones, T.B. and S. Chan, *Charge Relaxation in Partially Filled Vessels*, Journal of Electrostatics, Vol. 22, 1989.

Jones, Thomas B. and Jack L. King, *Powder Handling and Electrostatics: Understanding and Preventing Hazards*, Lewis, 1991.

Jones, Thomas B. and Kit-Ming Tang, *Charge Relaxation in Powder Beds*, Journal of Electrostatics, Vol. 19, 1987.

Kaiser, Kenneth L., *A Study of Free-Surface Electrohydrodynamics*, Ph.D. Thesis, Purdue University, 1989.

Kallman, Raymond, *Realities of Wrist Strap Monitoring Systems*, Electrostatic Overstress/Electrostatic Discharge Symposium Proceedings, 1994.

Keithley, *Low Level Measurements*, Keithley Test Instruments, 1993.

Kessler, LeAnn and W. Keith Fisher, *A Study of the Electrostatic Behavior of Carpets Containing Conductive Yarns*, Journal of Electrostatics, Vol. 39, 1997.

Khalifa, M., ed., *High-Voltage Engineering: Theory and Practice*, Marcel Dekker, 1990.

King, Ronald W. P., *Transmission Line Theory*, McGraw-Hill, 1955.

Koyler, John M. and Donald E. Watson, *ESD from A to Z: Electrostatic Discharge Control for Electronics*, Van Nostrand Reinhold, 1990.

Krevelen, D.W. Van, *Properties of Polymers*, Elsevier, 1990.

Kussy, Frank W. and Jack L. Warren, *Design Fundamentals for Low-Voltage Distribution and Control*, Marcel Dekker, 1987.

Larsen, Øystein, Janicke H. Hagen, and Kees van Wingerden, *Ignition of Dust Clouds by Brush Discharges in Oxygen Enriched Atmospheres*, Journal of Loss Prevention in the process industries, Vol. 14, 2001.

Lee, Kai Fong, *Principles of Antenna Theory*, John Wiley & Sons, 1984(b).

Lee, Thomas H., *The Design of CMOS Radio-Frequency Integrated Circuits*, Cambridge University Press, 1998.

Lerner, C.M., *Problems and Solutions in Electromagnetic Theory*, John-Wiley & Sons, 1985.

Lo, Y.T. and S.W. Lee, ed., *Antenna Handbook*, Van Nostrand Reinhold, 1988.

Lüttgens, Günter and Norman Wilson, *Electrostatic Hazards*, Butterworth-Heinemann, 1977.

Maissel, Leon I. and Reinhard Glang, *Handbook of Thin Film Technology*, McGraw-Hill, 1970.

Marshall, S.V., Richard E. DuBroff, and G.G. Skitek, *Electromagnetic Concepts and Applications*, Prentice-Hall, 1996.

Maurer, B., M. Glor, G. Lüttgens, and L. Post, *Hazards Associated with Propagating Brush Discharges on Flexible Intermediate Bulk Containers, Compounds and Coated Materials*, Electrostatics 1987, Institute of Physics Conference Series, No. 85, 1987.

Melcher, James R., *Continuum Electromechanics*, MIT Press, 1981.

Miner, Gayle F., *Lines and Electromagnetic Fields for Engineers*, Oxford University Press, 1996.

Mizuno, A., *Electrostatic Precipitation*, IEEE Transactions on Dielectrics and Electrical Insulation, Vol. 7, No. 5, October 2000.

Montrose, Mark I., *EMC and the Printed Circuit Board*, IEEE Press, 1999.

Montrose, Mark I., *Printed Circuit Board Techniques For EMC Compliance*, IEEE Press, 2000.

Moon, Parry and Domina Eberle Spencer, *Field Theory for Engineers*, D. Van Nostrand, 1961.

Moore, A.D., ed., *Electrostatics and Its Applications*, John Wiley & Sons, 1973.

Morse, Philip M. and Herman Feshbach, *Methods of Theoretical Physics*, McGraw-Hill, 1953.

Nagy, John, Austin R. Cooper, and Henry G. Dorsett, Jr., *Explosibility of Miscellaneous Dusts*, Bureau of Mines, Report of Investigation 7208, 1968.

Nagy, John, Henry G. Dorsett, Jr., and Austin R. Cooper, *Explosibility of Carbonaceous Dusts*, Bureau of Mines, Report of Investigations 6597, 1965.

Nanevicz, J.E., *Some Techniques for the Elimination of Corona Discharge Noise in Aircraft Antennas*, Proceedings of the IEEE, Vol. 52, No. 1, January 1964.

Nanevicz, Joseph E., *Static Charging and its Effects on Avionic Systems*, IEEE Transactions on Electromagnetic Compatibility, Vol. EMC-24, No. 2, May 1982.

National Association of Relay Manufacturers, *Engineer's Relay Handbook*, Hayden, 1960.

Nishiyama, Hitoshi and Mitsunobu Nakamura, *Capacitance of Disk Capacitors*, IEEE Transactions on Components, Hybrids, and Manufacturing Technology, Vol. 16, No. 3, May 1993.

Ott, Henry W., *Noise Reduction Techniques in Electronic Systems*, John Wiley & Sons, 1988.

Paul, Clayton R., *Introduction to Electromagnetic Compatibility*, John Wiley & Sons, 1992(b).

Peek, F.W., *Dielectric Phenomena in High Voltage Engineering*, McGraw-Hill, 1929.

Perls, Thomas A., *Electrical Noise from Instrument Cables Subjected to Shock and Vibration*, Journal of Applied Physics, Vol. 23, No. 6, June 1952.

Pidoll, Ulrich von, Helmut Krämer, and Heino Bothe, *Avoidance of Electrostatic Hazards during Refueling of Motorcars*, Journal of Electrostatics, Vol. 40–41, 1997.

Plastics for Electronics: Desk-top Data Bank, The International Plastics Selector, 1979.

Pratt, Thomas H., *Electrostatic Ignition Hazards*, Burgoynes, 1995.

Pucel, Robert A., Daniel J. Massé, and Curtis P. Hartwig, *Losses in Microstrip*, IEEE Transactions on Microwave Theory and Techniques, Vol. MTT-16, No. 6, June 1968.

Ramo, Simon, John R. Whinnery, and Theodore Van Duzer, *Fields and Waves in Communication Electronics*, John Wiley & Sons, 1965.

Rao, Nannapaneni Narayana, *Elements of Engineering Electromagnetics*, Prentice-Hall, 1994.

Ratz, Alfred G., *Triboelectric Noise*, Instrument Society of America Transactions, Vol. 9, No. 2, 1970.

Riggs, Olen L., Jr. and Carl E. Locke, *Anodic Protection*, Plenum Press, 1981.

Saums, Harry L. and Wesley W. Pendleton, *Materials for Electrical Insulating and Dielectric Functions*, Hayden, 1973.

Schneider, M.V., Bernard Glance, and W.F. Bodtmann, *Microwave and Millimeter Wave Hybrid Integrated Circuits for Radio Systems*, The Bell System Technical Journal, Vol. 48, No. 6, July–August 1969.

Schwab, Adolf J., *High-Voltage Measurement Techniques*, The MIT. Press, 1972.

Schweitzer, Philip A., ed., *Corrosion and Corrosion Protection Handbook*, Marcel Dekker, 1988.

Sclater, Neil, *Electrostatic Discharge Protection for Electronics*, TAB, 1990.

Secker, P.E. and J.N. Chubb, *Instrumentation for Electrostatic Measurements*, Journal of Electrostatics, Vol. 16, 1984.

Shen, L.C., S.A. Long, M.R. Allerding, and M.D. Walton, *Resonant Frequency of a Circular Disc, Printed-Circuit Antenna*, IEEE Transactions on Antennas and Propagation, Vol. AP-25, No. 4, July 1977.

Slade, Paul G., ed., *Electrical Contacts: Principles and Applications*, Marcel Dekker, 1999.

Smith, Douglas C., *Unusual Forms of ESD and Their Effects*, Electrostatic Overstress/Electrostatic Discharge Symposium Proceedings, 1999.

Smits, F.M., *Measurement of Sheet Resistivities with the Four-Point Probe*, The Bell System Technical Journal, No. 3, May 1958.

Smythe, William R., *Static and Dynamic Electricity*, McGraw-Hill, 1968.

Somerville, J.M., *The Electric Arc*, John Wiley & Sons, 1959.

Spangenberg, Karl R., *Vacuum Tubes*, McGraw-Hill, 1948.

Spencer, Ned A., *An Antenna for 30-50 MC Service Having Substantial Freedom from Noise Caused by Precipitation Static and Corona*, IRE Transactions on Vehicular Communications, VC-9, No. 2, August 1960.

Static Control Components, Inc., *Choosing the Right Static Bag*, Technical Bulletin P1, 1996.

Stevenson, William D., Jr., *Elements of Power System Analysis*, McGraw-Hill, 1975.

Strojny, Jan A., *Some Factors Influencing Electrostatic Discharge from a Human Body*, Journal of Electrostatics, Vol. 40–41, 1997.

Sze, S.M., *Physics of Semiconductor Devices*, John Wiley & Sons, 1981.

Takuma, T., M. Yashima, and T. Kawamoto, *Principle of Surface Charge Measurement for Thick Insulating Specimens*, IEEE Transactions on Dielectrics and Electrical Insulation, Vol. 5, No. 4, August 1998.

Taylor, D.M. and P.E. Secker, *Industrial Electrostatics: Fundamentals and Measurements*, Research Studies Press, 1994.

Taylor, D.M., *Measuring Techniques for Electrostatics*, Journal of Electrostatics, Vol. 51–52, 2001.

Thomson, J.J. and G.P. Thomson, *Conduction of Electricity Through Gases*, Vol. II, Dover, 1969.

Tippet, John C. and David C. Chang, *Characteristic Impedance of a Rectangular Coaxial Line with Offset Inner Conductor*, IEEE Transactions on Microwave Theory and Techniques, Vol. MTT-26, No. 11, November 1978.

Tolson, P., *Assessing the Safety of Electrically Powered Static Eliminators for Use in Flammable Atmospheres*, Journal of Electrostatics, Vol. 11, 1981.

Valdes, L.B., *Resistivity Measurements on Germanium for Transistors*, Proceedings of the IRE, February 1954.

Vosteen, William E., *A Review of Current Electrostatic Measurement Techniques and Their Limitations*, Monroe Electronics, 1984.

Walker, Charles S., *Capacitance, Inductance, and Crosstalk Analysis*, Artech House, 1990.

Watt, Arthur D., *VLF Radio Engineering*, Pergamon Press, 1967.

Weber, Ernst, *Electromagnetic Theory: Static Fields and Their Mapping*, Dover, 1965.

Weeks, W.L., *Antenna Engineering*, McGraw-Hill, 1968.

Wenner, Frank, *A Method of Measuring Earth Resistivity*, Bulletin of the Bureau of Standards, Vol. 12, No. 4, October 11, 1915.

White, Joseph F., *Microwave Semiconductor Engineering*, Van Nostrand Reinhold, 1982.

Williams, George M. and Thomas H. Pratt, *Characteristics and hazards of electrostatic discharges in air*, Journal of Loss Prevention in the Process Industries, Vol. 3, October 1990.

Wolff, Edward A., *Antenna Analysis*, John Wiley & Sons, 1967.

Wood, Jody W., Personal Communications, New England Wire Technologies, 2003.

Zahn, Markus, *Electromagnetic Field Theory: a problem solving approach*, Robert E. Krieger, 1979.

Milton Keynes UK
Ingram Content Group UK Ltd.
UKHW051947071024
449327UK00026B/2200